VOYAGER

PHOTOGRAPHS FROM HUMANITY'S GREATEST JOURNEY

VOYAGER

PHOTOGRAPHS FROM HUMANITY'S GREATEST JOURNEY

Jens Bezemer
Joel Meter
Simon Phillipson
Delano Steenmeijer
Ted Stryk

With a special essay from Dr. Garry Hunt

teNeues

SOMEWHERE, SOMETHING INCREDIBLE IS WAITING TO BE KNOWN

Carl Sagan, 1977

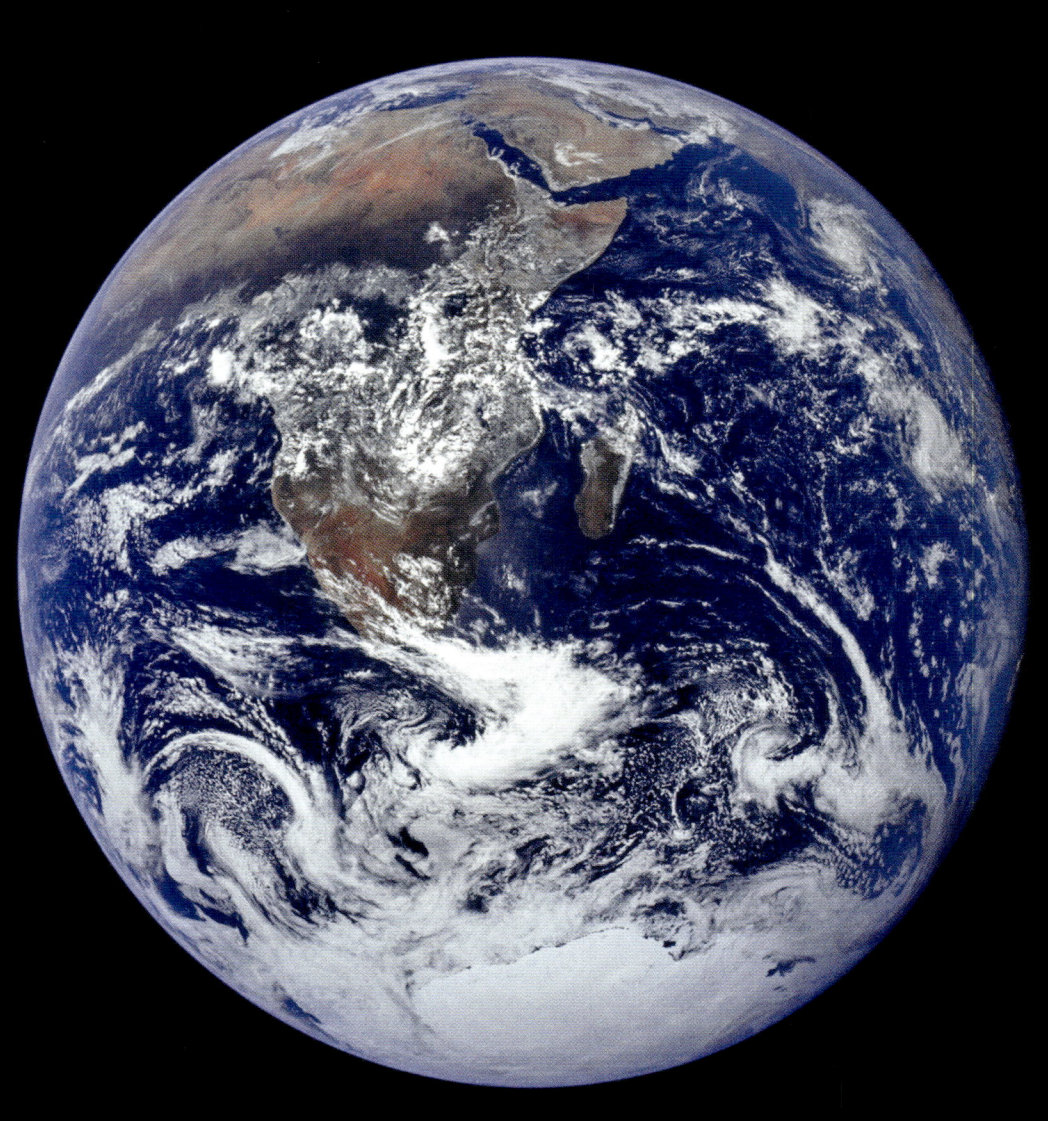

THE NEXT BEND IN THE ROAD

What does it mean to "know" something? People across different centuries, cultures, and countries often thought they had the world figured out. Nevertheless, we are a curious bunch. Our appetite for comprehension is rarely fulfilled. The urge to know has characterized humans for as long as we can remember. We want to see what is around the corner. We have to go past the next bend in the road. Not for the sake of doing so, but because we want to know what is there. The not knowing crushes something in the human spirit.

Although that curiosity has always been there, it takes on different starting points, forms, and agendas depending on the era and culture. If you asked some practical thinkers in Babylonia about the upper ontologies in artificial intelligence, they would probably give you a glass of water and ask you to lie down for a bit. The common thread, however, is that we expand our knowledge along the way. Not only do we expand it in our minds, but also in space. For the Chinese, the cosmos was an infinite space with floating pieces of condensed vapor, while the Europeans saw it as a celestial sphere made out of a substance called "quintessence." People thought the Earth was flat until a Greek mathematician in the third century BCE had an epiphany with two wooden sticks on a sunny day. Moreover, it is less than one single human lifetime ago that we did not know about other galaxies, about the universe expanding, and about the Big Bang.

The way we expand this knowledge is highly dependent on the way we think. Ancient Greek thought is built around wholeness and a circular understanding of time, while in modern times we have seen the birth of the subject and a more linear time perspective. Our kind of knowledge is derived, and our conclusions are drawn from our sensory experience. We know the things around us, insofar as we can sense the things around us.

The Voyager mission of NASA was, and still is, an ambitious attempt to make our Solar System, and that which is beyond, part of that sensory experience. Despite the fact that no humans are there along for the ride, Voyager carries our senses. Two robotic spacecraft embarked on a journey to expand our knowledge and answer the very same cosmic questions that curious children across centuries have asked themselves, their parents, and their teachers. Where do we come from? From what is the universe made? Are we alone?

Planetary scientists call this approach "remote sensing," whereby robotic sensors provide the data to expand our knowledge. Cameras as eyes. Drills, scoops and arms as fingers. Sampling probes as noses. Radio antennas as ears. Flying spacecraft and rovers on the ground form the "bodies" that carry around these capabilities for us, providing the pictures, the data, and the audio fragments from places far away. Nevertheless, make no mistake; these machines are not alive. "Don't anthropomorphize the spacecraft," says Voyager Project Manager John Casani. "They don't like it."

The impact of the images these spacecraft send back home cannot be overstated. Making use of an extraordinary alignment between the planets, Voyager 1 and Voyager 2 became responsible for the first detailed views of our Solar System beyond Mars. The giant planets Jupiter, Saturn, Uranus, and Neptune transformed, once and for all, from tiny specks of light to wondrous manifestations via images that gave writers, musicians, filmmakers and poets ample ammunition for decades of creativity.

Judged with the quality standards of today, the images are not that spectacular. With an average resolution of 800 x 800 pixels, it has been difficult to give them a home in print. At least not in a way that does justice to the images and what they represent to a contemporary audience. With this book, VOYAGER, we intend to turn this around. By enriching, processing, combining and treating these photographs with the tools of the 21st century, we embraced the opportunity to explore and revisit them in a new way. We hope that by showing you these images with the clarity of our time, we contribute in a tiny way to expanding the knowledge retrieved by the great men and women on the Voyager team.

The spacecraft are still ploughing through space and are now the farthest human-made objects from Earth. Although their cameras have been turned off, their communication and data are still to this day, expanding our knowledge. Both Voyager 1 and Voyager 2 have broken through the boundaries of our Solar System and entered interstellar space. Two sentinels destined to become artifacts from our home that will be continuing their exploration long after our species, our planet, and our Solar System are gone. This makes Voyager rightfully humanity's greatest journey. A journey to explore the truly unknown. However, as we hope to show you with this book, it became a journey to get to know ourselves as well.

Left: the iconic 'Blue Marble' photograph taken by the crew of Apollo 17 on December 7th, 1972.

VOYAGER

ORIGIN

Space History

Although Voyager to this day is still going strong, the world looked very different when the mission came into existence in the 60s and 70s. Those with any hunger for science and space did not have the luxurious abundance of the internet that offers the opportunity to feed that hunger with countless articles, informative pages, videos, movies and interviews. Let alone TV channels feeding your needs, ranging from The Discovery Channel to National Geographic, and from NASA TV to The History Channel. There was public broadcasting, and when space history was being made, you had to wait until they showed it to you in granular black-and-white images.

That space history, for some part, aligns with the history of the Cold War. The cosmic arm wrestle started on October 4th, 1957, when the Soviet Union launched the first artificial satellite to orbit Earth: Sputnik. After several years of competing in the development of Intercontinental Ballistic Missiles (ICBMs) to carry nuclear weapons, the design and launch of Sputnik kick-started the Space Race. Every 96 minutes, the satellite would orbit Earth, sending down radio beeps from its transmitter, reminding Americans and the rest of the world of the Soviet Union's risen power. A month later, when they sent a second one into the sky carrying the dog Laika, the Americans knew it was their turn.

Although they sent a rocket into space on January 31st, 1958, carrying a satellite called Explorer, the Americans actually owe a large portion of their expertise to another country. The team that achieved the feat largely consisted of German Rocket Engineers who had once developed ballistic missiles for Nazi Germany and were now working for the "other side." Under the direction of Wernher von Braun, the team further developed the German V-2 Rocket into a more powerful version called Juno, carrying the instruments to prove the existence of what we now call the Van Allen radiation belts around Earth. That same year, all the different space exploration activities decided to bundle their forces and consolidate into a new government agency, the National Aeronautics and Space Administration (NASA).

On April 12th, 1961, Yuri Gagarin had the honor to follow in the footsteps of the dog Laika. After take-off, the Soviet cosmonaut orbited around Earth in a historic journey of 108 minutes, becoming the first human in space. NASA took about three weeks to provide an answer when they hurled Alan Shepard into space. His flight however, was not an orbital flight, but rather a suborbital trajectory, which means that it went into space but did not go all the way around Earth. The result became 15 minutes of fame for the American astronaut. Still, with the first artificial satellite, the first dog, and the first human in space orbiting Earth, the Soviet Union was hitting milestone after milestone. However, they did not stop there. With the first object hitting the Moon in 1959 (Luna 2), the first spacewalk as well as the first woman in space (Valentina Tereshkova), the Soviet Union clearly were in the lead in this arm wrestle.

It was time for another, bigger milestone for the United States to sink their teeth in. On May 25th, 1961, American president John F. Kennedy gave it to them: "I believe that this nation should commit itself to achieving the goal, before the decade is out, of landing a man on the moon and returning him safely to Earth." The 60s became all about achieving that goal. With a program called Project Gemini, NASA started making real progress when astronauts started testing technology needed for future flights to the moon as well as their own ability to endure many days in spaceflight. Project Apollo became next in line, set out to orbit around the moon and the lunar surface. In 1969, the United States sent the first astronauts to the moon on Apollo 11, with Neil Armstrong setting foot on its surface and making "giant leaps."

Other Planets and Moons

In the background of these public historic and cultural events, the Soviets and Americans were planning to push even farther into the Solar System. Partly due to the realization that the Soviets were not ready for the technologically heavy weightlifting of putting a man on the moon, a switch was made by the Russians towards robotic missions to other planets and moons. Early successes were mixed with spectacular disasters. In December 1971, they put a spacecraft on Mars, but its transmission lasted for

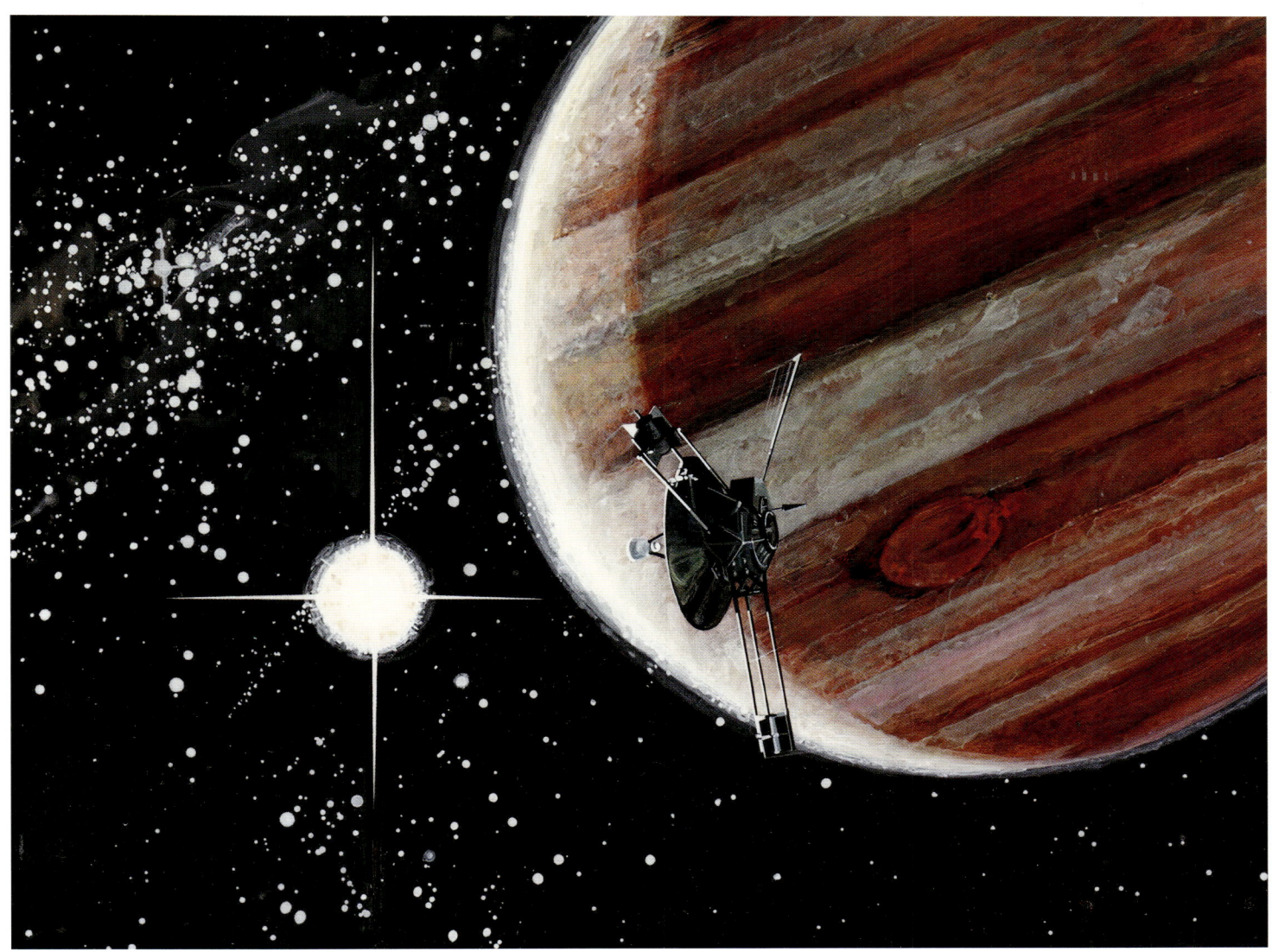

Artist Rick Guidice depicting Pioneer 10 passing Jupiter, 1973.

only 14.5 very expensive seconds. NASA on the other hand started their probe development in the 60s and throughout the 70s with a program called Mariner, which studied Venus, Mars, and Mercury. During the 70s, NASA also carried out Project Viking (originally also named Voyager but was cancelled and given a new name) which managed to land two probes on Mars that took photographs, examined the chemistry, and tested the dirt of the Martian surface.

But what about the four distant gas giants Jupiter, Saturn, Uranus, and Neptune? Shrouded in mystery, knowledge about the next four planets in line was limited. People knew they were mostly made up of hydrogen and helium, and some methane on the outer planets. Jupiter has four moons orbiting around it called Io, Europa, Ganymede, and Callisto,

of which everybody was highly convinced they were just battered ice balls. Saturn had rings and the major satellites were known, but that was about it. For centuries, people had been looking at these four planets through their telescopes as fuzzy blobs. In the early 1970s, NASA launched its first attempt to visit Jupiter and Saturn with the help of the Pioneer probes. Although they did critical groundwork for subsequent missions and were important for showing it was possible to travel through the asteroid belt, the Pioneers carried limited instrumentation and an elementary camera. Consequently, many of the mysteries remained mysteries. The real nature of Jupiter, Saturn, Uranus, and Neptune in terms of weather, distribution, mass or constitution was unknown.

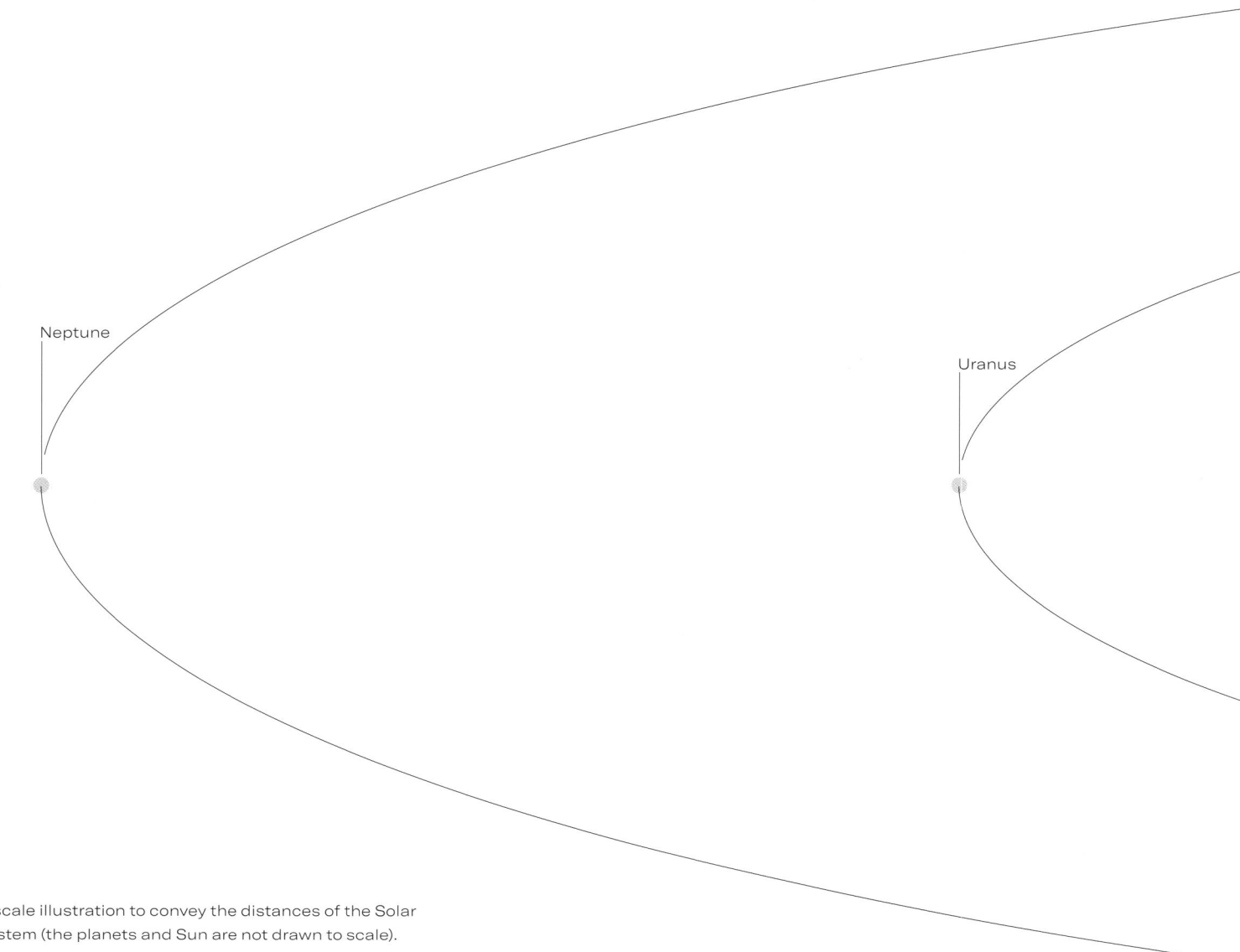

Neptune

Uranus

A scale illustration to convey the distances of the Solar System (the planets and Sun are not drawn to scale).

Gary Flandro and Gravity Assist

Visiting the four outer gas giants to capture detailed images and measurements seemed, quite literally, far-fetched. But then again, every major achievement probably does in the beginning. Often (and the cliché is true here) it starts with imagining it on a cocktail napkin or a flash of inspiration. That is exactly how humanity's greatest journey got started. During the mid-60s, the Jet Propulsion Laboratory (JPL) in Pasadena, California was working on using "gravity assist." In orbital mechanics, gravity assist, also known as a swingby, is the use of relative movement and gravity of a planet (or other astronomical object) to alter the speed and path of a spacecraft. Could this technique also be used for outer Solar System missions?

A young graduate student named Gary Flandro, studying instabilities in rocket combustion and working part-time at JPL to study aerodynamics and the trajectory of missiles, was asked to dive into that question. With a keen interest in the history of the academic field of celestial mechanics, Flandro quickly found out that the idea of gravity assist already had existed since the 1800s and was used by pioneers such as Urbain Le Verrier to figure out deviating comet flybys near Jupiter. On the shoulders of such giants, the main goal became finding out whether visiting the four outer planets with the use of gravity assist was possible, while still allowing a spacecraft to carry a significant amount of mass for all the necessary instruments.

The Perfect Alignment

So what was the flash of inspiration that Flandro had? In the spring of 1965, it occurred to him that gravity assist could perhaps be responsible for not just one slingshot, but a multitude of them. If the outer planets were aligned just right, this would permit a spacecraft to be propelled to more than one gas giant after passing by Jupiter.

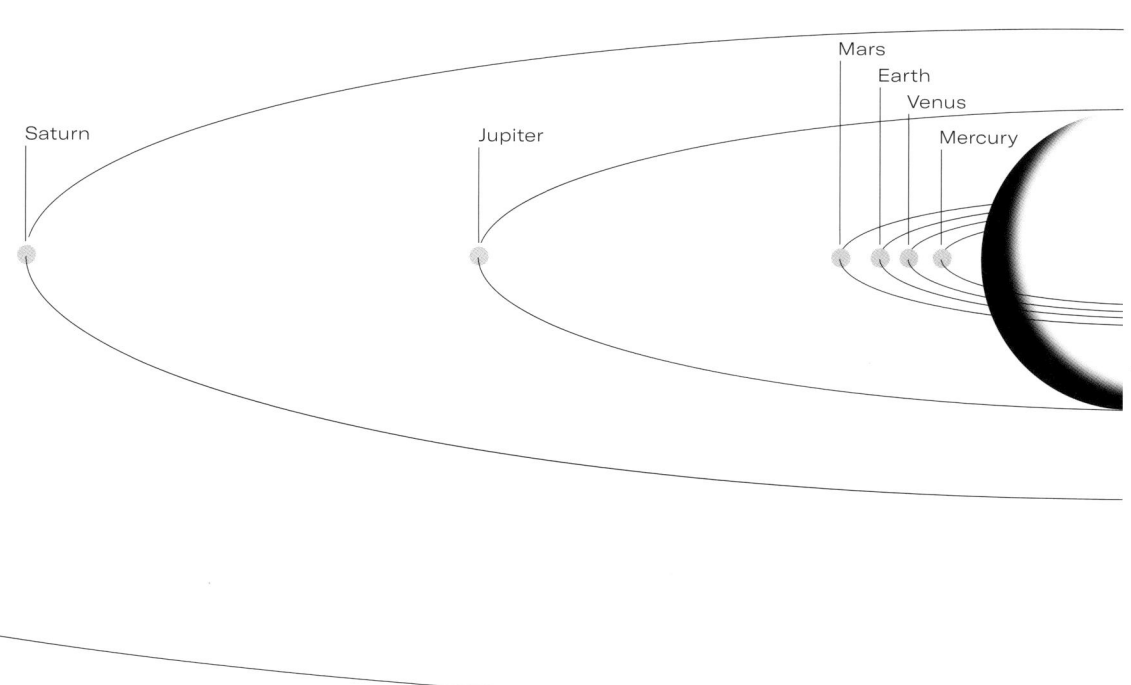

Saturn

Jupiter

Mars
Earth
Venus
Mercury

His research showed that such an alignment takes place about once in every 176 years. The last time it happened, exploration was still done with wooden ships, and the next alignment was expected to be in the 1980s. With a little bit of back planning, Flandro found out that if it was launched somewhere in the 1970s, on a single trajectory with the shortest possible trip time, it would be possible for a spacecraft to visit Jupiter, Saturn, Uranus, and Neptune, or Jupiter, Saturn and Pluto. Coined as the "Grand Tour," this became a once-in-two-lifetime's opportunity.

Not everyone, however, shared this sense of opportunity. Already when Flandro wrote up his findings and published them in Acta Astronautica in 1966, the response was quite bleak, with people openly mocking the idea. NASA was in the middle of the enormous feat of putting people on the moon, and now we wanted to send a spacecraft into space that has to operate for multiple decades? The longevity was unfathomable for many at the time. Still, Flandro and JPL were now on a quest

and proceeded to form a proposal. Formed in 1966, this included four spacecraft, two of which would fly past Jupiter, Saturn, and Pluto, and a second pair that would take on Jupiter, Uranus, and Neptune. The combined cost was estimated at $1 billion.

Again, there was plenty of discouragement to go around. Both NASA as well as the administration of President Nixon were baffled by the price tag and disapproved of the proposal. By that time, every month $2 billion was being spent on a war in Vietnam that was growing in political importance by the day. With NASA's already slim annual budget of $5+ billion being squeezed further by Congress, and all its programs competing for a gradually dwindling amount of resources, this highly expensive Grand Tour did not stand a chance. Luckily, the door was left open a tiny bit, with the JPL Grand Tour mission team being told that if they could come up with something "a little less grandiose," NASA and the administration would consider it. So they started to work it out.

MJS-77

Although they immediately decided to scale back from four to two spacecraft, an equally important decision was made to primarily use technology that had already been developed for the previously mentioned Mariner series. This 10-mission program, conducted by NASA and JPL, launched a series of robotic interplanetary probes from 1962 to 1973. All Mariner spacecraft were based on a hexagonal or octagonal "bus", which housed all of the electronics, and to which all components were attached, such as antennas, cameras, propulsion and power sources. At the time of the Grand Tour reshuffle of 1972, the first nine Mariner probes, with varying degrees of success, had focused on close-up observations and measurements of Venus and Mars. Mariner 10, which was the first to use a gravity assist trajectory, was planned to launch in 1973. Accelerating as it entered the gravitational influence of Venus, Mariner 10 was to be flung by the planet's gravity onto a slightly different course to reach Mercury. Mariner 11 and Mariner 12 were still in an infant stage and up for grabs.

What resulted from this was called "Mariner Jupiter Saturn 1977" or MJS-77. A trimmed-down version, designed specifically as something "a little less grandiose." Mission manager Harris "Bud" Schurmeier and his team scrapped two spacecraft, took off the atmospheric entry probes of the remaining two, and scaled the trajectory back to only include flybys of Jupiter and Saturn. All that cutting back had a significant impact on the price tag as well, which had now dropped to a more workable $250 million. The story goes that this proposal had NASA hooked immediately, only leaving the man in the Oval Office to be persuaded. As the NASA Administrator went to the President, he said "The last time the planets were lined up like that, Thomas Jefferson was sitting at your desk, and he blew it." The Nixon administration approved of the scaled back proposal, making the mission a formal fact.

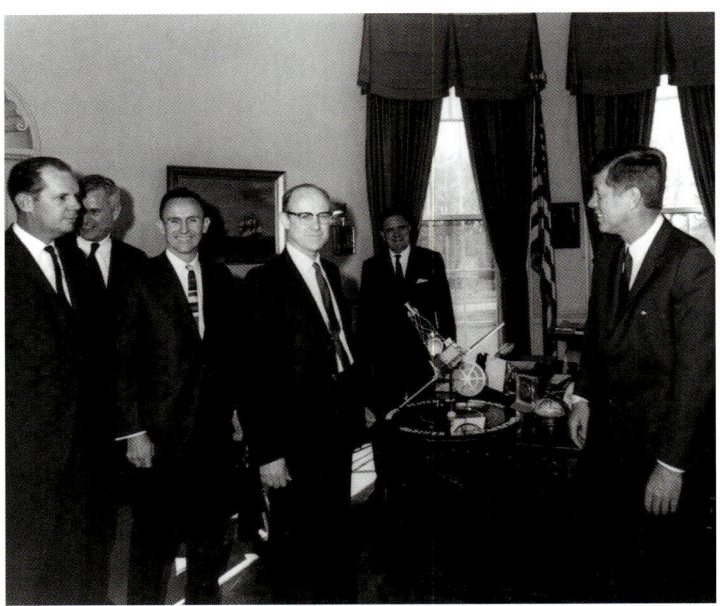

Director of NASA's Jet Propulsion Laboratory, Dr. William H. Pickering (center), presents a model of a Mariner spacecraft to President John F. Kennedy in 1961.

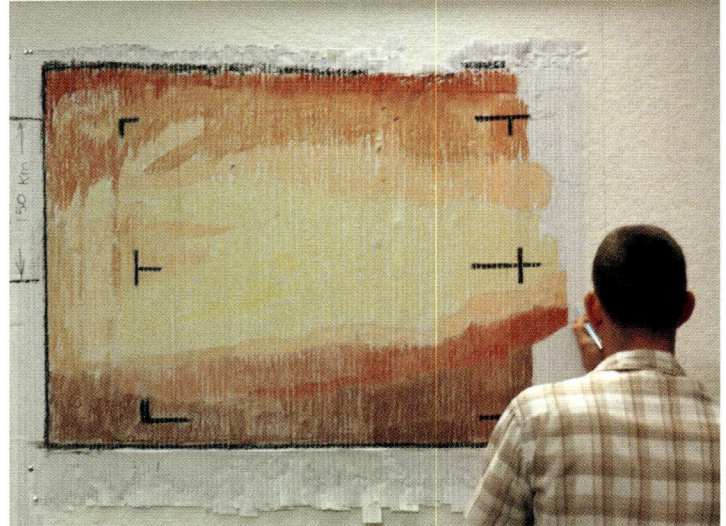

The first 'image' of Mars sent from Mariner 4 in July 1965. The imaging team at JPL couldn't wait for the official processed image. Instead they placed the data strips side by side and hand-colored the data numbers like a 'paint by numbers' picture.

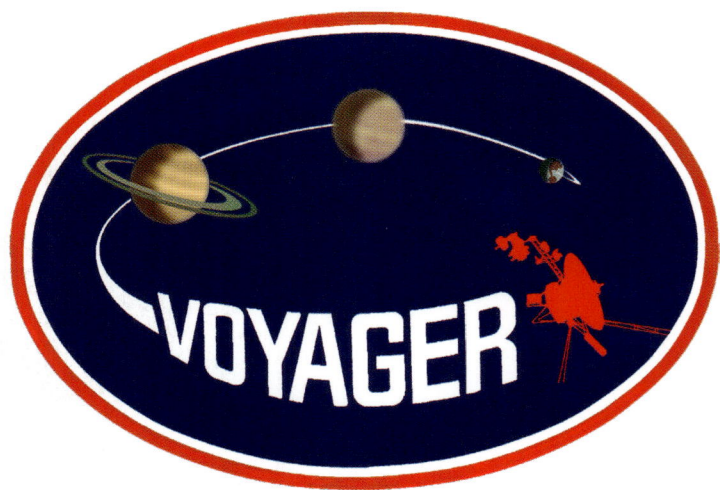

Voyager mission badge depicting only the Earth, Jupiter and Saturn.

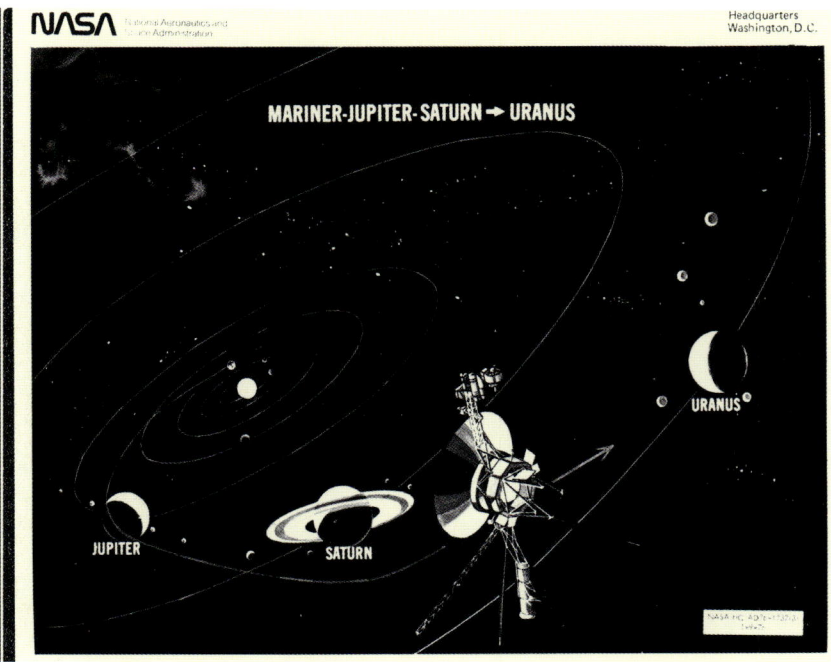

Cover image from a Mariner – Jupiter – Saturn – Uranus document showing the planetary slingshot technique.

The Name

Even a few months before the launch in 1977, the mission was still not being called "Voyager". This name carries a complicated history. Everyone involved with the mission at JPL generally agreed that the name MJS-77 was not the sexiest name ever given to an audacious mission of sorts. John Casani, the mission Project Manager at the time, said "Who the hell cares about what year we launched the mission? We need a crisp name!" For this reason, a lab-wide contest was organized to give the mission a better name. A case of champagne was delivered as a prize to the sender of the winning name Voyager. Nevertheless, it was not an easy pick, because it was a name that carried some emotional baggage throughout the corridors of JPL. Sometime before, another project also called Voyager had been proposed and was in development for years, until it was scrapped. Although the mission eventually did see daylight when it resurfaced as Viking, a mission to Mars, picking Voyager from the list of options felt like possibly jinxing it. Did the name carry bad karma? With just a few months to go until the launch, the scientists and engineers remembered that they were scientists and engineers; their decisions should be made on rational grounds. The name was chosen, turning Mariner 11 and Mariner 12 into Voyager 1 and Voyager 2.

A Mission within a Mission

With 11 scientific instruments and 11 different teams, the project started and its success was determined at getting one spacecraft past Saturn. At least, that is the version on paper. The scientists and mission managers had not forgotten about Gary Flandro and the Grand Tour. Right from the get-go, a clear determination arose to do everything in their power to reach for all of the outer planets. Ed Stone, project scientist and, the true father of the mission, recalls that "The idea of getting to Uranus and Neptune was being pursued, but quietly, partly because nobody wanted the mission to be considered a failure if we did not survive past Saturn!" So, it became a mission within a mission. It made the big challenge for the years ahead towards the launch twofold: figure out the perfect trajectory and build a spacecraft that has the capability to fly far and last forever.

TRAJECTORY

Flandro's Footsteps

As the project began, JPL started the onerous assignment to implement the Grand Tour mission opportunity within the standard trajectories. This took people to follow in the footsteps of Gary Flandro, and pick up the work where he left off. Navigation expert Charlie Kohlhase, together with Paul Penzo and Joe Beerer, began running thousands of trajectories through the computers. Their search was aimed at an itinerary that would not only yield the best photography and scientific measurement, but also bring the Voyagers to the four gas giants and their moons. "We searched through 10,000 different gravity assist flight path options from Earth to the outer planets," says Kohlhase.

The biggest concern was taking into account the level of propellant that was needed to correct for navigational errors during the flybys. Take Jupiter, for example. The enormous mass of the planet could flip the trajectory by about 90° when a spacecraft flies past. Taking the swingby too close or too far off the perfect path would deflect the Voyagers either too much or too little, resulting in a loss of propellant that might make reaching Saturn impossible. The final propellant estimate looked so small, that Kohlhase began to question its accuracy, giving himself some bad dreams along the way: "I dreamed that somehow we did something wrong and that we'd missed some terms, and the actual demands were bigger and the spacecraft just couldn't get to Saturn." Of course, the positions of the planets and their moons were of great

importance. However, unlike any other mission before, there was a lot more to take into consideration when calculating the trajectories of the Voyagers. The team had to take precautions to minimize any conceivable environmental risk, like near-Jupiter radiation levels or near-Saturn ring particle hazards, for example. Both risks had the potential to throw a monkey wrench in the works. Solar conjunction was avoided since it was known that, if the Sun was between the spacecraft and the Earth, the communication would be disrupted. Even small arrival time adjustments were taken into consideration to cause key scientific milestones to occur when radio telescopes on Earth were well positioned to receive the transmissions. The number of conditions that were applied exceeded any prior research done to identify the feasibility of multi-planet opportunities, making the level of detail and thoroughness unprecedented.

Titan

An important driver for the trajectories was already determined before the project received an "OK go" in 1972. Back when the team was preparing to sell the Jupiter-Saturn mission, a need for something more exciting was felt. How could they make it scientifically more interesting and rewarding? Although reaching the planets was the primary objective for the engineers and mission planners, the scientists who were designing the onboard instruments were easily as interested in the moons of Jupiter and Saturn as they were in the planets themselves. Moreover, when it comes to being scientifically interesting, it does not get more

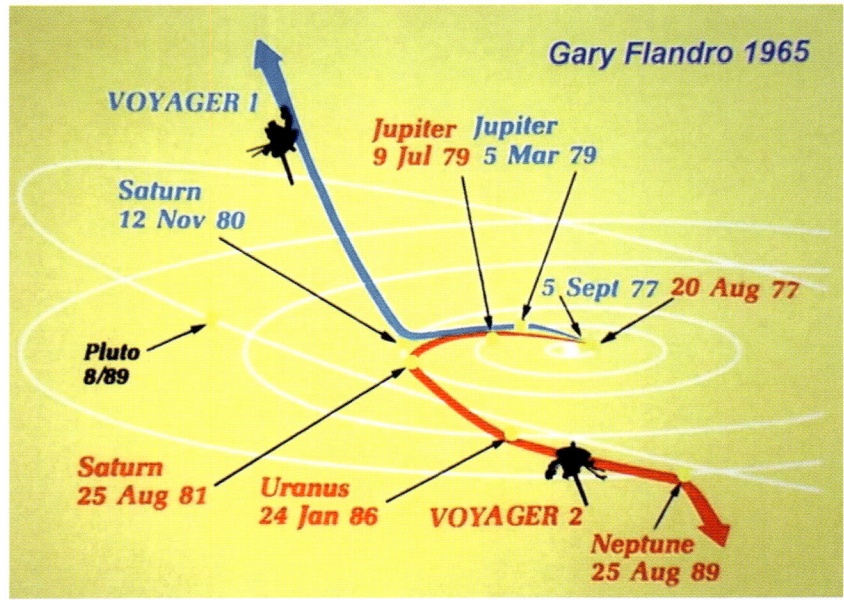

Original diagram showing the upward trajectory of Voyager 1 at Titan, which would mean missing the possible Pluto flyby.

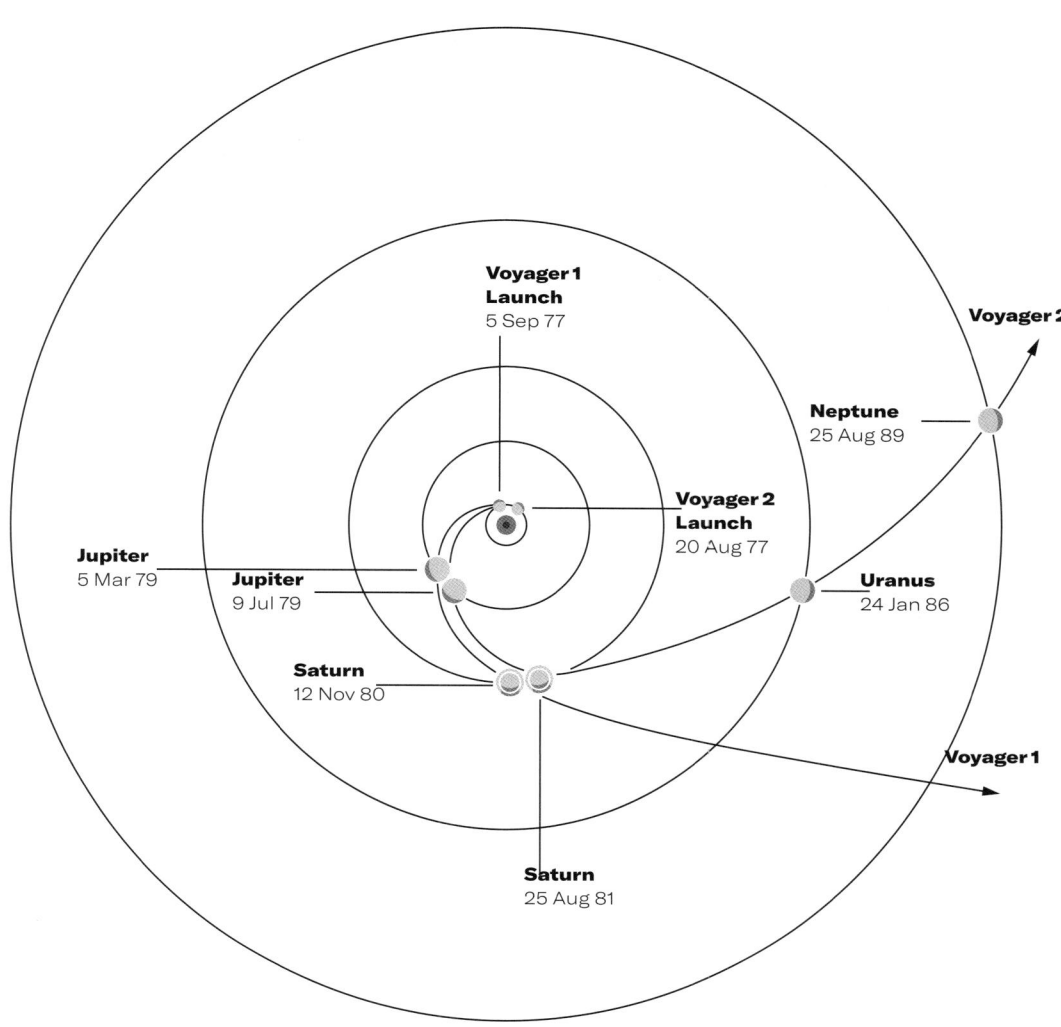

The trajectories that allowed both Voyager spacecraft to visit the four gas giants and exit the Solar System. Also stated is the date of each spacecraft's closest encounter with the planet.

interesting than Saturn's moon Titan. Known to be shrouded in a dense nitrogen atmosphere like Earth, this moon has long promised the scientific community potential clues to the chemistry of Earth, way back when life emerged some 3.8 billion years ago. A Titan flyby had the appeal they were looking for at the time.

Therefore, it was clear that for the first flight they would concentrate on getting Jupiter data, Saturn data and Titan data. To reach Titan, the first spacecraft would need to be deflected upward by its encounter with Saturn, out of the elliptical plane upon which all the planets orbit the Sun. If this were to be the case, and a close flyby to Titan would be made, the team knew this would be at the cost of a Pluto visit. However, given its scientific relevance, nobody at JPL had to think twice about making that sacrifice.

The Final Trajectory

In the end, the team came up with two routes designed to take the spacecraft to Jupiter and Saturn. Voyager 1 was initially targeted to pass by Jupiter with

a close flyby of its moon Io, and then follow its route to Saturn, which included a close flyby of its moon Titan. The trajectory of Voyager 2 did include Jupiter with all its moons, but headed for Saturn without a close flyby of Titan. Voyager 1 would therefore navigate a slightly shorter path to the first two gas giants as its twin probe. Even though Voyager 1 was set to launch three weeks after Voyager 2, by the time the two spacecraft were ploughing through the main asteroid belt between Mars and Jupiter, Voyager 1 would pass by Voyager 2 and take the lead.

Saturn, as we now know, was not bound to be the last stop on humanity's greatest journey. The mission within the mission was very much taken into account. The choice to include a Titan flyby made it impossible for Voyager 1 to also head for the Ice Giants, or Pluto, which had also been considered. However, Voyager 2 was given a pathway that made it the chosen one to aim for Uranus and Neptune. With their perseverance, attention to detail and determination, Kohlhase and his team made sure that when the Voyager mission was to launch in 1977, Flandro's dream of the Grand Tour was very much alive and kicking.

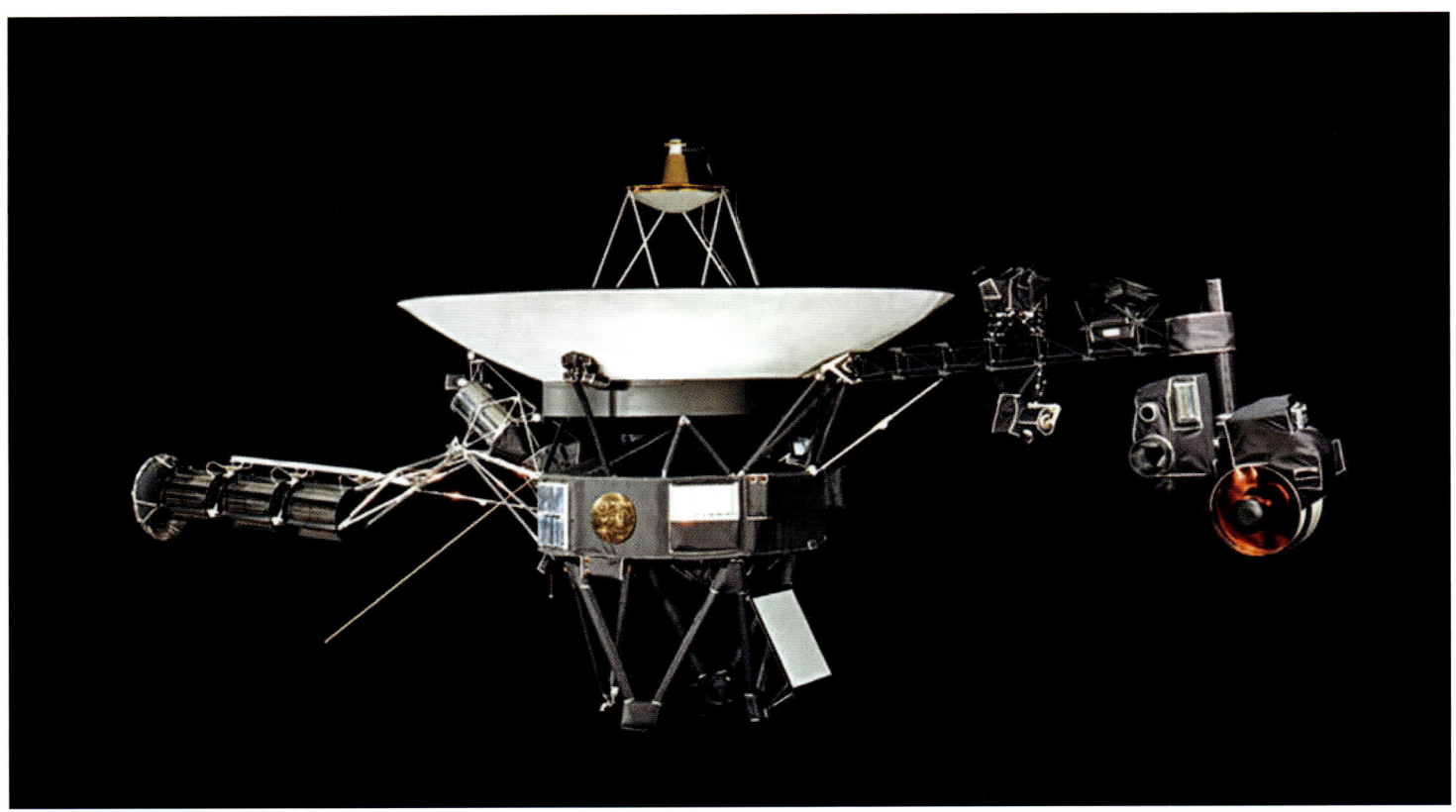

SPACECRAFT

Introduction

The Voyager spacecraft is quite an odd-looking creature. It is clear that it did not come out from the styling garages of Enzo Ferrari, nor did it take inspiration from the slick and seamless spacecraft seen in Stanley Kubrick's 2001: A Space Odyssey, released in 1968, nine years before the launch of both Voyagers. Its design however does have a strange beauty, perfectly fitted for the role of interplanetary exploration, capable of traveling the extreme distance that no other human-made object has traveled before. It is an insect-like probe, with long limbs that carry the scientific devices and cameras as our eyes and ears in space. It also carries the protruding antennas and satellite dishes to keep Voyager in contact with Earth.

The Voyager Program was conceived and commissioned before Neil Armstrong had stepped foot on the Moon back in 1969. In addition, as the mission funding was reduced and the mission goals of the Grand Tour being shortened, the design team at JPL in Pasadena, California, had to build an affordable vehicle that would be capable of traveling to the Saturnian System. The head of JPL at the time, William Pickering, suggested that the Voyager engineers revisit one of JPL's most successful spacecraft designs.

He was talking about the Ranger Lunar probes, which were designed in the late 50s by JPL for the purpose of taking exploratory photographs of the lunar surface to be used for the forthcoming Apollo missions. The Rangers were of a simple and robust construction that could easily be retrofitted for new missions. The Mariner probes and Viking Orbiters were all spin-offs from its earlier design. Pickering instructed the Voyager team to use these earlier models, and to design and reconfigure them for its deep space journey. A nuclear-powered generator replaced the original Mariner's solar panels, which would not provide enough power to support the low-light conditions seen at Saturn. The engineers would add a long extension boom that would house the scientific instruments needed for the voyage.

Although the probe was commissioned and built for a trip to Saturn, the engineers secretly made sure that the craft they were to build could continue the journey deeper into space, capable of making the trip to Neptune and beyond. The build and construction of the Voyager probes have been compared to the tradition of Isambard Kingdom Brunel, whose engineering feats were built to be fit for purpose, long after their primary use or lifespan had passed.

The solid construction and ingenious way the spacecraft were built made sure it could survive the extreme distances it would need to travel. The technology was designed and programmed in a way that made sure that any potential errors or faults could be self-autonomously detected. Moreover, the number of additional fail-safes created to make sure the engineers could fix the craft remotely and remain operational were also put into place.

Navigation

A trip across the Solar System to end up swinging by Neptune's atmosphere by a few thousand miles requires some razor-sharp navigation and tracking capabilities. As JPL describes on their website, "The Voyager delivery accuracy at Neptune of 62 miles (100 km), divided by the trip distance or arc length traveled of 4,429,508,808 miles (7,128,603,456 km), is equivalent to the feat of sinking a 2,260 mile (3,630 km) golf putt, assuming that the golfer can make a few illegal fine adjustments while the ball is rolling across this incredibly long green."

Voyager would send out and receive constant, precisely timed radio signals to and from Earth. Voyager navigators would be able to measure the time it took for these signals to make a round trip to accurately calculate the probe's location within the vast openness of space. If the navigators noticed that the craft was wandering off course, or a new flyby route was required, they could make these "illegal adjustments" by firing a series of small thrusters that would make fine, tiny corrections to put the probe back on track.

Other devices, such as the Canopus Star Tracker were fitted on board to make sure that the attitude and craft orientation were correct. This instrument would search for Canopus, one of the brightest stars in our night sky and calculate Voyager's orientation in relation to the star's position.

Data Communication

The Voyager team knew that the mission would only be successful if it could maintain clear and constant communication with the spacecraft along the 4.4 billion miles (7.1 billion km) long journey to the outer Solar System. The scientific data, as well as its location and the craft's operational status was relayed back to Earth via the dominant 12-foot (3.7 m) brilliant white satellite dish, which was mounted on the Voyager craft to broadcast the information back. The dish would also receive the trajectory corrections, adjustment data, imaging sequences for each of the flybys, as well as any hardware and software reprogramming if any faults or errors developed on board.

Radio signals sent from Earth were sent through the S-band of the microwave communication spectrum. Voyager would normally communicate back to Earth via the X-band of radio frequencies. The craft could also communicate back via the S-band frequency in case a problem was to develop with the high gain X-band signals. The "beam width" spread of the X-band broadcasts from the Voyager satellite dish was just 0.5° of arc. However, this signal that would emit from the 12 foot (3.7 m) dish would spread out nearly 5.6 million miles (9 million km) wide after 620 million miles (1 billion km) of travel.

These signals had to travel unfathomably vast distances, and as Voyager traveled farther away, the signal strength became much weaker. Major upgrades were made to the series of huge satellite dishes that were positioned around the globe from NASA's Deep Space Network, tasked with sending and receiving the communications from both Voyagers.

As the spacecraft traveled behind a planet or one of its moons, all radio communications were blocked out, and all data would need to be recorded and stored, and communicated back to Earth once the craft was in signal range again. When the spacecraft were built, there were no "flash drives" or memory disks. The most advanced technology at the time to store data was the simple 8-track tape. Something similar to the original tapes that made it possible to listen to your own music on those lengthy road trips. The tape player/recorder had just enough memory to store around one hundred pictures.

Science

What do we want to know about our distant neighbors? This was the question given to Voyager's Principal Investigators (PIs). This collective of scientists was tasked with submitting the questions they wanted answered about the four planets to which the probe would travel. This list of questions they collated allowed for the scientists and engineers to know, and to prioritize, which scientific instruments would be of the most value to carry on board a spacecraft with limited space.

Since the scientists would all want to have "their" instrument pointing to a particular target as the probe was flying by during each encounter, the biggest challenge was not fitting all the instruments onto the craft, but working out how they could all operate simultaneously. This gave mission planners the complex task of juggling around the instrument sequences and make sure that the maximum amount of data could be brought back to Earth.

In the end, ten scientific instruments were placed on board. Besides this, an additional experiment that would measure the atmospheres of the planets was integrated within Voyager's main radio communications transmitter. Some of the key Voyager instruments were installed on the "scan platform" situated on the end of one of Voyager's booms. The scan platform could move and point to a particular target with incredible pinpoint accuracy. It was crucial that the platform could point the imaging camera to its subject within one-tenth of a degree. This amazing feat of engineering may sound extreme, but was completely necessary when you are trying to take a photograph of a tiny section of a planet tens of thousands of miles away, while traveling over 35,000 miles per hour (56,000 km/h). To ensure that the images brought back were not blurred by motion, they had to engineer the scan platform to compensate for the spacecraft's movements. The scan platform had to steadily move thirty times slower than the hour hand on a clock.

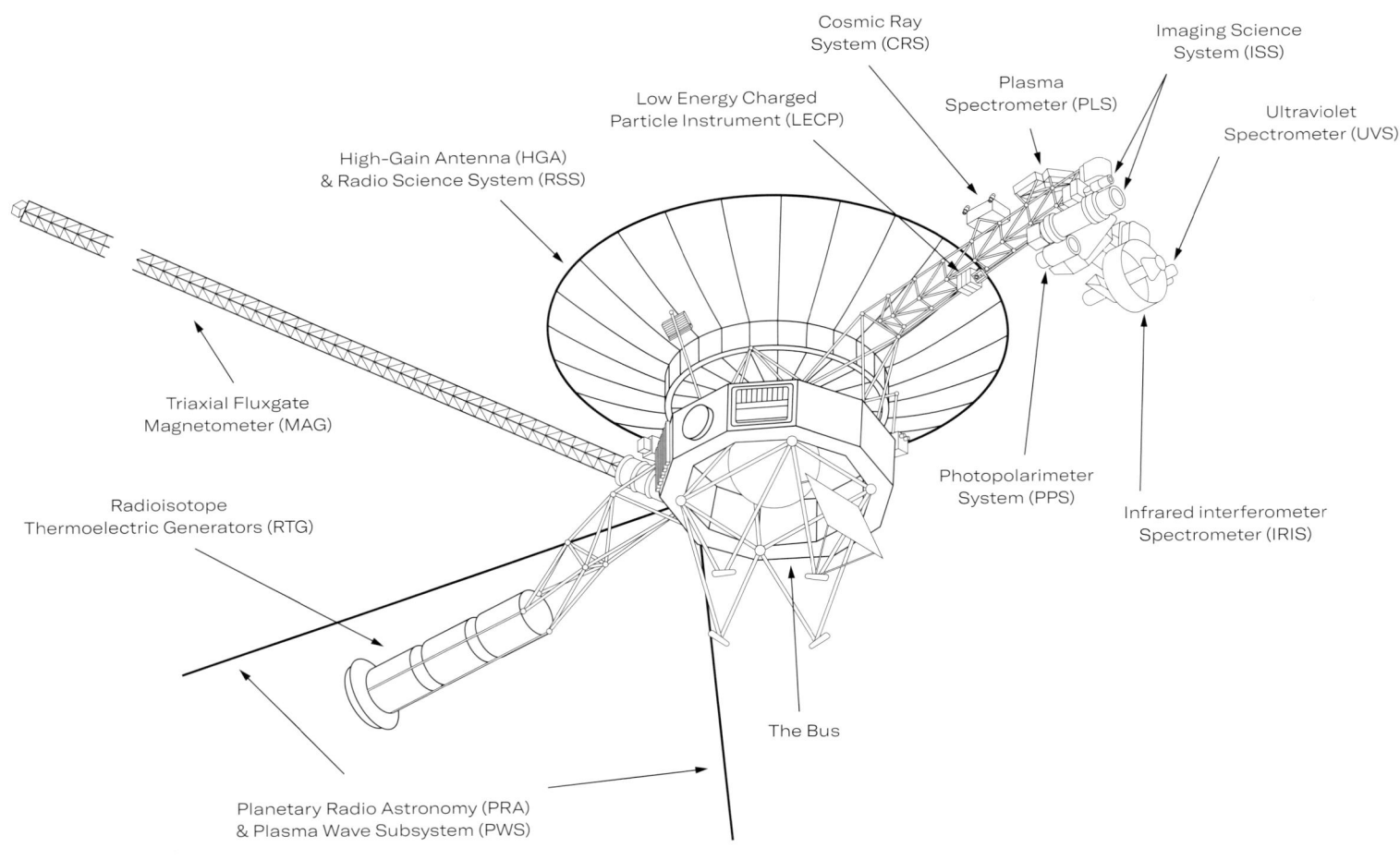

Cosmic Ray
System (CRS)

Low Energy Charged
Particle Instrument (LECP)

Plasma
Spectrometer (PLS)

Imaging Science
System (ISS)

Ultraviolet
Spectrometer (UVS)

High-Gain Antenna (HGA)
& Radio Science System (RSS)

Triaxial Fluxgate
Magnetometer (MAG)

Radioisotope
Thermoelectric Generators (RTG)

Photopolarimeter
System (PPS)

Infrared interferometer
Spectrometer (IRIS)

The Bus

Planetary Radio Astronomy (PRA)
& Plasma Wave Subsystem (PWS)

Spacecraft Design

Each of the Voyagers was constructed from around 65,000 parts, each weighing 1,704 pounds (773 kg). The scientific instruments mounted onto the probe made up one-seventh of the probe's mass. The "Bus," a ten-sided shape made of ten compartments, would contain a number of science experiments, radio transmitters and computer hardware. A backup of each device was installed in case of hardware failures occurring during the mission. The higher than anticipated radiation levels from Jupiter recorded by the Pioneer probes meant that the Voyager builders had to add extra metal shielding around the bus to stop the lethal radiation from damaging the internal hardware.

The Voyager probes are three-axis stabilized spacecraft, which means that Voyager flew in a fixed, steady position. This method of stabilization was an important design requirement as it meant that the camera could be held in a steady fashion and allowed much higher-quality images to be captured. The earlier Pioneer craft were spin-stabilized, which meant that the probe was constantly rotating, using a gyroscopic action to make sure that the heading and direction of the

craft remained on course. A rotating spacecraft however makes it much more challenging to take a photograph when the camera is always moving.

On board Voyager, three computer systems and software were custom-built for the project, which were some of the most powerful and innovative technologies of the time. The Computer Command System (CCS) was responsible for controlling the overall mission systems. The Flight Data System (FDS) controls the flight of the spacecraft and the Attitude and Articulation Control System (AACS) controls the spacecraft orientation, making sure that the high-gain antenna (HGA) is pointing towards Earth for communication signals. It would also control the attitude maneuvers and the positioning of the scan platform in order to achieve the targeted science objectives.

If the AACS had to make a change in trajectory, or the HGA to be realigned with Earth, it would ignite the small hydrazine thrusters that were mounted on the side of the bus. Hydrazine is a monopropellant fuel, not needing another chemical to mix with to ignite. The limited volume of the hydrazine tanks meant that the Voyager team had to be very careful not to overuse the thruster fuel during the mission.

High-Gain Antenna (HGA)

This antenna transmitted data back to Earth through both the X-band and S-band frequency channels. The X-band channel sent science and engineering data. The S-band data transmissions were used to communicate engineering data, as well as the operational status and health of the probe.

Imaging Science System (ISS)

The eyes of Voyager were a system consisting of two cameras. One with a narrow-angle lens, and the other with a wide-angle lens. It had the power to clearly photograph a newspaper headline from a distance of 0.62 mile (1 km) due to the resolution of the narrow angle camera. The vidicon detector would capture black-and-white images through different color filters mounted on a filter wheel, allowing the imaging scientists back home to recreate the color images just like the ones that make this book.

Planetary Radio Astronomy (PRA) and Plasma Wave Subsystem (PWS)

Two separate experiments fitted onto the two long V-shaped antennas. The PRA and PWS detected the various radio signals emitted by Jupiter and Saturn to gain a greater understanding of the planets' dynamic energy output.

Infrared Interferometer Spectrometer (IRIS)

Investigated the global and local atmospheric composition, and the energy balance between the solar energy absorbed into the planet against the amount of internal energy radiated from the planets and their moons. The varying temperatures at different atmospheric heights were also recorded as well as the thermal properties and the composition of Saturn's rings.

Ultraviolet Spectrometer (UVS)

A sensitive light meter that searched and detected specific colors of ultraviolet light that certain chemical elements and compounds give off. By detecting these specific ranges of UV light, it determined when particular ions and atoms were present. The UVS was the longest operating instrument on both spacecraft, as the others were gradually switched off over the mission duration.

Radioisotope Thermoelectric Generators (RTG)

Three nuclear-powered generator units supplied Voyager 1 and Voyager 2 with the electricity it would need to keep its instruments and computer systems fitted on board during the 4 billion plus mile (6.4 billion km) trip across the Solar System. Each RTG unit had a plutonium core; the decaying isotope particles released heat energy, which was converted by the thermoelectric converter into electricity.

Low Energy Charged Particle Instrument (LECP)

The differences in energy fluxes and other properties of ions and electrons, such as interplanetary and solar energetic particles, were measured by this particle measuring system.

Cosmic Ray System (CRS)

This system looked for the very energetic particles in plasma often found in the extreme radiation fields such as Jupiter. The CRS also searched for other extremely high-energy particles that are emitted from other stars. This data revealed information on the energy content, acceleration process, and the life history and dynamics of cosmic rays in the galaxy.

Radio Science System (RSS)

This system utilized the crafts telecommunications to determine the physical properties of the ionospheres, atmospheres, mass, gravity fields and densities of the gas planets and their moons, as well as detecting and measuring the ring system around Saturn.

Photopolarimeter System (PPS)

This system was constructed with a 6-inch (15.2 cm) compact telescope with a series of analyzers and spectral waveband filters that would be able to scan each of the planet's cloud surfaces for composition and texture. The telescope could also provide density and atmospheric scattering data at all four planets. The PPS would also scour the planets for any evidence of lightning strikes and auroras.

The Bus

The chassis of the spacecraft, encased within the ten-sided box, in separate compartment bays each housing the various engineering subsystems and computers. The hardware for the radio transmitters were contained in bay one, for example. The centerline of the bus formed the Z-axis. The probe was designed to roll about this axis by firing the small thrusters placed on various edges of the bus frame.

Triaxial Fluxgate Magnetometer (MAG)

This system recorded the magnetic fields of the planets, moons and planetary rings. It also discovered what influence the solar winds would have on each of the planets' magnetospheres. Lastly, it measured the properties of the interstellar magnetic field once the Voyagers flew beyond the range of solar influence.

Plasma Spectrometer (PLS)

This device investigated the macroscopic change in the properties of plasma ions as the spacecraft would travel through the Solar System. The PLS was the first device to detect the heliopause boundary and register the first observations of plasma from interstellar space.

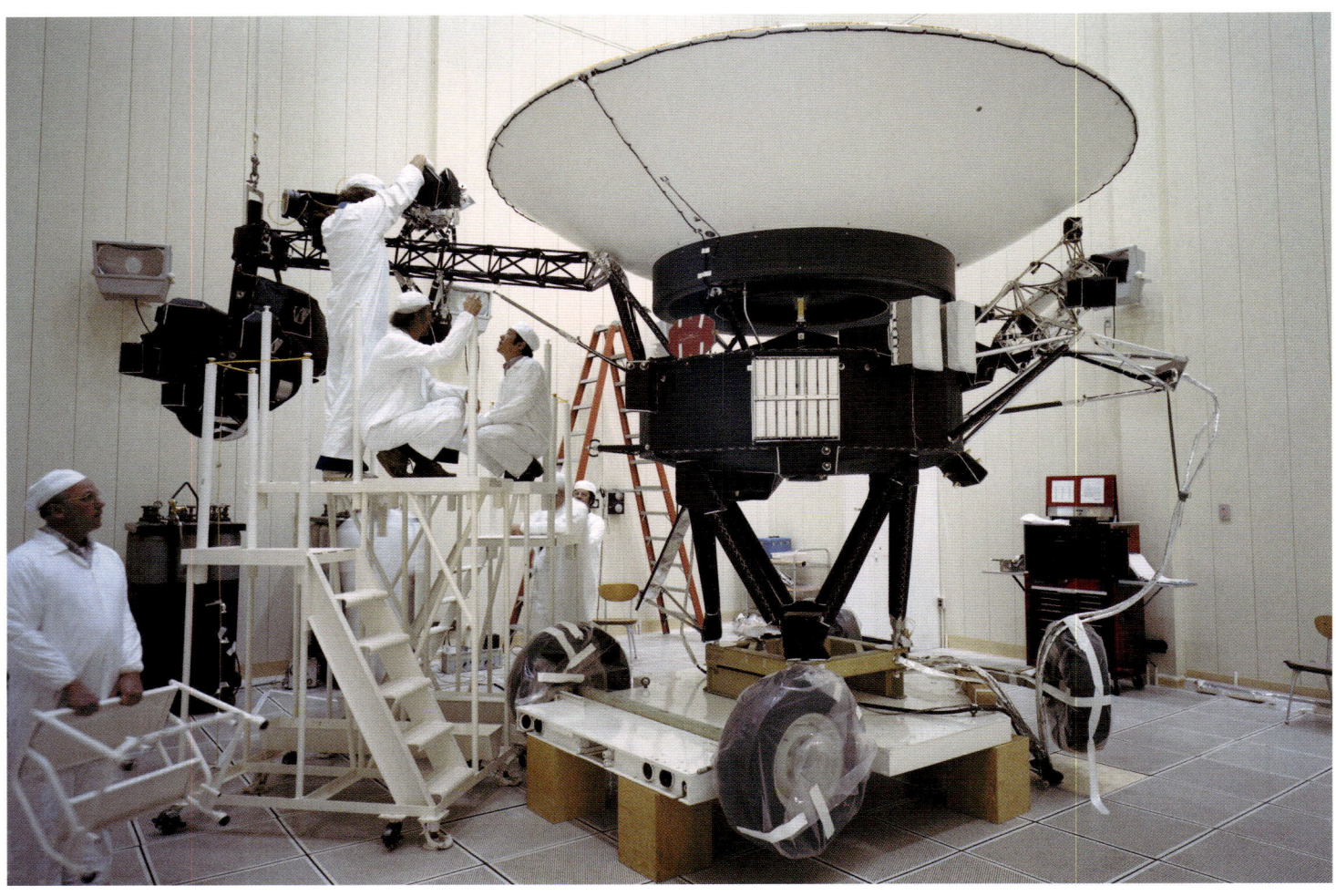

Build

Voyager was designed and built by JPL in Pasadena, California. As part of NASA, JPL has been at the center of robotic space exploration since the earliest beginnings of our ventures into space. Three Voyager spacecraft were made inside Building 179, the iconic "High Bay" spacecraft assembly area. This was the same facility where the Ranger, Mariner, Viking, Galileo, and Cassini spacecraft were built.

There were no robots or help from 3D printers to build the most difficult and technologically advanced spacecraft that JPL had built up to that point. It took over 1,500 engineers and more than five years to design and build, not to mention the $200 million budget spent to get the two probes onto the launch pad. A diverse range of expertise was called upon from all over the world to build the spacecraft, from thermal electrical systems and software engineering to material scientists, planetary and space physicians, human resource managers, welders, wire bundle routers, and machine tool handlers. The project also consisted of subcontractors and equipment suppliers from all over the USA, down to universities around the world, whose staff and students built the many instruments to perform the scientific experiments during the voyage. Just to name a few.

Three spacecraft were constructed. VGR77-1 was assigned as the Proof Test Model (PTM). This craft was the one used to run tests in simulated deep space conditions, to make sure the design, construction, and durability could be able to withstand the journey. VGR77-2 was the second built and was earmarked to become Voyager 2, with VGR77-3 being assigned the role of Voyager 1.

Although the PTM was tested more rigorously than the two flight models, a series of tests were run to make sure all the craft were built to requirements. Individual instruments and the entire spacecraft were put on vibration tables to simulate violent shaking, far stronger than what the spacecraft would endure during takeoff.

As the spacecraft construction was finished in Pasadena, VGR77-2, and slightly later, the VGR77-3 were sent in parts to Cape Canaveral, Florida. From there, NASA crews reassembled and tested the two spacecraft again. The last step in the process was to fold the camera and instrument boom down alongside the bus. With the satellite dish facing up, a metal capsule was lowered to encase the probe, making it ready to be mated with the rocket that would launch it out of Earth's atmosphere.

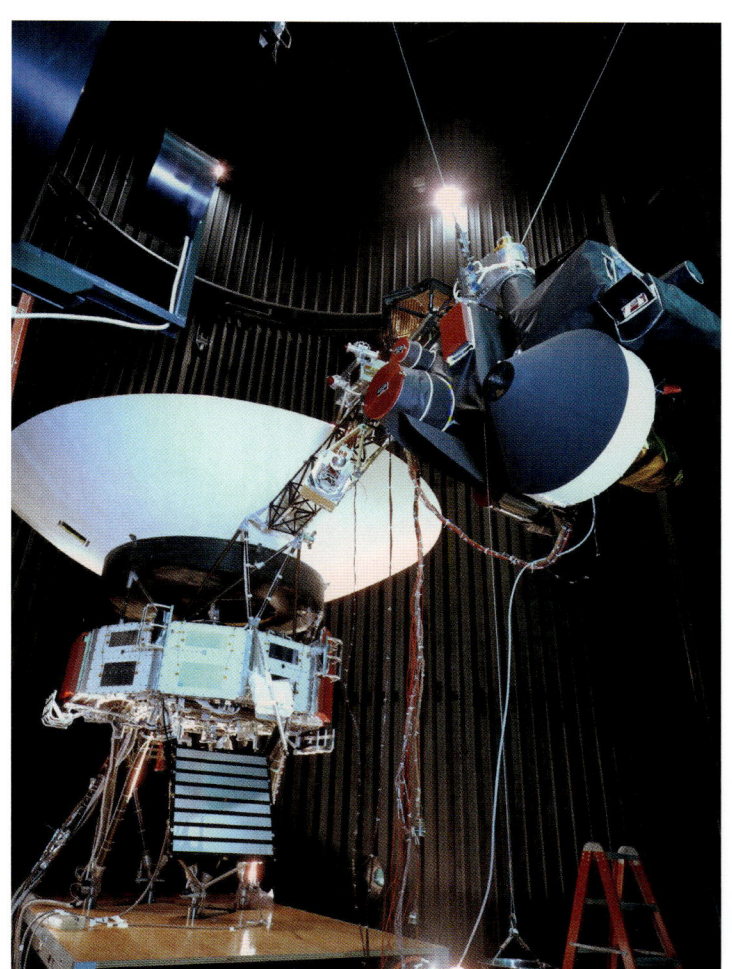

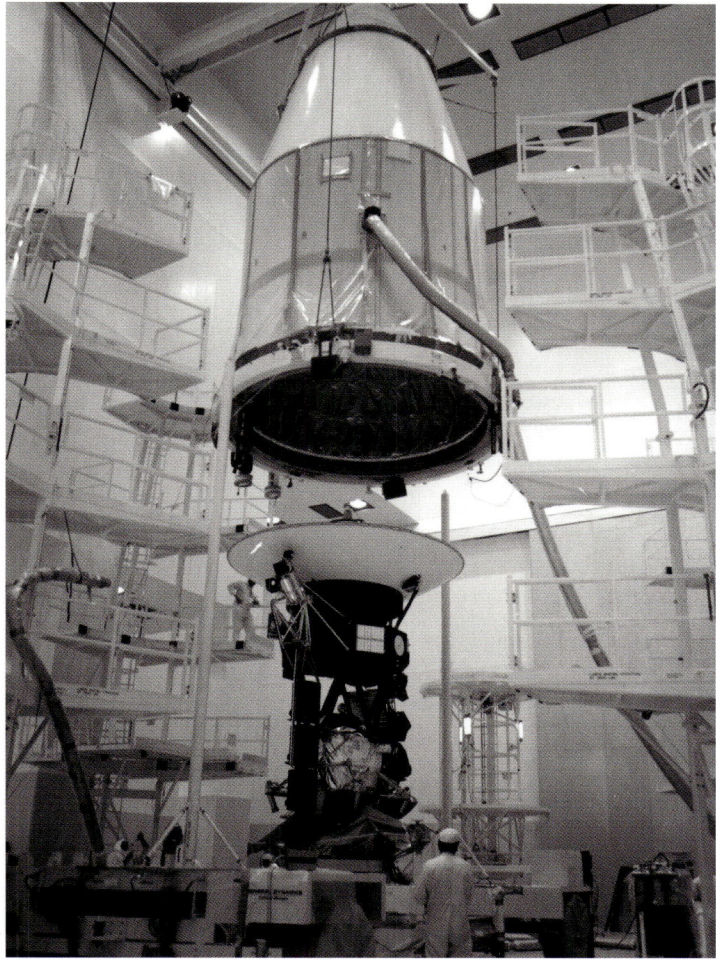

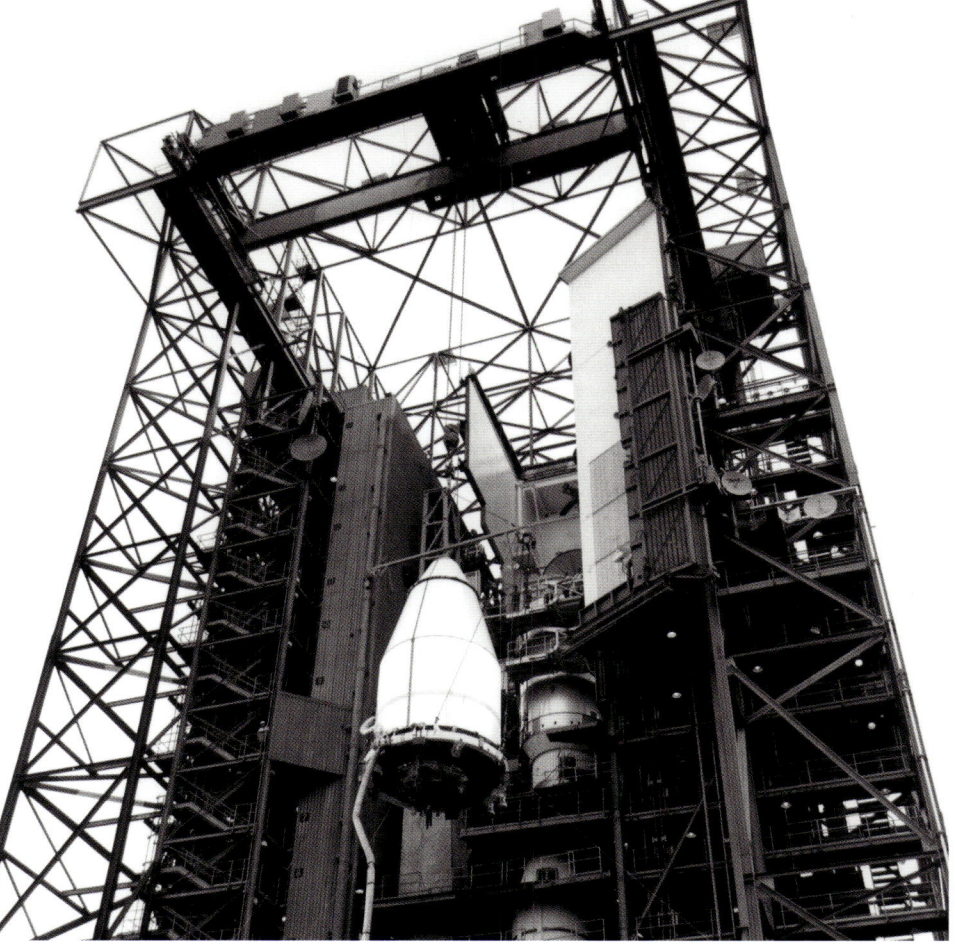

Left: engineers testing the Voyager 2
spacecraft on March 23rd, 1977.
Above: Voyager proof test model,
which did not fly in space.
Right: Voyager 2 spacecraft during
the encapsulation stage at NASA's Kennedy
Space Center in Cape Canaveral, Florida.
Below: Voyager 2 being mounted atop the
Titan/Centaur-7 launch vehicle.

PICTURE THIS

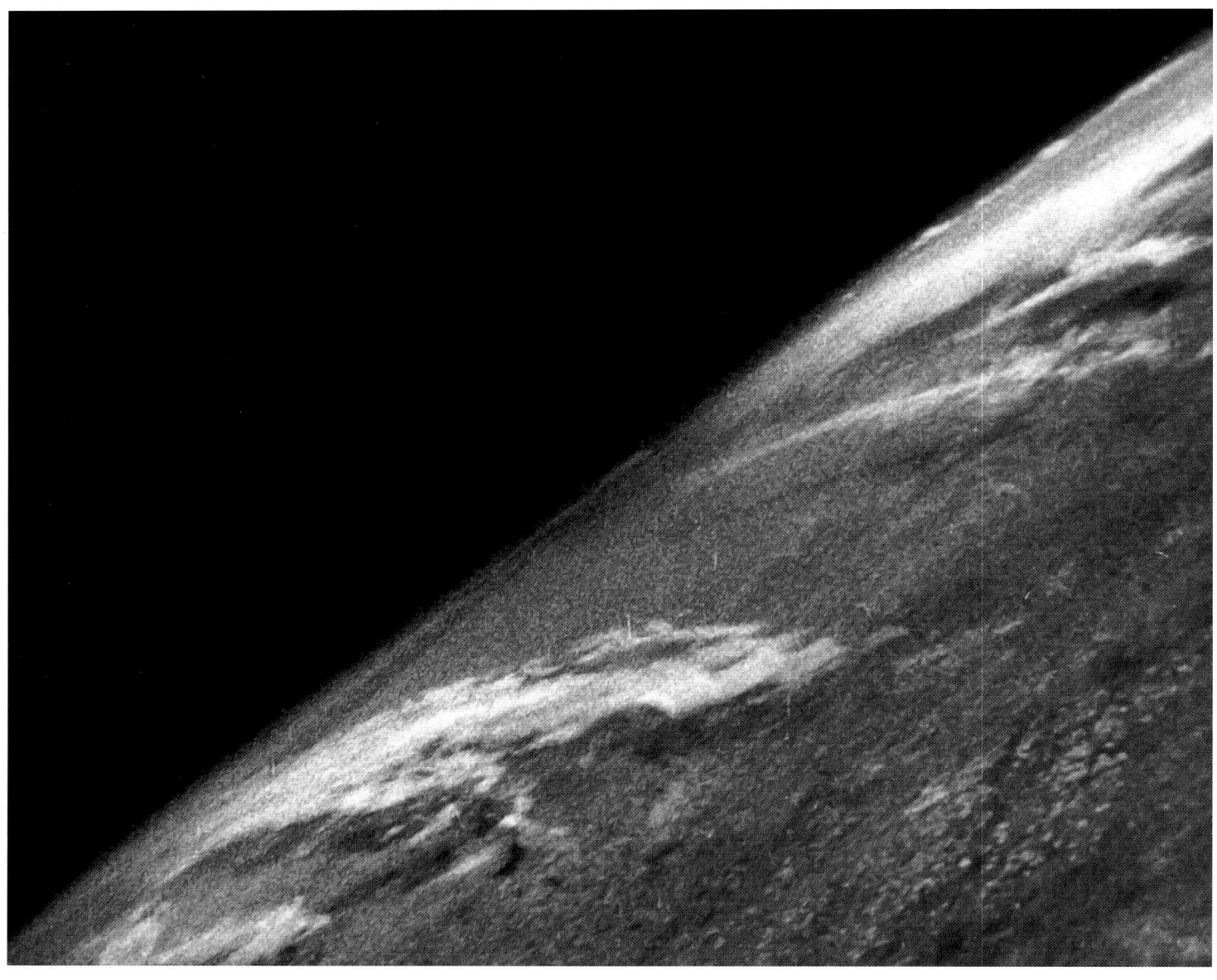

The first photo taken from space was captured on October 24th, 1946 from a sub-orbital US-launched V-2 rocket at White Sands Missile Range. The highest altitude (65 miles, 105 km) was 5 times higher than any picture taken before.

THE POWER

An Appealing Image

What is it exactly that makes an image powerful? Scientists have been on a search for the answer to that question, leading to interesting insights. Apparently, it is a matter of memory and processing. People generally remember 10% of what they hear, 20% of what they read and 80% of what they see. Of all the information we take into our brain, 90% is visual. We can understand what an image means in 13 milliseconds, less than the blink of an eye, and we process visual input 60,000 times faster than text. The power of a picture, so it seems, is built into our sensory experience.

A single image can tell a story to billions of people. It transcends language as well as physical barriers, and it requires no prior knowledge of the subject being portrayed. By reaching out to so many people, images have played a crucial role in documenting the history of humankind. From cave paintings to engravings, and from woven illustrations to the first photograph, visual communication throughout the ages has evolved immensely.

Advertising

The development of advertising has made sure that people are constantly surrounded by imagery. This started in the beginning of the 20th century when black-and-white photography started being used for print advertising in the local and national newspapers. Pictures were seen as a preferable communication form because they performed the cardinal principles of advertising: attract attention, arouse interest and create desire. Fast-forward to today and we see everything in brilliant color and high definition. From glossy magazine pages to product packaging, and from sides of moving buses to the screens of our smartphones, eye-catching images are the key to stop people in their tracks or halt their news feed scrolling.

Chances are that in one of those occasions, whether it is on the wrapping plastic of a candy bar or a magazine ad, a wild and distant planet is portrayed, like Saturn and its rings. Although we take the fact that we know what these planets look like for granted today, portraying them would not have been possible without the Voyager mission and the images they captured for the first time. Their impact was huge. When it comes to the power of a picture, planets are almost a category in itself.

Picture a Planet

For a very long time, portraying a planet visually was a guessing game. Although we may see a big blue marble in a split second when someone says "Earth", knowledge about the size and shape of our planet was not that obvious to the 100 billion people that have lived before us.

Although Pythagoras is generally acknowledged to be the first to recognize that the Earth is spherical, it was his colleague Eratosthenes that proved it 250 years later by performing one of the most famous scientific experiments of all time. At noon, he placed a stick in the city of Syene, that did not cast a shadow because the Sun was directly overhead, while at the exact same moment a stick was placed in the more northern city of Alexandria, which did cast a short shadow. From this, he deduced the only possible explanation, that the Earth was spherical.

The Impact

2200 years later, in the era of space travel today, we can actually climb on a spacecraft, leave our planet, turn around and have a look. This was done for the first time in the late 1940s, not by ourselves but by a camera. Suborbital German V-2-Rockets, which had been captured by the US Army after World War II, were shot into space from the White Sands Missile Range in New Mexico, resulting in grainy photographs showing the curvature of the Earth.

It was about two decades later, on August 23rd, 1966, that NASA's Lunar Orbiter 1 spacecraft took the first truly global-scale photograph of our planet, showing a magical black-and-white Earth rising behind the horizon of the moon. With its portrayal in LIFE Magazine and NASA handing them out as posters, the photograph became a huge public relation hit. Two years later, the famous Apollo 8 Earthrise photograph has been credited to inspire Earth Day in 1970 as well as propel the modern environmental movement into the mainstream by illustrating how finite our resources truly are from a different perspective. In just a couple of years' time, it was becoming clear what impact this kind of pictures could have.

Dr. Garry Hunt

For Voyager, given the sheer impact these pictures could have, this made the flybys and the resulting images quite a thrill. With every image that would come in at JPL, history was changed. No one knows this better than Garry Hunt. Based at University College London and the only resident British scientist on the mission, Hunt was selected by NASA as one of the nine scientists that formed the Imaging team. To give us a first-hand account of what it was like when these images would roll in and to share the meaning and value of these photographs, Hunt would be the right person for the job. That is why we asked him to do exactly that for this book. Luckily, he said "Yes."

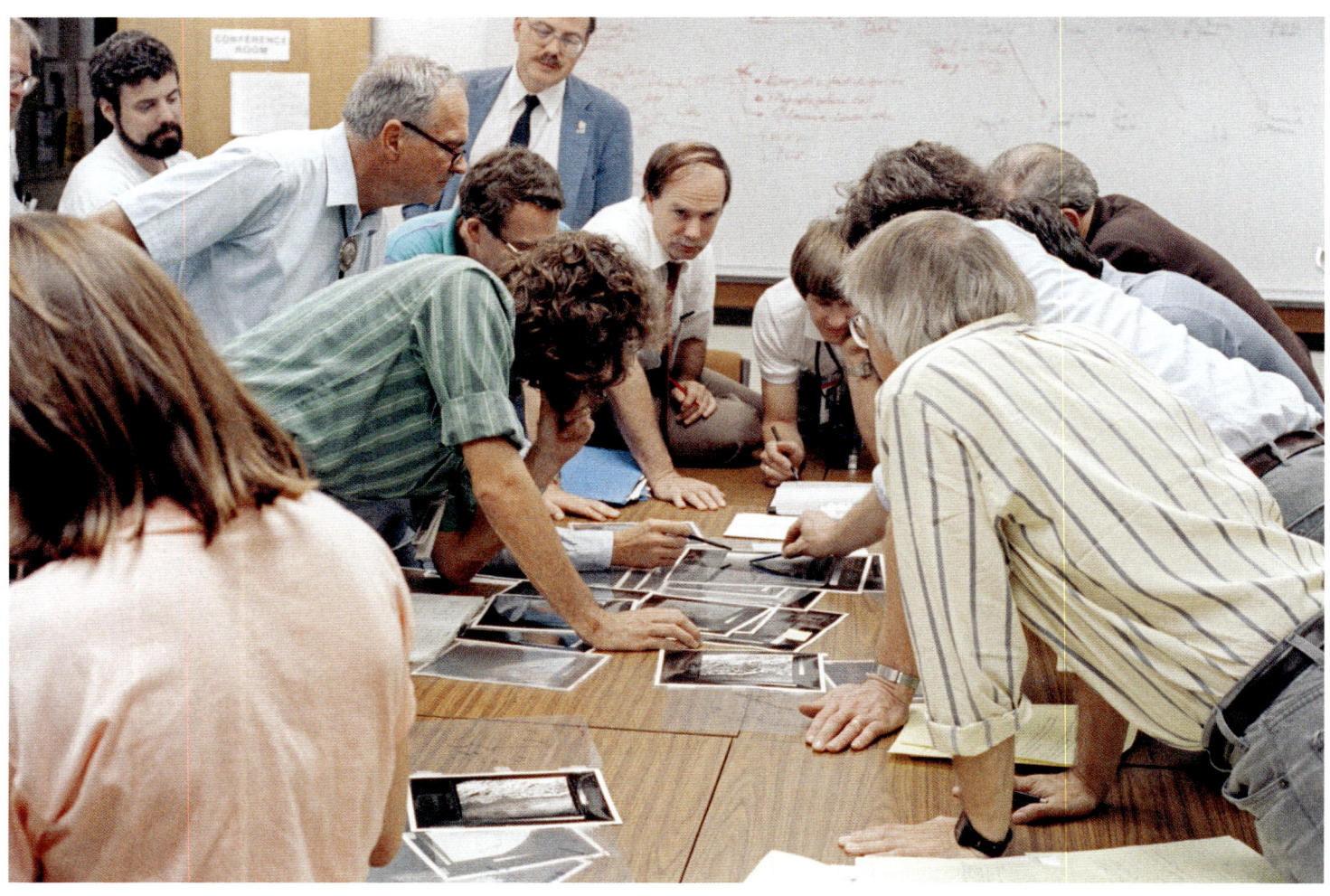

Members of the Voyager science team reviewing recently
photographed images of Neptune's moon Triton in August 1989.

THE EYES OF THE BEHOLDER
BY DR. GARRY HUNT

Pieces of Art

"A picture is worth a thousand words", and the never-ending Voyager mission has taken these words to their fullest extent. This mission transformed the four giant planets, Jupiter, Saturn, Uranus and Neptune, with their rings and satellite systems, which previously had been no more than tiny specks of light in the night sky, into worlds of their own with close-up views from a spacecraft camera.

The thousands and thousands of pictures are certainly wonderful pieces of art, beautiful in their own right. They have been displayed everywhere: in art galleries, on chocolate boxes, on sweatshirts, in advertising, on TV programs, and even on giant roadside advertising billboards. The list is endless. They can be seen in black-and-white, artistically portrayed in every color, combined as collages, some pretending to be real, others clearly false.

These pictures are for everyone to enjoy. They are items for discussion wherever they are displayed.

A Scientific Discovery

Still, the Voyager Imaging Cameras were not just designed to be an artist's palette. The mission was developed to explore the outer Solar System and provide new insight into these planets and satellites. It is a quest to understand their origin, how they have developed, and the origin of the Solar System as a whole. Exploring the planets in our Solar System also provides a unique opportunity to examine the processes that shaped our distant planetary neighbors as well as providing natural laboratories to investigate terrestrial processes under very different conditions from those currently found on Earth.

Hence, the Voyager images were really part of a very detailed science investigation of the outer planets.

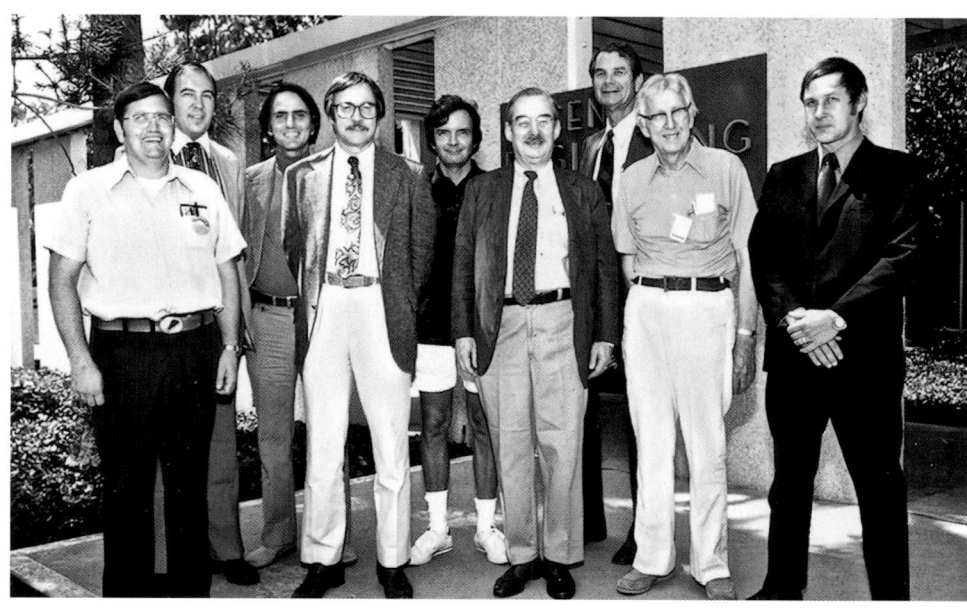

The Voyager imaging team (back row: Garry Hunt, Carl Sagan, Toby Owen, Merton Davies, front row: Ed Danielson, Brad Smith, Allan Cook III, Verner Suomi, Larry Soderblom).

The results and countless new discoveries that continue to be made decades later have changed our understanding of these distant bodies and have rewritten the astronomy textbooks. The cameras are the eyes of the spacecraft, with a critical and essential role in ensuring the correct trajectories are followed. To do so, every encounter and measurement sequence was carefully planned and transmitted to the spacecraft and carried out with precision. A simple, infinitesimal error in the spacecraft's position, particularly early in the mission, would be greatly magnified as the spacecraft progressed further and further away.

Learning from the Past

When snapping a picture in space you immediately think of the Apollo astronauts with their hand-held Hasselblad cameras, or the current astronauts on the International Space Station with their digital SLR camera systems. However, the Voyager Imaging System was nothing like either of these. Voyager is a robot, and the pictures are the result of the spacecraft systems following precise instructions received from the Earth.

For a mission on a path moving further and further into the unknown depths of our Solar System, the Voyager camera system had to be extremely reliable. That was why it was based on tested and certified instrumentation using a modified version of the slow scan vidicon cameras that had been used on previous Mariner Missions. Special improvements were made to the overall signal to noise ratio compared with the Mariner 10 design and at the same time retained as much as possible of the previous proven camera system characteristics to minimize the cost and risk.

Rich Plans, Rich Team

A major concern was the lifetime of the mission. When the mission activities started in 1972, the plan was for the spacecraft to reach Jupiter and Saturn by 1982, a mere ten years. Originally, NASA had no formal plans for an extended mission to include the more distant worlds of Uranus and Neptune, which would require the Imaging System and all the other instruments to continue functioning until 1990 at least. Of course, all the scientists and engineers involved had other ideas! It is important to remember that we are now in this digital world of the 21st century, but here we are discussing a mission that used technology from half a century ago. While this may seem like the distant dark ages, these pictures and the continuing scientific studies using them are proof of its success.

Previously the scientists selected by NASA for the Imaging Investigations had all been astronomers. However, this time it was different. In 1972, NASA selected from the thousands of proposals submitted by scientists throughout the world, nine people as the core team. Eight came from the US and one from the UK (me), but more importantly, four were astronomers, one a geologist, two meteorologists, a geodesist and an engineer. This reflected the wide range of interdisciplinary scientific studies to be conducted during the encounters. Being selected for the Voyager team changed my life, made a huge impact on my research studies as an atmospheric scientist and has added a great deal to my subsequent career. Being a member of this major NASA mission provided a unique opportunity to apply my knowledge of the behavior of atmospheres and their weather systems to worlds very different from the Earth. As the mission progressed from the Jupiter encounter in 1979, additional scientists were added to work with the flood of data.

The first five years were extremely exciting and hectic as we prepared the camera system for the mission launch. An immovable date. I worked as part of the imaging team to develop the camera system, to select the most suitable camera filters, and to develop the best observations to understand the distant objects. In particular, I always made sure we had the most suitable observations to study any atmospheric features. This involved regular meetings with all of the team members. These were the pre-internet days. This meant endless journeys from the UK to locations all over the US, while my US colleagues had just short trips. At the same time, back at home in the UK, I was also creating a world class imaging processing system at University College London. This way, my colleagues and I could analyze the data directly at home.

A Challenging Task

Taking pictures from a robot spacecraft is not as straightforward as the simple point and shoot camera system we are familiar with today. There are considerable differences in the types of observation to be made depending on the specific scientific question being investigated. Some are concerned with the movement of the atmospheric cloud systems and the structure of the atmosphere, which requires repeated pictures of the same position for a period of time. This involves the ability for us to view motions of the entire planet, local motions, and interactions between individual cloud features and their development and dissipation. Moreover, it also required the ability to search these distant objects for features unique to their environment, just like a weather forecaster. However, the observations of the moons were quite different. Until now, the moons were little more than specks of light seen in the best telescopic observations. This was the first opportunity to map their surfaces and carry out the first detailed geological studies of these objects. This required us to make high-resolution pictures, through different filters of the same regions. Only then we could create maps of the entire object, and with the ability of the camera system quickly make further observations of newly discovered areas displaying unexpected features. On this voyage into the unknown, we could not make the mistake of thinking all the moons were likely to be very similar.

The first important step was to generate a timeline of the observations to be made at each encounter, which had to incorporate every possible objective in a conflict free and flexible manner. The centuries of telescope observations of Jupiter and Saturn had provided us some basic information of their visible appearance. Little was known about their satellites beyond the existence of a few moons, while even less was known about the environments of Uranus and Neptune. Throughout the mission, new discoveries were being made and theories proposed, which required new observations to be included in order to examine these new facts and ideas. During the encounters, we found unexpected features, which demanded further investigations. New observational sequences had to immediately be included into a rapidly updated timeline. Somehow, these requirements were added to the robot spacecraft to instruct the onboard instruments how to perform. This was not a simple matter, as the spacecraft were moving away from the Earth at a great speed,

while the one-way communication time from Earth to Jupiter ranged from 35 to 45 minutes, while communication to Neptune took over 4 hours.

The task of the cameras was not limited to just producing pictures for the imaging experiment, they were also providing pictures to support measurements by other instruments mounted on the scan platform. The cameras were the eyes of the spacecraft too, to pinpoint its location by taking regular pictures of well-known stars. During an encounter, pictures were taken of various moons against backgrounds of stars with known positions. This helped to determine the spacecraft location and the time the picture was taken. The pictures were then transmitted to the Earth and JPL via the Deep Space Network of Antennas directly, or first stored on the tape recorders if the Earth and spacecraft were not in line of sight.

Drinking Out of a Fire Hose

Each Voyager encounter was incredibly exciting. The photographs would reveal themselves on control-room monitors, pixel by pixel. When a planet was getting closer, say for example Jupiter, every next picture was the largest picture ever taken of this giant planet. As the spacecraft moved nearer and nearer, the individual planet was giving us a first glimpse of its distant world. This acceleration as we were approaching, and the quick accumulation of incoming data, became a thrilling experience. I remember we referred to it as "drinking out of a fire hose"; you are trying to take a little sip, and this torrent of images is coming out.

This observatory phase was a spectacular time for me, seeing the weather systems grow larger and larger with increasing complexity and beauty. I was able to analyze these data while the geologists waited for the spacecraft to get closer to the satellites. The ever-changing faces of Jupiter and Saturn with their own interacting cloud systems, driven by the powerful weather systems. The Great Red Spot of Jupiter, that whirlpool of emotion, was now seen in detail, interacting with neighboring storms, scientifically meaningful and aesthetically beautiful. Suddenly those tiny specks of light, the previously known satellites seen only through telescopes, became worlds of their own interlaced with countless new features.

A Dark Direction

For the Voyager Mission, this was meant to be the successful accomplishment of Part 1. Actually, it was the start of another big step for Voyager 2, on a pathway to Uranus and Neptune. A trajectory carefully selected from 10,000 possible routes. Extending the mission sounds straightforward. The journey as far as Jupiter and Saturn had been a success for the science teams and spacecraft engineers. Nevertheless, after a very successful Saturn encounter, an unexpected problem arose. The scan platform would not move. Would this

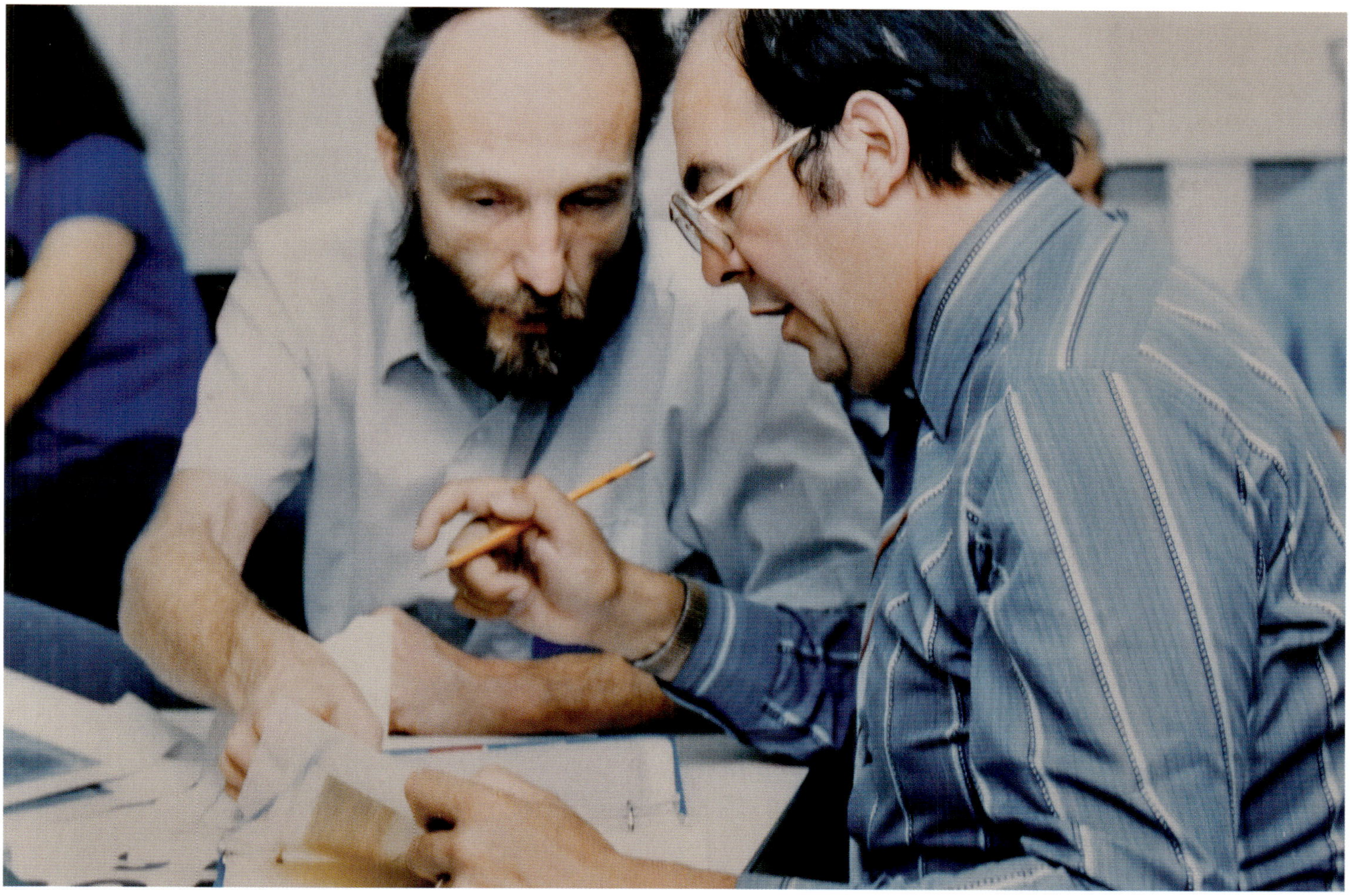

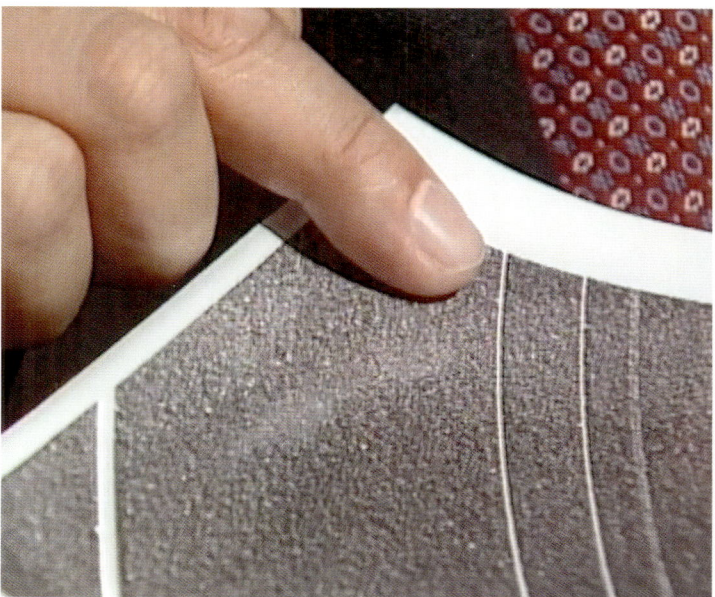

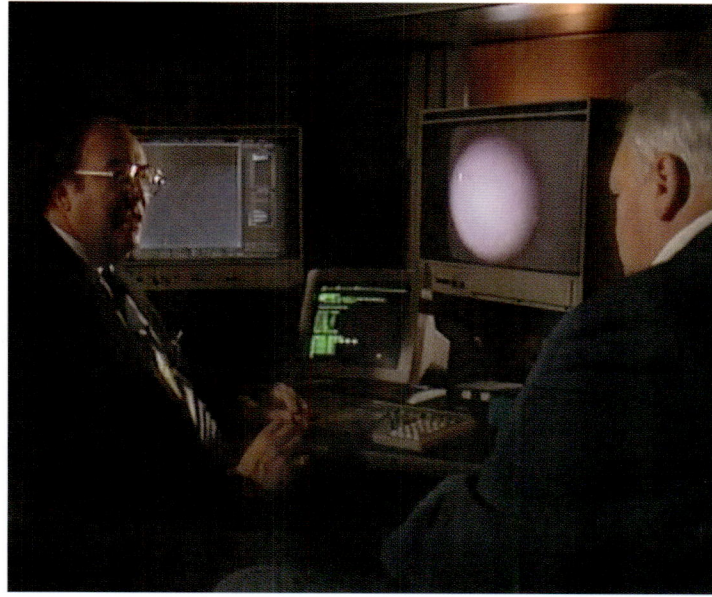

Top: Garry Hunt and Andy Ingersoll (CalTech) discussing the features in Jupiter's atmosphere in 1979. Bottom left: Garry highlighting the faint ring system found in the images taken from the Uranus flyby. Bottom right: Garry discussing the processing of Uranus images at JPL with Sir Patrick Moore for the BBC television program 'Sky at Night'.

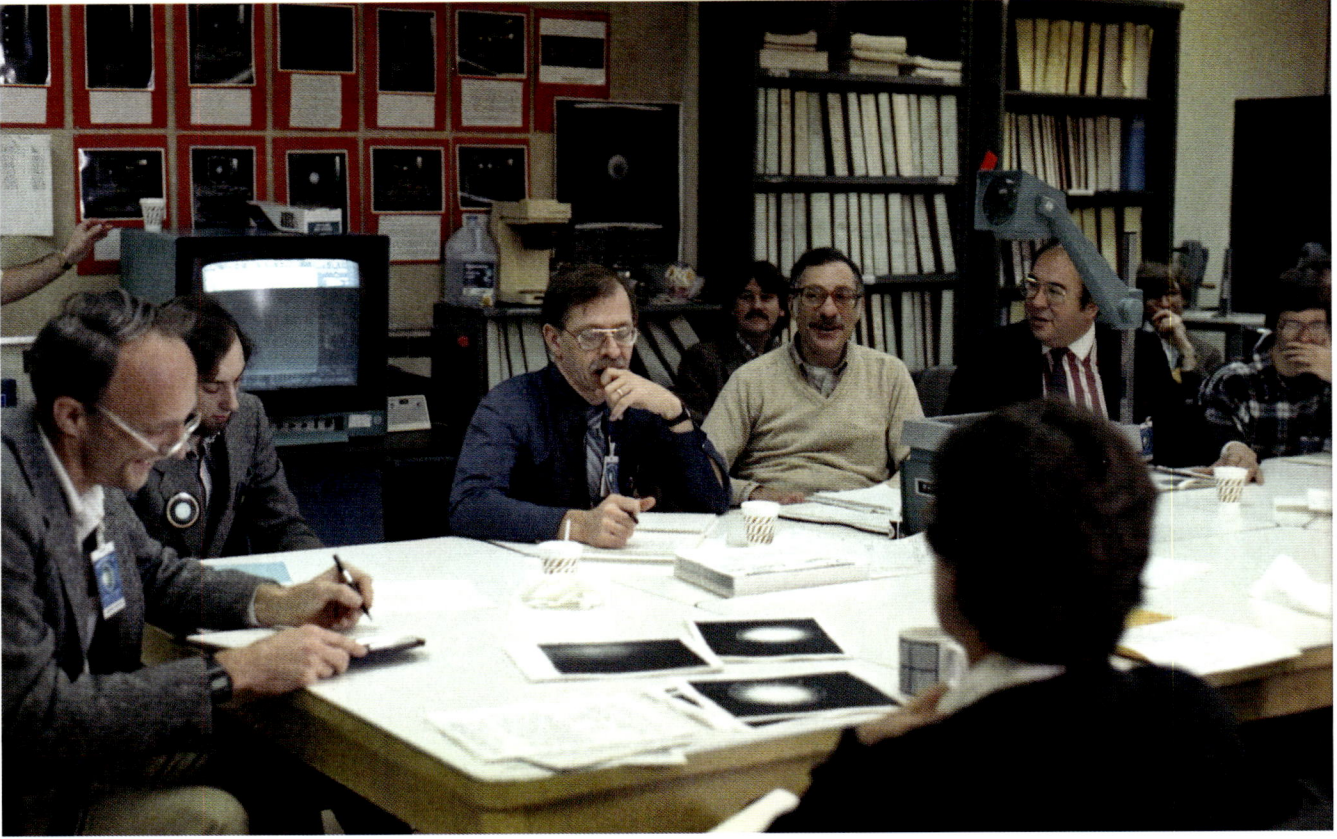

The Voyager Imaging Atmosphere Groups discussing the Uranus atmosphere pictures.

prevent the spacecraft reaching Uranus and Neptune? Of course, we know the answer as the mission was saved by the brilliance of the JPL engineers who would, after days of careful analysis and many sleepless nights, overcome the problem by commanding the scan platform to move in smaller and slower steps, and it worked.

Traveling further and further away from the Earth has the additional problem of the lower light levels when taking the pictures. At Neptune, the light levels are 900 times fainter than those on Earth. This is like trying to take a picture in a darkened room without a flash. Simply increasing the exposure times for each Voyager image was not a sensible option. In addition, the spacecraft is not completely stable and passing moving targets at great speed increases the dangers of smeared images. The spacecraft engineers once more came to the rescue and reconfigured the spacecraft operations between Saturn and Uranus so it would move more smoothly than before, helping to speed up the rate at which pictures were taken. This was equivalent to making Voyager 30 times steadier than the hour hand on a clock. A further development was the use of a new procedure to compress the data more efficiently, thereby introducing a technique to account for the large amounts of dark space and low contrast features in the image that helps to speed up the time to transmit images from Neptune to Earth. The successful accomplishment of these encounters with Uranus and Neptune is a result of the brilliant supporting activities of the flight staff and engineers of JPL.

Timeless Beauty

The Voyager images are all outstandingly beautiful and pushed our scientific knowledge of planetary systems further. It has also raised a wealth of new questions that remain to be answered by future generations of scientists and new missions to these outer reaches of our Solar System.

It is impossible for me to choose a favorite picture. I adore them all. Every day as I look at the many Voyager pictures in my house, I am constantly reminded of the brilliance of the Voyager Mission and especially the camera systems that have brought us so much new knowledge of the outer Solar System, while also posing still more major questions. I am very proud to have played a part in making these pictures that have helped rewrite the astronomical textbooks that you can now enjoy.

Decades later, these pictures are still being studied, resulting in even more discoveries about these distant planets and satellites. The scientific intrigue stimulated by these iconic images continues. The beauty of the Voyager pictures is timeless, and they are now here for you to enjoy in higher quality and resolution than ever before.

Dr. Garry Hunt
Voyager Imaging Team (1972-90)
West Wimbledon, London,
April 17th, 2020

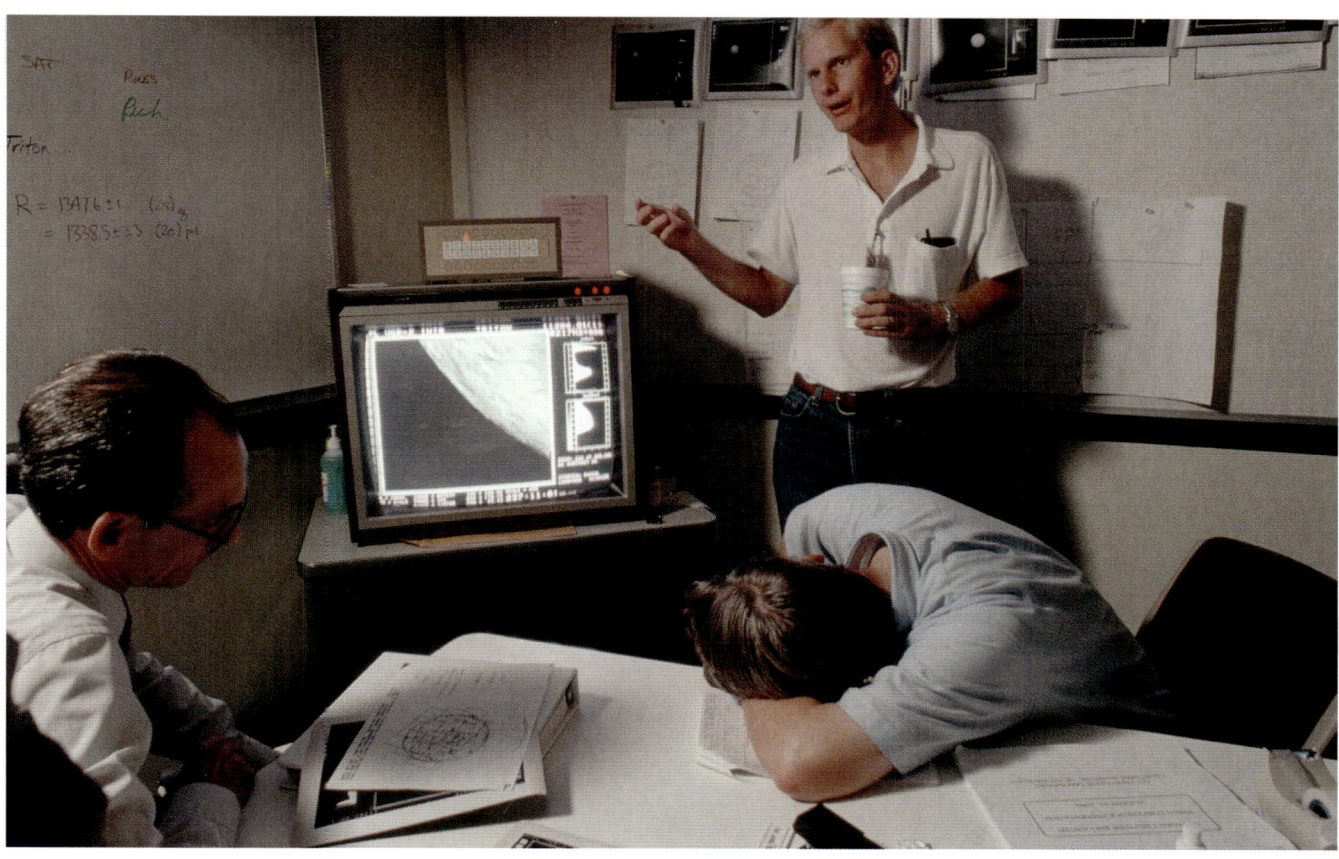

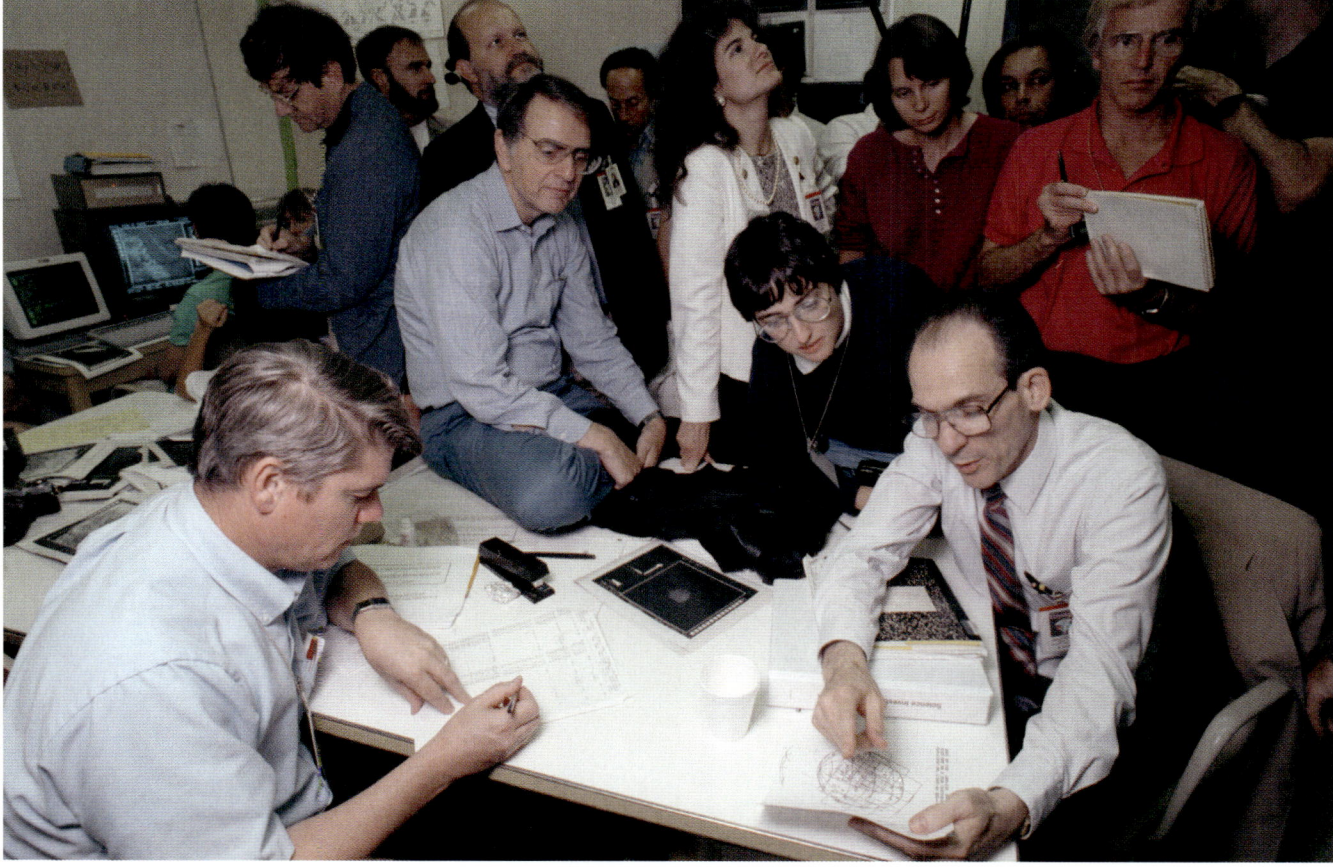

Above: scientists at NASA's Jet Propulsion
Laboratory examining fresh images of Neptune's
moon Triton. Below: Ed Stone, Carl Sagan and
other members of the Voyager Imaging team,
discussing images of Triton.

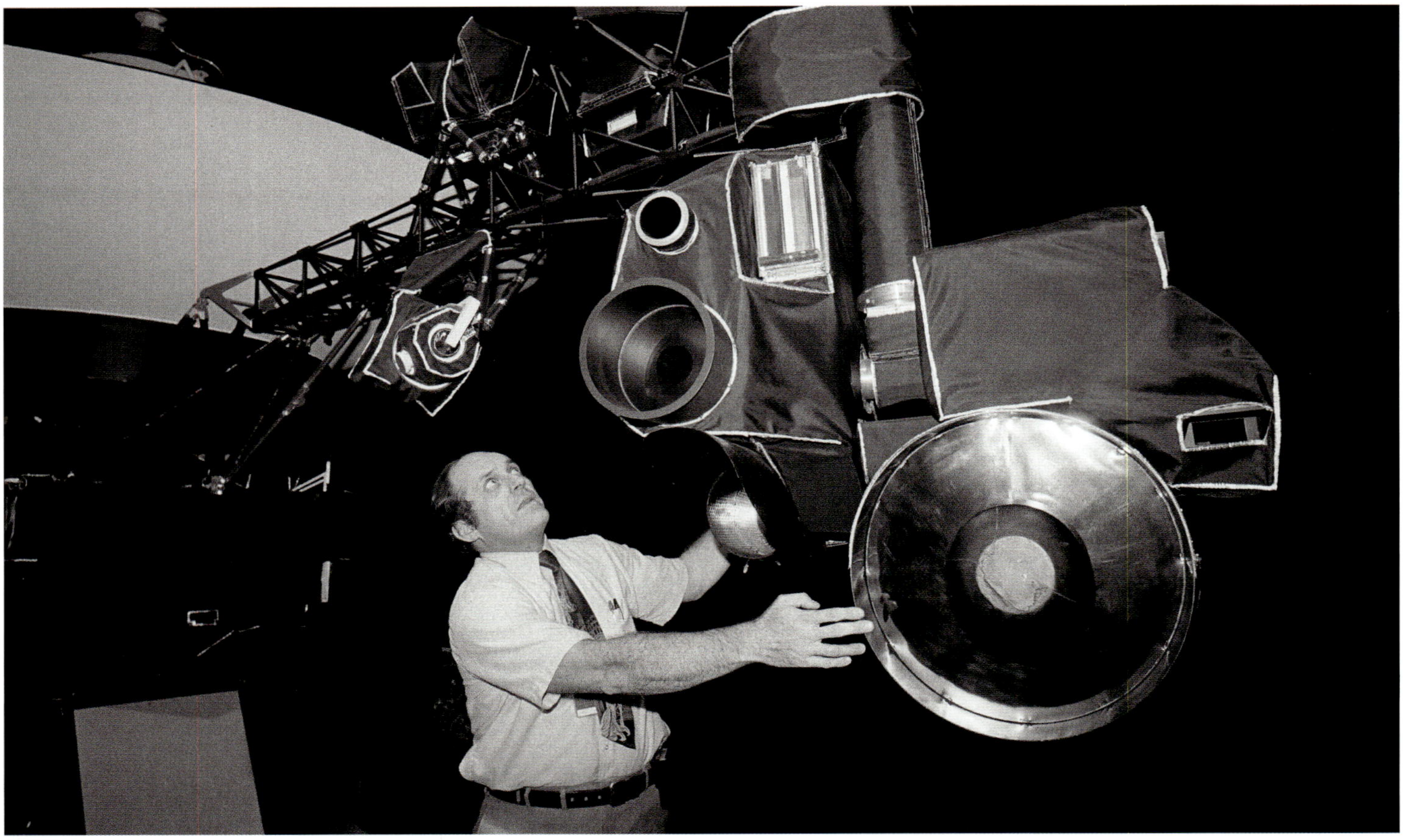

Voyager 2 mission Director Dick Laeser holds the cameras and scan platform on the Voyager mock-up model during a news briefing at the JPL in Pasadena, August 26th, 1981.

THE CAMERA

Introduction

The iconic images of Jupiter, Saturn, Uranus and Neptune that filled children's schoolbooks and showcased on special live television broadcasts around the globe, had a significant and lasting impression on us. Using just two cameras, the Voyager spacecraft allowed us to experience these distant and mysterious worlds in detailed images that no human had ever seen before.

The camera systems fitted to Voyager that captured these beautiful images were an upgrade of the similar vidicon cameras that were used on JPL's Mariner spacecraft that launched a few years earlier. Although the vidicon cameras were not the most advanced cameras available to the engineers at the time, the proven success from the Mariner missions meant the camera was a sensible and reliable choice. The two television vidicon cameras were upgraded. The changes to the sensors and other parts made the camera work better for the lower light conditions at Uranus and Neptune.

The three-axis stabilization of Voyager meant that the camera remained steady during the flight. This allowed the camera to take much higher resolution and complete images of Jupiter, Saturn and the other planet systems. This was a big improvement compared to the Pioneer 10 and 11's camera — which was mounted on a spinning spacecraft — which could only capture low resolution images, one pixel at a time, using a scanning photometer. Jupiter's Great Red Spot for example could only be captured entirely in one frame with the rudimentary Pioneer camera. In comparison the Voyager camera, at the highest resolution, would require taking forty pictures, mosaicked together to show the complete full image of the Great Red Spot.

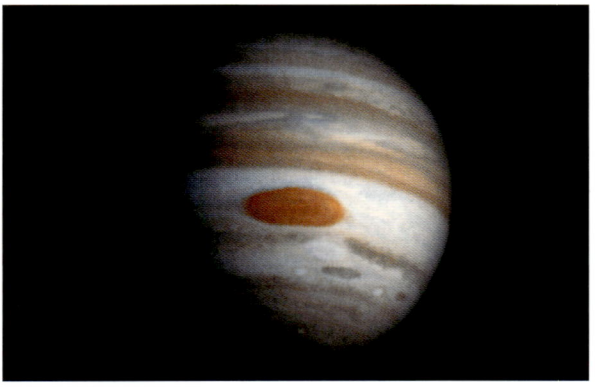

Jupiter captured by Pioneer 11 in December, 1974. An example of the low resolution image quality that the Pioneer spacecraft could capture.

The Camera

The ISS is made of two television-type cameras that shoot only black-and-white images. An eight-color-filter wheel was mounted on the front of the lenses to allow the imaging scientists back home to overlay the different color filter images and convert the black-and-white images to color. One has a low-resolution 200 mm wide-angle lens with an aperture of f/3, and the other camera uses a higher resolution 1500 mm narrow-angle f/8.5 lens. The resolution on the narrow-angle lens was capable of reading a newspaper headline from a distance of 0.62 miles (1 km). The majority of valuable photographic data was from the narrow-angled camera due to the greater resolution and detail the camera could capture. The wide-angle camera had more of a supporting role, providing complete views of the planet or moon, which would help the imaging team process and stitch the high-resolution mosaic images together more easily. Moreover, it would also shoot continuous photographic surveillance of the entire planet systems when the spacecraft would be making its closest flybys.

The two cameras are both slow-scan vidicon designs produced by the General Electrodynamics Corporation in Dallas, Texas. The imaging electronics were built by the Xerox Corporation in Pasadena, California. The cameras use a 1-inch selenium-sulfur vidicon that would turn the optical image into a digital image. Light would pass through the focusing lens, the filter wheel, and then through the shutter. The focused and filtered light hits the faceplate sensor, which would scan the image one line at a time. It measures the brightness of the fragment of light that struck the frame and converts it into a digital pixel. The active frame area of the camera is 0.48 x 0.48 inch (11.14 x 11.14 mm) and fitted 800 scan lines, each with 800 pixels on each line, with 640,000 pixels

in total to make up the image. The image resolution is incredibly low by today's standards, and is one of the reasons why a number of images taken by the Voyager cameras are of low quality. The maximum resolution and image quality depended on the trajectory of the spacecraft, was about 1 to 1.5 miles (1.6 to 2.6 km) per pixel with the closest flybys.

The camera had a color filter wheel fitted behind the lens. The filter would allow the imaging scientists to create color images from the black-and-white camera. The filter wheels on the two cameras contained clear, violet, blue, green, orange, sodium, and methane filters. In order to make a colored image, the camera would shoot the same image a number of times, each time using a different color filter. The imaging team would then process the images once they were received, coloring each of the different filtered images and then layering the images on top of each other; these would create the full color image.

The high speed of the spacecraft and the low-light conditions meant that the risk of blurred, "image smeared" images was high, especially as the spacecraft traveled further away from the Sun. The lower light conditions at Uranus and Neptune made it particularly difficult to photograph quality imagery. Longer exposure times had to be used during these encounters. The shutter had to be left open considerably longer than the original maximum setting of 15.360 seconds used at Jupiter and Saturn. To help reduce the risk of image smear in low-light conditions, the team had a few ideas in mind to incorporate into the image sequencing commands. Firstly, the camera's pre-amplifiers were built to lower system noise and to incorporate a high-gain state, and secondly a genius trick was pulled off by slewing the camera in a smooth motion to compensate for the velocity of the spacecraft.

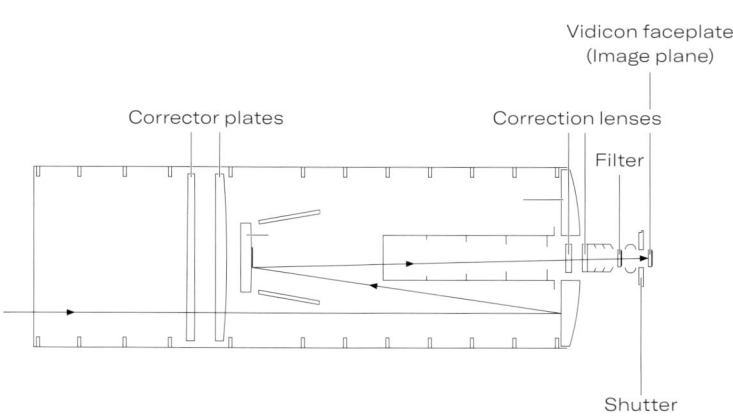

Sideways diagram of the narrow-angle camera.

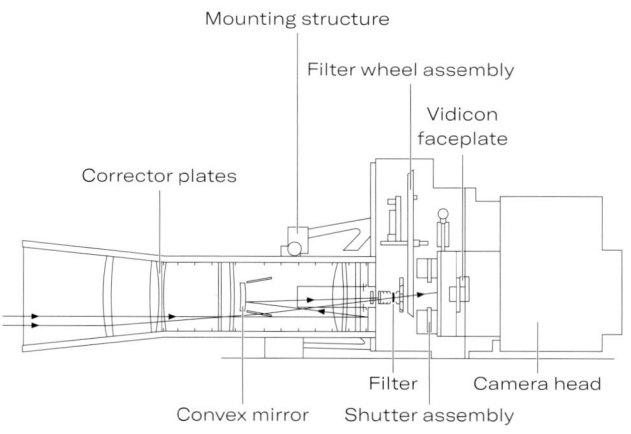

Sideways diagram of the wide-angle camera.

Targets of Opportunity

Scientists and astronomers from around the globe would submit requests on what kind of photographs they would like Voyager to capture during its encounters with the four planet systems. It would be the task of Voyager's Principal Investigators (PIs) to determine what "targets of opportunity" should be captured by the camera.

The general camera objectives would aim to capture the global distribution and movement of cloud patterns on Jupiter and Saturn. The color and composition in photographs would also help to give an indication to how their atmospheres were built up. Geology studies were another major target of opportunity. With the camera looking out for impact craters, volcanic activity, polar caps, and variations in topography, just to name a few. Photographing the planet's rings with the Sun's light behind them during occultation maneuvers was also another important target of opportunity, as these images would allow astronomers to accurately determine the composition and size of a planet's ring system.

Camera Programming

Unlike the other onboard science instruments on Voyager, the camera was not autonomous. The Flight Data Subsystem (FDS) would handle and process the camera commands and sequences that were programmed by the Imaging Sequencing Team back at JPL. These commands were being constantly reprogrammed and uploaded via radio signal from the Deep Space Network dishes to the spacecraft.

Image sequencing was a tough, stressful job. The consequence of making mistakes during the closest approach of the planet system was huge. The margin for error during these "one time only" opportunities to photograph simply was not there. The job required people who were incredibly attentive and concerned about accuracy and detail.

The team would have to figure out simple sets of time-stamped instructions for the camera to follow. For example, an instruction could be: "Point the camera to the left at a certain time, take twenty photographs with the narrow-angled camera, using a range of different color filters, wait 15 minutes and then repeat by taking another twenty photographs." These commands were thoroughly tested, reviewed and critiqued by the full imaging sequence team before being uploaded to the craft. The Imaging Sequencing Team would also have to estimate the exposure times of the camera. This is quite the challenge when you are trying to photograph a place that has never been seen before. If the surface of a

moon was icier and more reflective than anticipated, a long exposure time would let too much light into the camera, and you would end up with a completely burnt out image. On the other hand, if the rock color was darker, and the exposure time was too short, you would end up with a useless black image. In order to make educated and calculated estimates for exposure times and other camera setting parameters, the team would use whatever information they could gather, ranging from Earth based telescope observations, theoretical calculations, data brought back from the Pioneer mission, as well as learnings from Voyager 1's earlier encounters.

Deep Space Network

The need to be in constant contact with the Voyager probes was essential to make sure the craft was still heading in the right direction to upload and receive new mission instructions (such as updated image sequencing), to change course during a particular flyby, or for the Voyager team back home to receive all the precious data and images that the spacecraft was sent to collect. All this is easier said than done when you are sending radio signals to a tiny probe that is billions of miles away.

Responsible for the important task of keeping in touch with the Voyager craft was NASA's Deep Space Network (DSN). The DSN is a network of three giant radio telescope facilities managed by JPL that are positioned strategically around the globe to allow for constant coverage as the Earth spins around on its axis. The three facilities are roughly equally spaced around the world from each other, about 120 degrees apart in longitude. One site is situated near Canberra, Australia, another at Goldstone, near Barstow, California, and the third just outside of Madrid, Spain.

Each of the facilities had three separate antennas, one 210-foot (64 m) diameter dish and two smaller 85-foot (26 m) dishes. The nine large dishes were needed because of the extremely weak radio signals being sent from the distant Voyager probes. For example, when Voyager had just passed Jupiter, Voyager's 23-watt radio transmitter sent a signal back to Earth that was about a hundred-millionth as powerful as the battery in your cell phone. By the time that Voyager 2 had travelled to Neptune, the signal was now five hundred times weaker than what was broadcast at Jupiter. It takes over 20 hours now for a radio signal to travel from Voyager 1 back to Earth.

All the received data and radio signals collected around the world were sent directly to the Mission Control and Computing Center at JPL in Pasadena, California. The office was responsible for all the

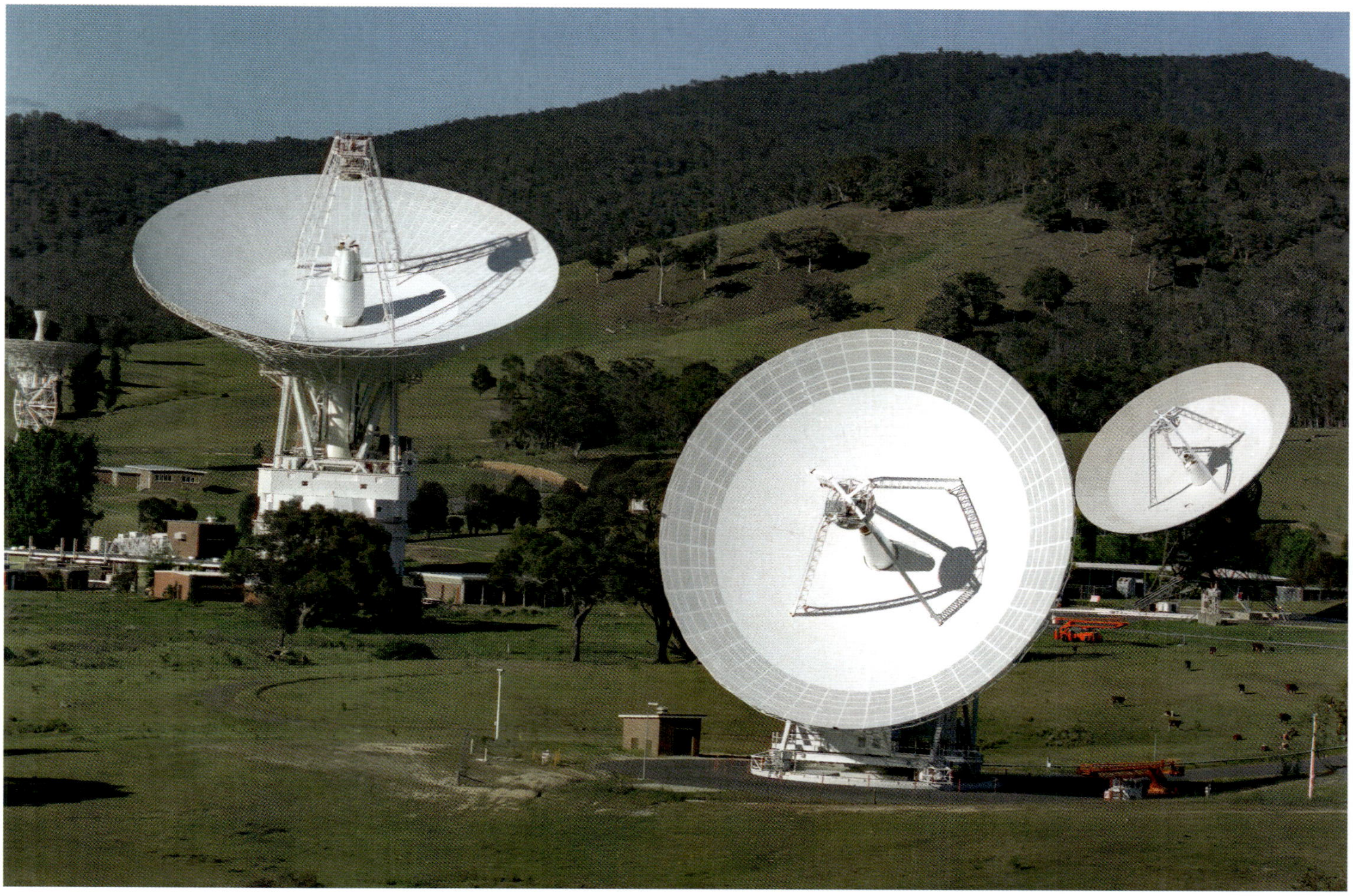

Three of the five large antennas at the Canberra Deep Space Communication Complex.

tracking, commanding and processing of all the data and communications from the spacecraft along with a spacecraft monitoring station in Florida. The data received from the different radio telescopes in Australia, Spain, and Florida was sent instantly via the NASCOM network of underwater ocean and overland cables back to Pasadena.

For example, if the spacecraft was out of signal range when the craft was passing by the dark side of the planet, the data would be stored on a digital 8-track tape drive, because it was not possible to directly broadcast back to Earth at that moment. Each tape drive could store around 67 MB of data. The tape would then play back the data and broadcast back to Earth when Voyager's dish was back in contact with the DSN. By the time Voyager 2 had completed its encounter with Neptune, around five trillion bits of data were sent back to Earth by both Voyager spacecraft. This amount of data would roughly fit on around seven thousand music CDs.

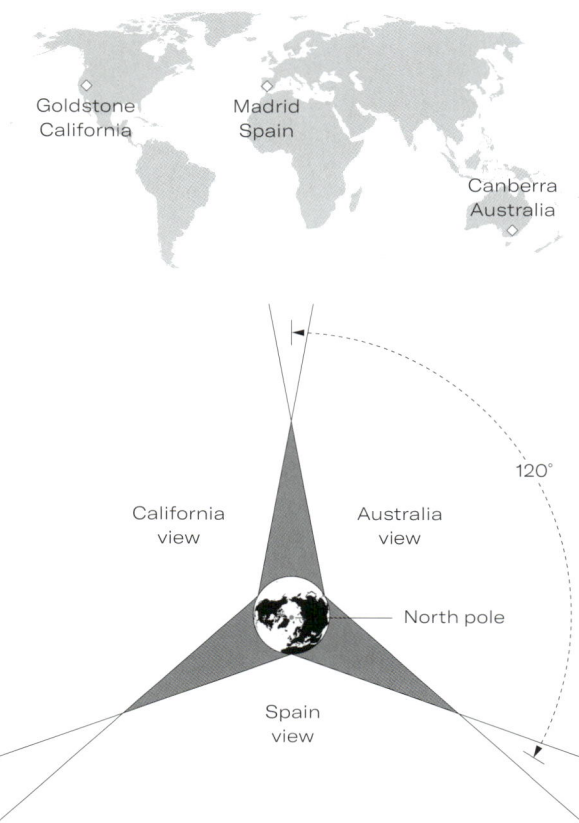

Goldstone
California

Madrid
Spain

Canberra
Australia

120°

California
view

Australia
view

North pole

Spain
view

Locations and distribution of the Deep Space Network stations around the globe.

FROM BINARY TO BEAUTIFUL

Breathing New Life

The dataset that makes up the collection of photographs that both Voyager craft took represents what is still one of the best datasets for studying the outer Solar System, and, in the case of Uranus and Neptune, will be only close-up dataset for at least another two decades. Thus it is imperative that we extract everything we possibly can out of these precious images.

Luckily, with the latest imaging software, mixed with a little bit of AI, and, most importantly, with Ted Stryk's expert background knowledge with the Voyager dataset, we have been able to breathe new life into some of the most important photographs captured in the Solar System into this book. Over the next few paragraphs, we show some of the few steps needed to convert those binary data signals that were streamed over the billions of miles back to Earth, to create the photographs that are featured in this book.

1970s Cameras

The Voyager imaging system is vintage early 1970s technology. Instead of using a CCD chip like a modern digital camera, they utilized vidicon tubes, which were used in early television cameras. CCD chips were brand new at the time, and although they were considered, it was deemed too risky to use them in a mission designed to last for many years. The vidicon sensors present several challenges. First, they are covered in calibration markings, called "reseau." Over time, vidicon tubes warp, and the dots on the images were used to correct the warping in the images, as this would be a problem when using them for mapping.

Additionally, vidicon is notorious for after-images, which means a bit of the previous image can be seen in the next picture. To solve this, the Voyager team used brute force, installing a flash bulb. Between each picture, the Voyagers would take an unrecorded picture of the bulb, which would saturate the vidicon tubes. Thus the after-image was pure white, which could be easily subtracted from the following image.

Image Smear Correction

Many of the images are smeared due to the motion of the spacecraft or camera pointing issues, so this has to be corrected. Some was done on board, using the scan platform to track the object during the exposure. The rest must be corrected via computer processing. For underexposed images and to maximize resolution in distant images, the component images, taken through various filters, are stacked and averaged, and then, if there is enough data to produce a color image, the color image is draped over the stack.

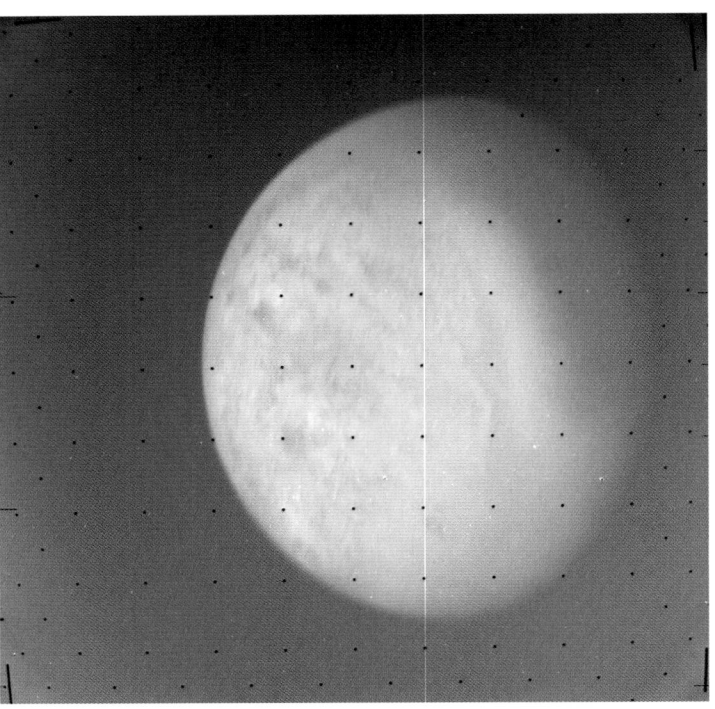

The raw dataset image of Triton above is a good example of the visible reseau markings (the numerous black dots) show how the vidicon camera's image has warped. The light background is the legacy result of the flash bulb, wiping clean the image ready for the next image.

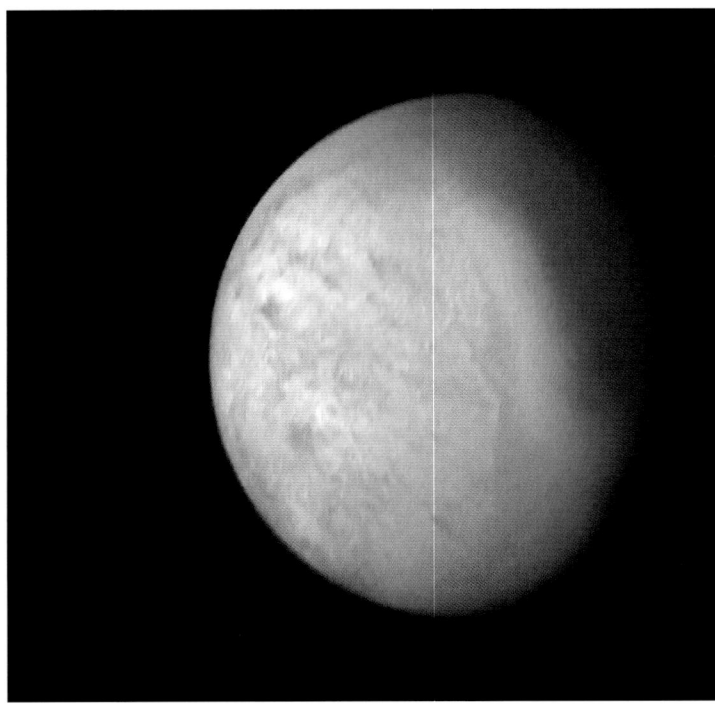

The processed image above shows how we have corrected this warp, and taken out the flash bulb legacy. The image has also been smear corrected via computer processing. Each of the color filter images is smear-corrected and then stacked on top of each to create the finished image. The corrected image of this particular image of Triton can be seen at the bottom of page 267.

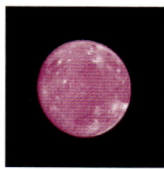 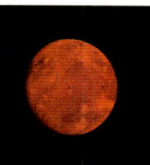 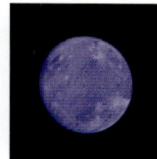 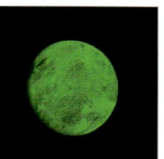

Getting Color from a Black-and-white Camera

The Voyagers, like most planetary missions, carried black-and-white cameras along with a filter wheel. They would take images through various color filters, which could then be combined to make color images. This is better for scientific analysis, because it allows more precise measurement of color differences. On modern missions, color is produced using red, green, and blue filters. However, the vidicon tubes were blind to true red light, so the red component is more orange than red. Occasionally, other filters, such as yellow and violet, need to be used due to the lack of availability of images from the other filters. The wide-angle cameras also carried some special filters to search for specific atmospheric components (mainly methane). The four filtered images (right) are placed on top of each other, blending the colors together creates the color image as in the example of Ganymede on the right.

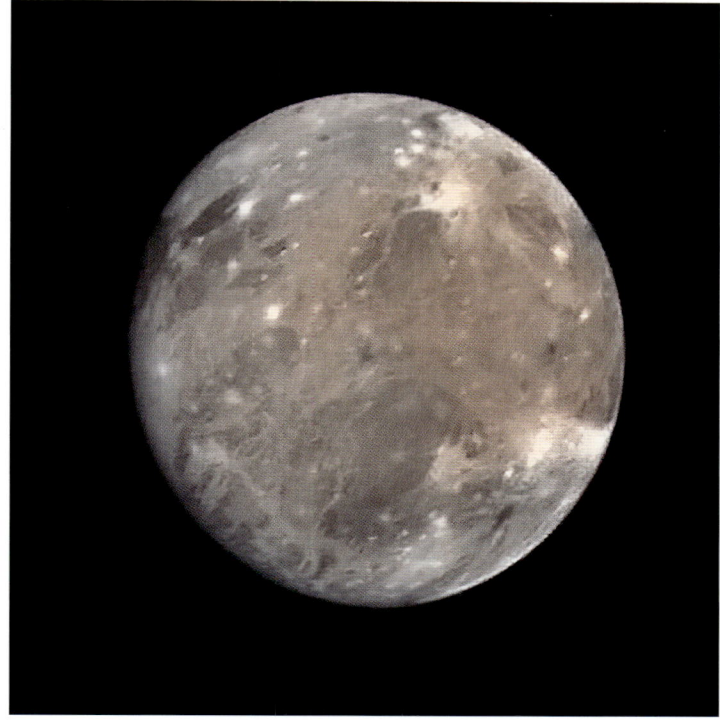

Puzzle Pieces

When the target was too big for the camera's field of view, the cameras took mosaics of images designed to be stitched together to form the complete image. Below, to the right, you can see the nine individual shots that Voyager 1 took of Callisto. It took the camera 46 minutes and 25 seconds to capture all nine of the images. Directly below you can see how the nine images are aligned and joined up to make the complete 'mosaic' image of Callisto. Further processing, such as correcting and warping the image to create the spherical shape of the moon.

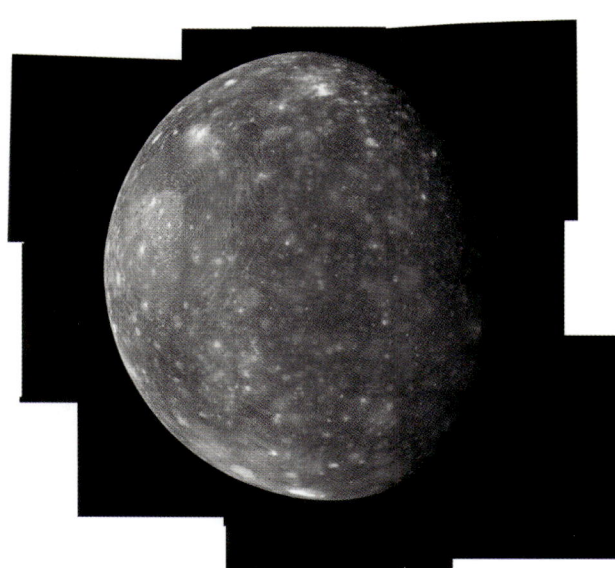

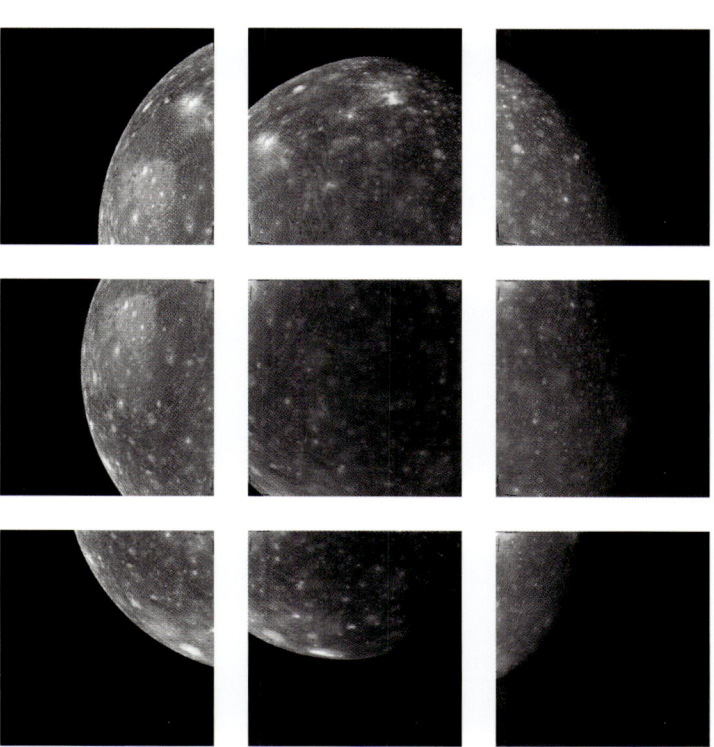

The nine separate shots of Callisto can be merged together to create a complete global view of the moon.

THERE IS NO END

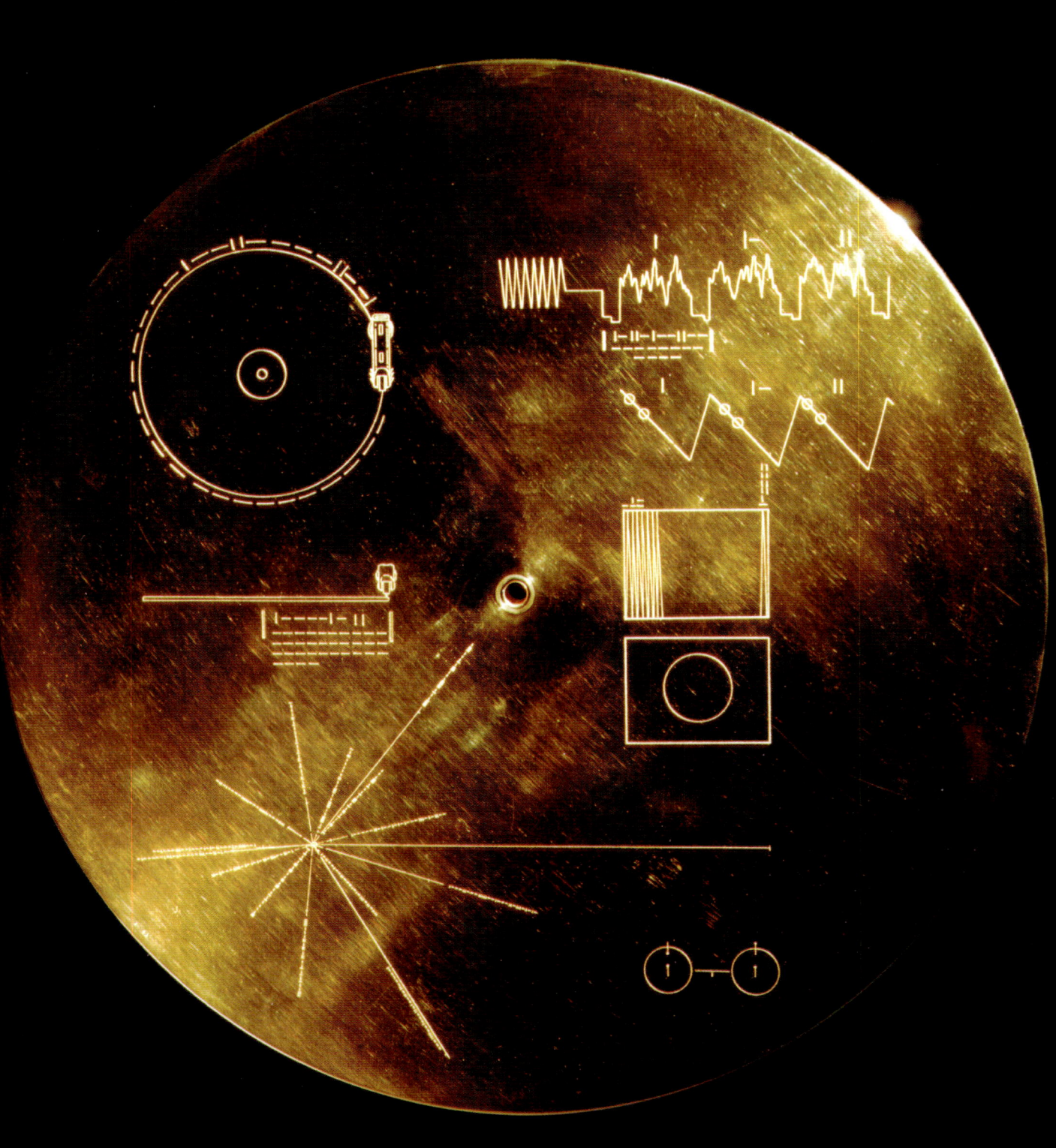

THE IDEA

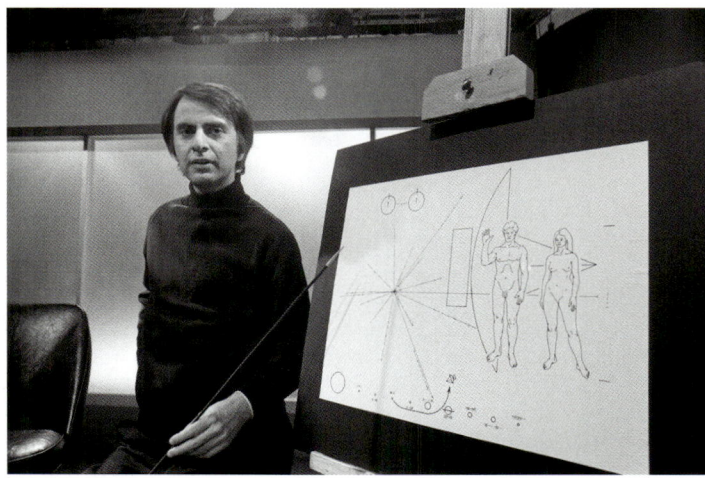

The 12-inch gold-plated copper disk that contained the sounds and images curated to give an insight into the diversity of life and culture found on planet Earth. Left: the Golden Record cover.

Dr. Carl Sagan in January 1974, standing next to an illustration similar to what was displayed on the Voyager Golden Record.

A Message in a Bottle

After the trajectories for Voyager 1 and Voyager 2 were chosen, so was their fate. Since the spacecraft would travel with incredible speed due to the gravity assist, escape velocity from our Solar System became inevitable. There is an irreversible and poetic truth to the fact that both Voyager spacecraft were destined to exit our Solar System: they are never coming home. Long after the scientific part of the mission ends, these human artifacts will be wandering the vastness of space in silence, snapshots of our species that wander the galaxy as time capsules. Just when the spacecraft were wrapped up and being shipped from JPL to Florida, an intriguing last-minute question came to mind. If we are going to throw a bottle into the ocean, why don't we put a message into it?

A question that leads to more questions, so it seemed. Astronomers, science-fiction writers and philosophers have pondered the challenge extensively. How do we communicate with space aliens? Moreover, do we need a special language for it? Throughout history, humans have shown an exceptional skill in decoding language. From linguists deciphering ancient Greek language to to Alan Turing cracking the Nazis' Enigma machine codes, trying to understand what our fellow people are uttering is something we do quite well. For those in the Voyager team that contemplated the message, this led to a strong belief that other intelligent life must be able to conduct a similar level of decoding. Besides, this time we were actually going to try to make ourselves understood.

Carl Sagan

Luckily, the Voyager imaging team had a crew-member that not only became one of the best-known scientists of the late 20th century, he was also the astronomer who made the study of extraterrestrial life credible. His name was Carl Sagan, and he would become responsible for transforming Voyager from a pure scientific mission into a cultural and artistic one. After being asked whether he would be willing to undertake the message for the spacecraft, Sagan without a blink of an eye confirmed and told the team it would cost "about 25,000 bucks" and started assembling a team. Two Cornell astronomers and one of their wives, a New York City science writer and an American expatriate artist living in Toronto were gathered to construct a universal message that transcends particular interest and speaks to the heart of what we are about as people of Earth.

For some, this sounded a bit too "Walt Disney" for their taste. What did we know about the recipients of the message? Instead of sending out our hopes, some were having fears. In their viewpoint,

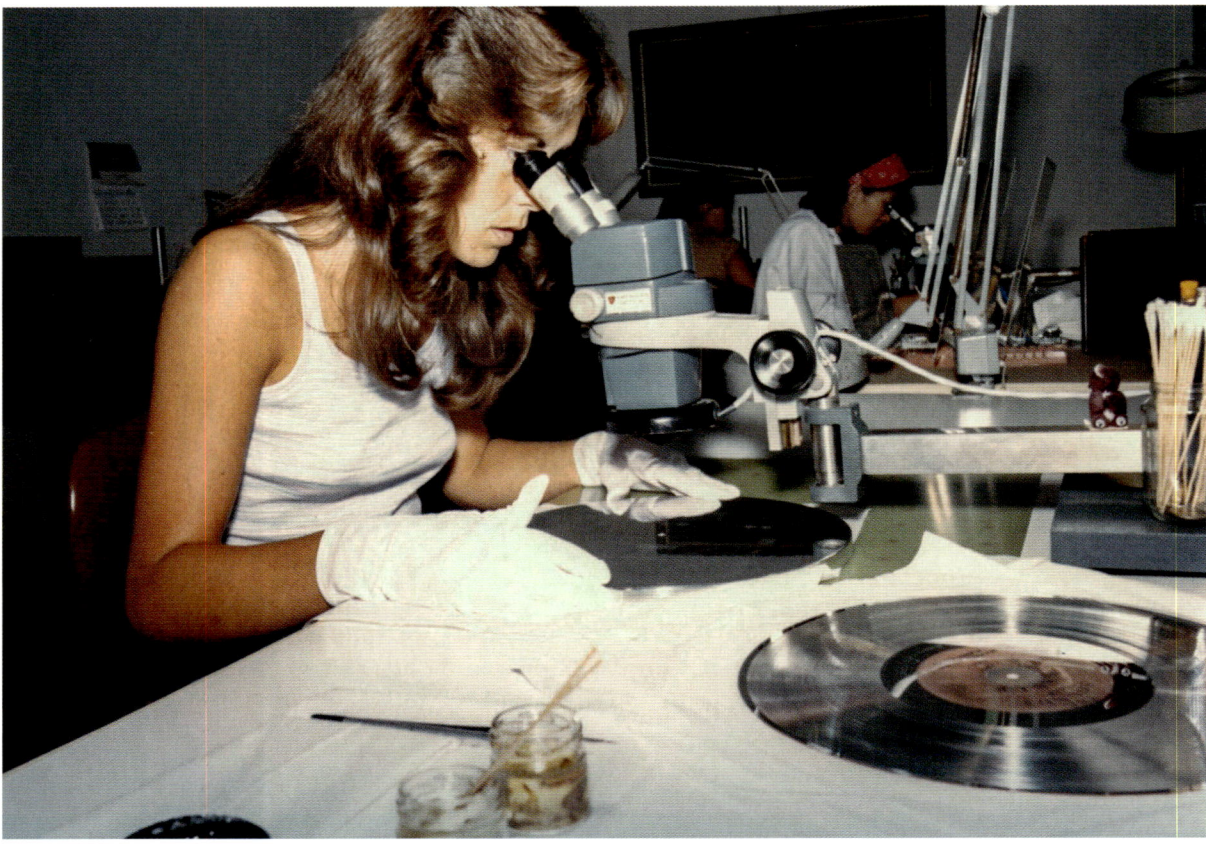

An employee inspecting the engraving on the nickel plated mother record at the James G. Lee Record Processing Center in Gardena, California, who were commissioned to cut and gold plate eight Voyager records.

sending a message to outer space could pose a grave danger. What if they were hunters? Or interstellar colonialists? We would be indicating our presence to beings whose intentions we do not know. Throughout the 70s this view had been shared a couple of times, especially in relation to the Search for Extraterrestrial Intelligence (SETI) projects that aimed to communicate with extraterrestrial life by using radio waves. In contrast, the culture on the Voyager team seemed much more optimistic and fearless. Besides, the debate seemed overdue anyway. TV broadcasts, military radars and spacecraft communication result in electromagnetic waves being shot into space on a daily basis. In the eyes of the Voyager team, it was time to turn this rustle into something more relevant.

The Pioneer Plaques

For Carl Sagan, this meant taking his previous attempt at an interstellar message to the next level. In the early 70s the NASA Pioneer 10 and Pioneer 11 spacecraft were hurled into space and became the first man-made objects to be accelerated beyond the Sun's escape velocity. As pathfinders for the future Voyager missions, the Pioneers executed critical preparatory work. They did initial science scouting and demonstrated technologies and celestial navigation methods, like showing how to get through the Main Asteroid Belt and the plane of Saturn's rings without a scratch. These spacecraft were the first ever to move beyond Mars and set course to slip out of our Solar System. Science journalists Eric Burgess and author Richard Hoagland publicly challenged humanity not to miss the opportunity to send these

emissaries into infinity without a message. Carl Sagan, his wife (artist and writer Linda Salzman) and astronomer Frank Drake accepted their challenge.

With only three weeks to finish the job, the trio came up with the idea to create gold-anodized aluminum plaques that could stick to the spacecraft and survive for a long period. The message that was etched into the plaque consisted of rudimentary drawings and markings based on physics and astronomy. Instead of communicating to our alien neighbors about our civilization or our motivation for the mission, the plaques were solely intended to carry an indication of the locale, epoch and nature of the senders. The message was not hailed as a great achievement, but looking back, the biggest criticism can also be regarded as a sign of the times. Besides rightfully calling out the numerous assumptions made in the symbols and the communication, much commotion was directed at the depiction of a naked woman and man in a line drawing. "Interstellar Pornography" was the label used. Looking back, the people that were involved recognized their message could have been better, and with Voyager came the opportunity to set things right.

The Mother of all Concept Albums

The Golden Record was born with a simple suggestion from Frank Drake. This widely known and respected astronomer is a pioneer when it comes to communicating with possible extraterrestrial life. From beaming short greetings towards a number of

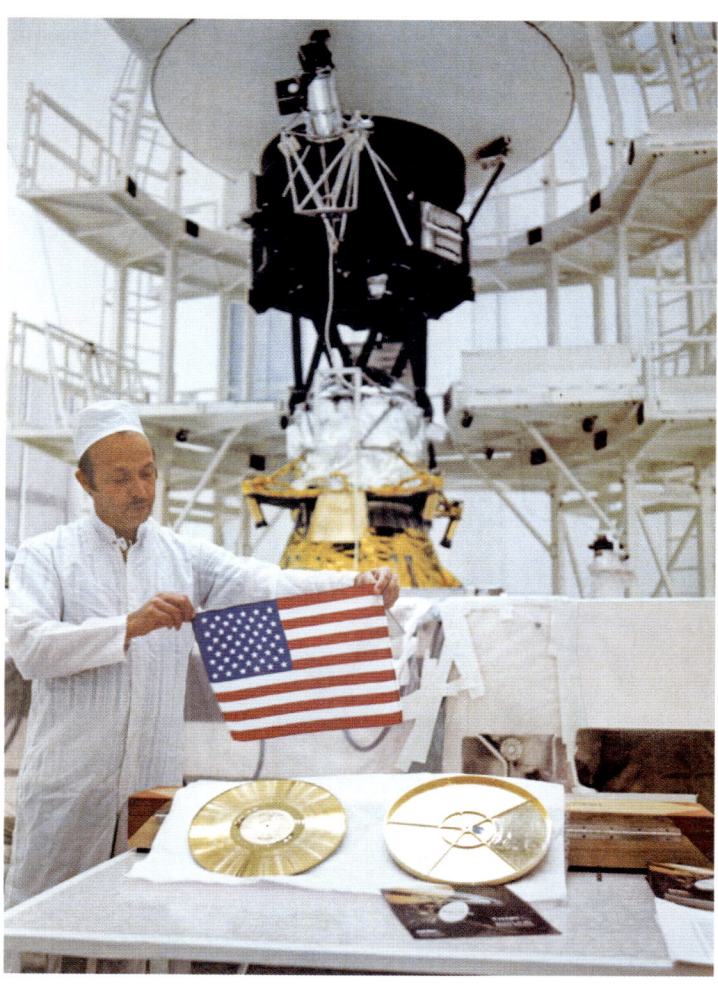

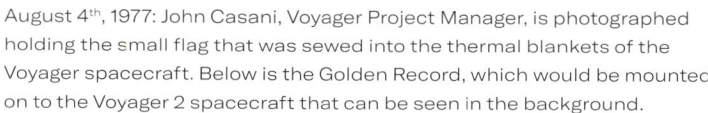

August 4th, 1977: John Casani, Voyager Project Manager, is photographed holding the small flag that was sewed into the thermal blankets of the Voyager spacecraft. Below is the Golden Record, which would be mounted on to the Voyager 2 spacecraft that can be seen in the background.

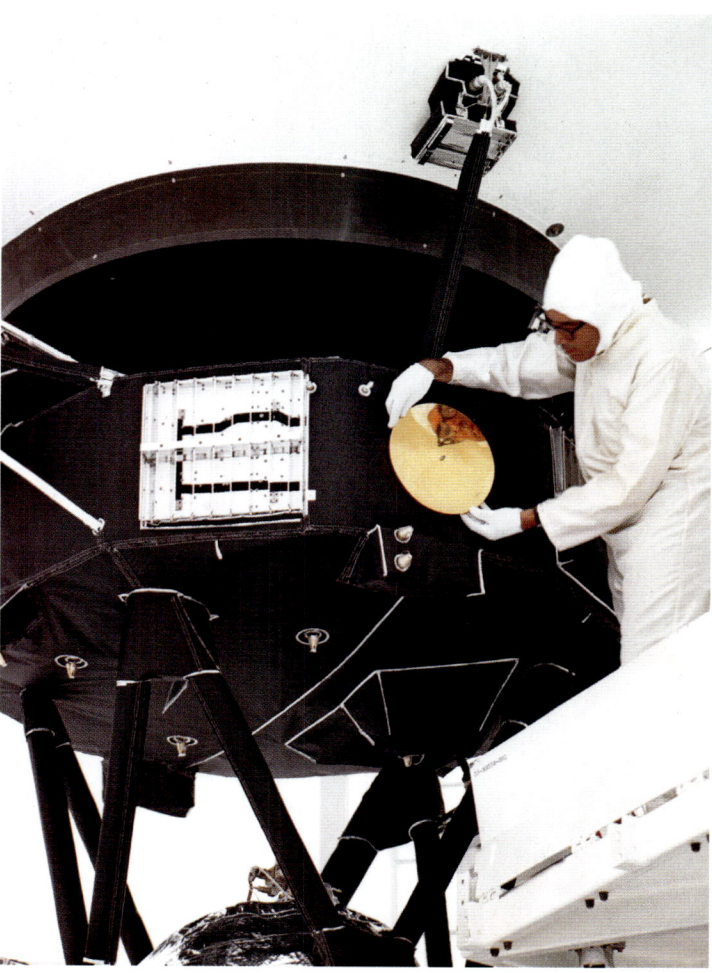

The Golden Record being mounted onto the Voyager 1 craft.

nearby stars using a radio telescope in the 60s, to coming up with the "Drake Equation" to calculate the number of intelligent civilizations, Drake was always a front-runner when it came to starting alien conversations. For Voyager, he was set on a celebration. How great would it be if, on the 100th anniversary of Thomas Edison's development of the phonograph record, we could launch the Voyager spacecraft carrying an LP record? According to Drake, an LP would take the same amount of weight and space as a plaque would. This changed everything. In an era where concept albums by Pink Floyd, David Bowie and Jethro Tull were topping the charts, the Golden Record was set to become the most ambitious of them all.

The perpetuating nature of storing information by etching it in the grooves of a disk was a near perfect solution. What better way to store sound and digital representations for billions of years? The big leap this took, in comparison to previous attempts like the Pioneer plaques, was the possibility of sending music. If the aim were to communicate a message that speaks to the heart and the hopes of what we are about as people on Earth, only information and pictures would not do the trick. Music has the capability to invoke emotions with an unmatched potency. For Carl Sagan and his team, the hope was that it could convey the human spirit and tap into the feelings of advanced life forms from a different time and place.

A Convenient Rush Job

To make the record of all records even more challenging, the total production time for the Golden Record was six weeks. A deadline that could not be toyed with. The alignment of the outer planets made sure that the launch window for Voyager was set. There was a once-in-176-years gravity assist at stake. Naturally, waiting for an LP to finish was not on the list of reasons to postpone the take-off. Members on the team say that Sagan did not really see the short time frame as an inconvenience. Given the budget restraints and cutbacks, stretching the Golden Record process would increase the danger of it being sidetracked in congressional oversight. A last-minute done deal however was much easier to give approval on. Therefore, the team buckled down in an intense and close collaboration to answer one question: How do you figure out a way to explain the world to aliens in 42 days?

THE CONTENTS

Living Up to an Ideal Version

We live in times of social media. Every day billions of people are choosing what kind of content about themselves they feel comfortable sharing with the rest of the world. It is often said these people place a filter on their own lives, displaying an almost picture-perfect version of themselves. Criticism of this phenomenon often uses labels like "fake" or "inauthentic". However, what if these people are actually ambitious instead? What if they are setting up a version of themselves and taking on the challenge to start living up to it?

It was this kind of thinking that fifty years ago, (when the only kind of "social media" was the official yearbook people received at the end of a school year), was on the minds of Carl Sagan and his team. Selecting which images, sounds and music to eternalize into the grooves of the Golden Record was no simple task. What version of our planet and ourselves do we portray? Do we celebrate humankind with its greatest achievements and focus on the inspiring side of humanity? Or... do we choose a more balanced representation that also takes into account the dark and brutal side of humanity we have displayed throughout the ages?

Carl Sagan and his team chose the former. Although they are undeniably a part of human nature, topics like revenge, greed, war, genocide, and injustice were left out of the Golden Record selection. The decision was not made out of dishonesty, but for the simple reason that it is not the part of us they wanted to send into space. It might very well be the case these records will outlive humanity and become the last artifacts of our existence. It felt more appropriate for the team to highlight the brighter side of who we are. Besides, what kind of message are we sending if we depict our capability to build atomic bombs and destroy those around us? Instead, Sagan saw this record as an opportunity and hoped for the positive influence it eventually had. Jon Lomberg, the Voyager Records Design Director and artist of the Record's now iconic cover drawing, states that, "It was like asking ourselves whether we could live up to the Voyager record? This is us at our best, or at least not at our worst. This approach made the record aspirational."

The Record

Before the question of the content could be delved into deeper, the team first had to find a way to construct the record. The main difference between a normal record and the Golden Record is that the latter was made from gold-anodized copper to make it last longer. Still, along the way, the metal is anticipated to face radiation that could corrode it or high-speed micrometeorites that most likely pit and gouge it. To protect them, the record was therefore wrapped inside a gold-plated aluminum casing.

Another challenging question they needed to answer was one that a modern teenager would probably also ask when given an LP record, "How do you play this?" If the record is ever to be found, chances are high the aliens in question did not have a lot of analog listening devices in their basement. The team needed to figure out a way to actually play the record. That is why they decided to include a stylus and a cartridge in the package of the record. Subsequently a drawing on the cover was made that shows the method by which the stylus is to be placed on the record.

"Earth, 1977"

However, you can only make a first impression once, so the record explanation is not the only aspect the team wanted to include on the cover. Just like any handwritten letter, postal card or modern email, this message needed a signal as to where in the world it came from. Next to the drawing, they wanted to place a map that indicates where the message that these space aliens were holding was launched: "Earth, 1977". This led back to those same challenging questions: how do you communicate with space aliens and do you need a special language for it? Unless their technology is so advanced that they study us from afar, human language or conventions are not part of their curriculum.

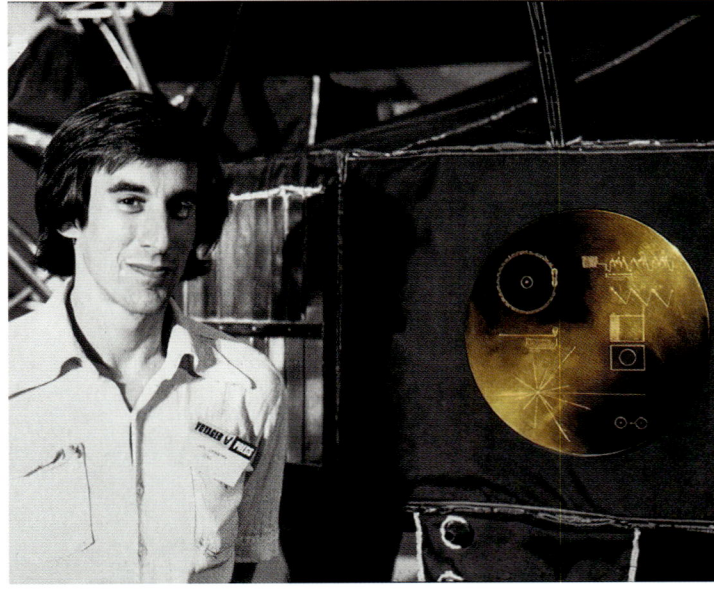

Jon Lomberg, the Voyager Record's Design Director and artist, standing next to the Golden Record's now iconic cover drawing that he created.

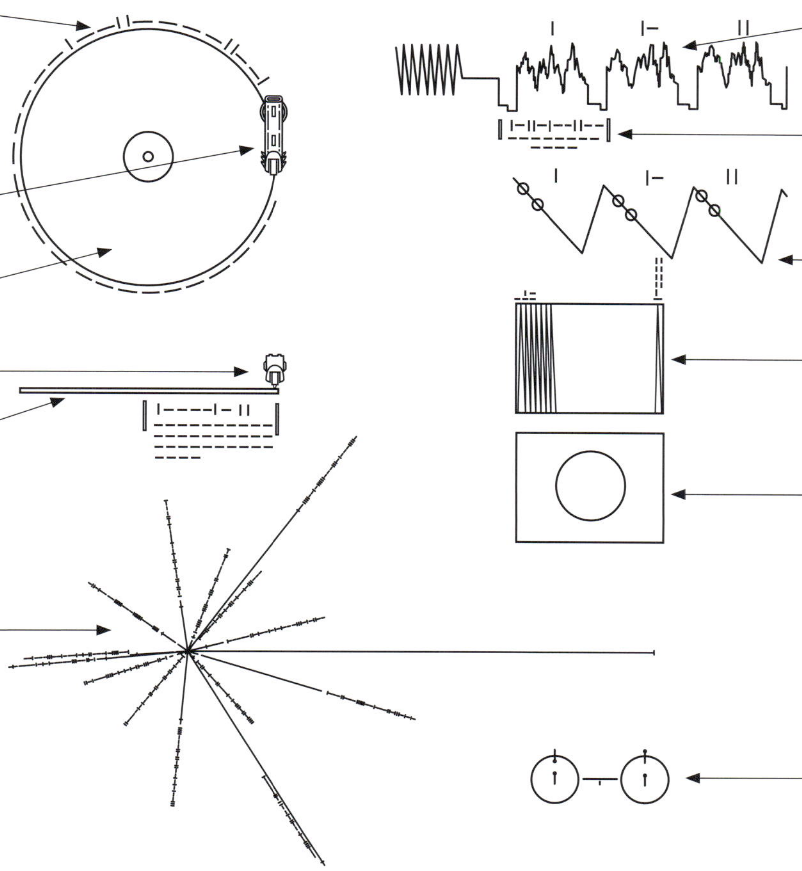

Binary code defining proper speed
(3.6 seconds/Rotation) to turn
the record (I=Binary 1, ━=Binary 0)
expressed in 0.70 x 10^{-9} seconds,
the time period associated with
the fundamental transition
of the hydrogen atom

Outline of cartridge with
stylus to play record
(furnished on spacecraft)

Pictorial plan view of record

Elevation view of cartridge

Elevation view of record
One side = ~1 hour

This diagram defines the location
of our sun utilizing 14 pulsars of
known directions from our Sun.
The binary code defines the
frequency of the pulses

General appearance of wave
form of ideo signals found
on the recording

Binary code tells time of
the scan (~8 ms)

Scan triggering

Video image frame showing
direction of scan. Binary code
indicates time of each scan
sweep (512 vertical lines per
complete picture)

If properly decoded,
the first image that
appears will be a circle

This diagram illustrates the two
lowest states of the hydrogen
atom. The vertical lines with the
dots indicate the spin moments
of the proton and electron.
The transition time from one
state to the code defines the
frequency of the pulses. The other
provides the fundamental clock
reference used in all the cover
diagrams and decoded pictures

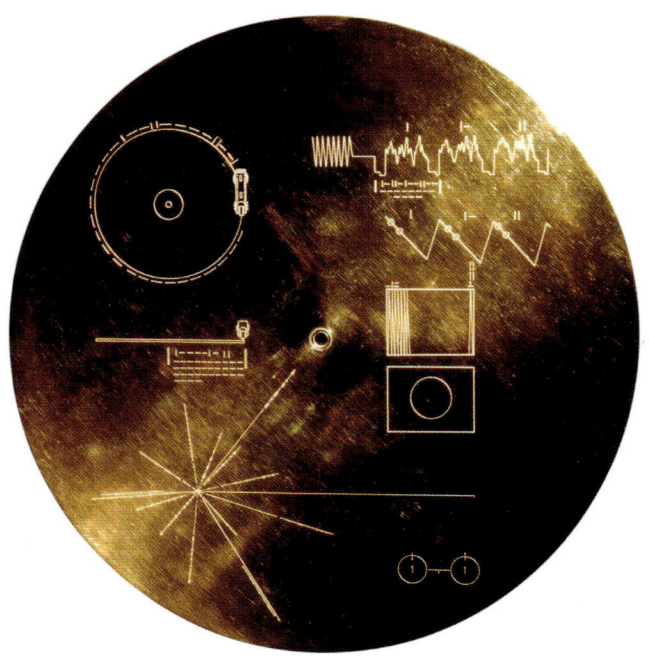

Just like with the clever conception of the LP-record, Frank Drake stepped up with his wealth of experience in extraterrestrial communication to find a solution. In order to create a map with which the beings could orientate themselves, the locations were given relative to a number of prominent pulsars. These pulsars are rapidly rotating stars that each have their own distinct frequency, which made it possible to not only convey our location, but also our time. This answered the challenge on what they wanted to say, but still did not cover how they could say it. For this, they turned to an abundant and plain simple element in the universe: hydrogen.

Hydrogen as a Language

Hydrogen consists of just one proton and one electron. What if, Drake pondered, you could use the transition time it takes one hydrogen atom to turn between its two lowest-energy states and apply that as your fundamental time scale? This 0.70 billionths of a second would be the basis for the representation of time on the Golden Record. The pulsar frequencies could now be indicated as "hydrogen time".

The highly basic character of hydrogen had Drake and the team convinced that intelligent life would be able to decode the information. Assumption was bliss however, since naturally there was no way of knowing for certain. In order to follow through with all the content, they had to presume that the aliens would understand atomic physics and recognize hydrogen, would use vision and hearing of some kind to take in the pictures and the music, and would be able to perform deductive reasoning and see causality of events to be able to appreciate and understand what these crazy disks were telling them.

The Pictures

If so, what would they be telling them? Overall, the Golden Record contains 116 pictures, 27 musical pieces and a catalogue of sounds and voice recordings. The pictures can be categorized into two different types, namely photographs and diagrams. The pictures refer to one another and enrich our understanding by doing so. Together, they form a series that tells a story. The selection for the pictures was purely motivated by the goal of conveying the uniqueness of life on earth. Any aesthetic appeal that came along with the images was seen as a bonus. Extraterrestrial head games had to be played by the team during the selection procedure, in order to put themselves in the position of the aliens: how would you look at this picture without the unconscious context of walking, eating, living and breathing on Planet Earth?

The pictures follow a path the same way knowledge follows a path in the education of a child. It creates the building blocks to increasingly understand more about the world in all its facets. For the Golden Record the first image became the calibration circle, which was followed by images conveying scientific knowledge. These had to explain how our numbers work and define ideas of distance and mass, using examples of the hydrogen atom and wavelengths of radiation that are emitted when hydrogen changes states. This in turn will facilitate the explanation of how many hydrogen atoms make up a human body.

The photographs in turn had to depict the uniqueness of life on Earth. Life in the form of trees, animals, seascapes and mountains, as well as people eating, learning, dancing, running, socializing, dressing and building. The level of detail that went into choosing and explaining the pictures is astonishing. A great example of this is the stroboscopic picture of a gymnast performing her routine on the balance bar. It includes the exercise's timescale of five seconds from start to finish, indicating people on Earth move around in matters of seconds as opposed to milliseconds or years. A picture of an ocean is not a beautiful scenic natural portrayal. Instead, aliens with an understanding of optics and fluid mechanics can understand it, to use the colors and patterns in the sky and the ocean to deduce properties like the pressure or chemical composition. Not one picture

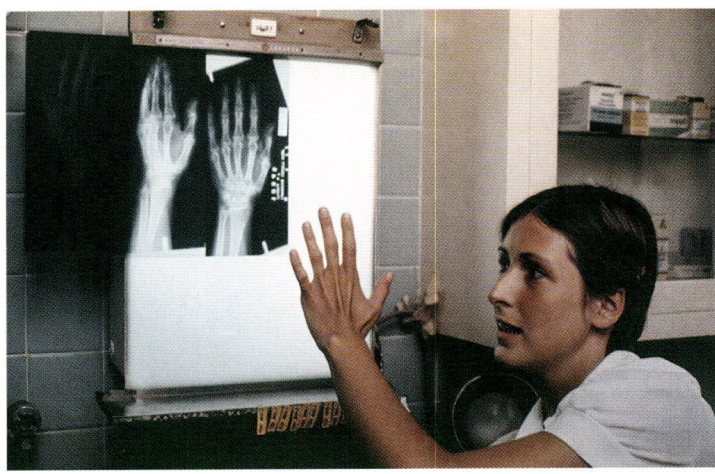

This image, portraying an X-ray of a hand, celebrates the human hand, which was key to the success of the human race's evolution. The X-ray image further visualises our development of technology.

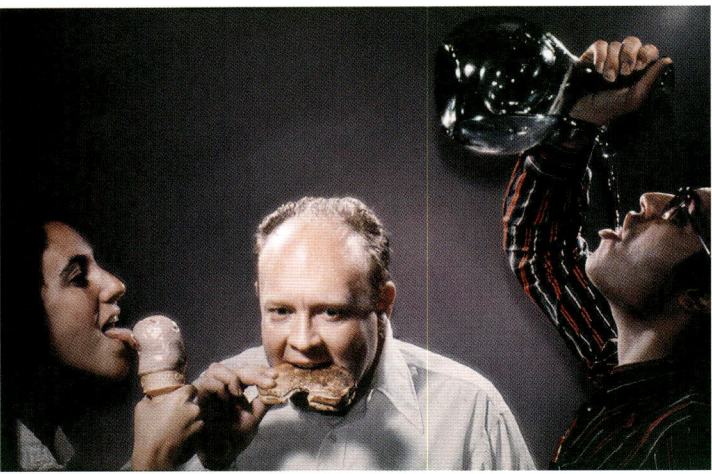

Demonstration of licking, eating and drinking: this photograph provided a clear image of how humans consume food and liquid.

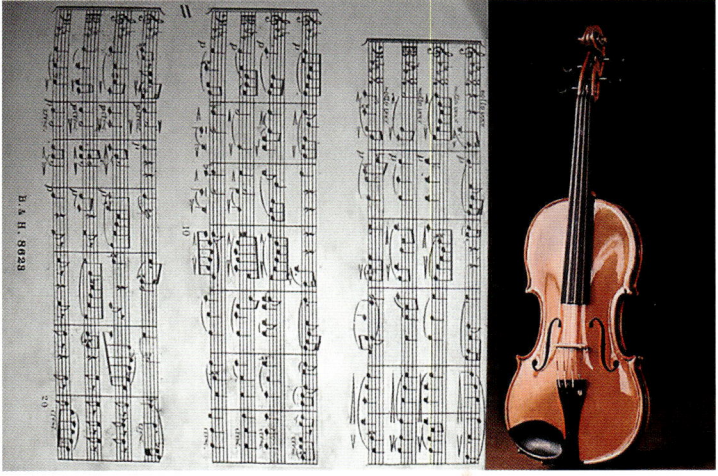

The image of a musical score and violin was selected to transform the general image selection from a 'photo album' to a loaded message. The combination of the score and instrument was to demonstrate what music is, which was further supported by the matching audio track found on the disk.

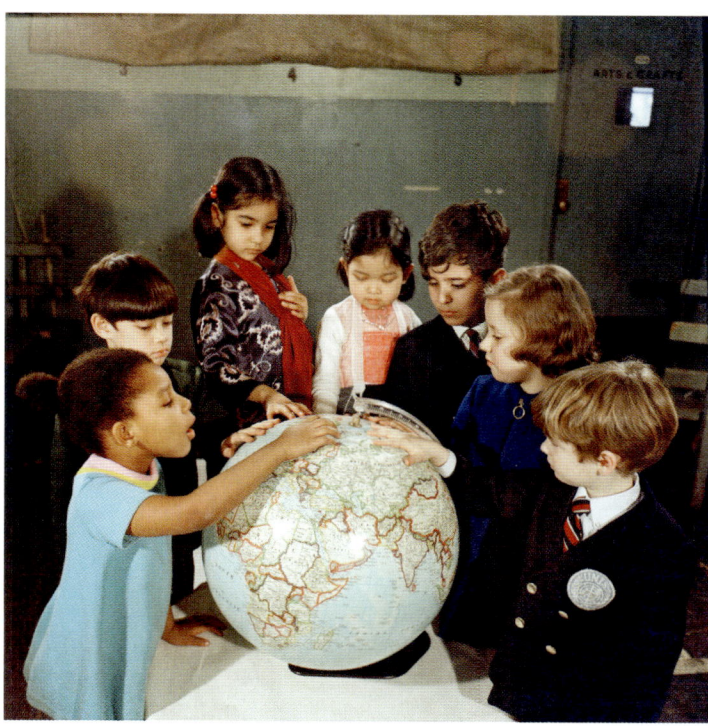

Children with globe: a group of children from the UN international school, the globe in the middle of the image shows the individual, man-made country borders.

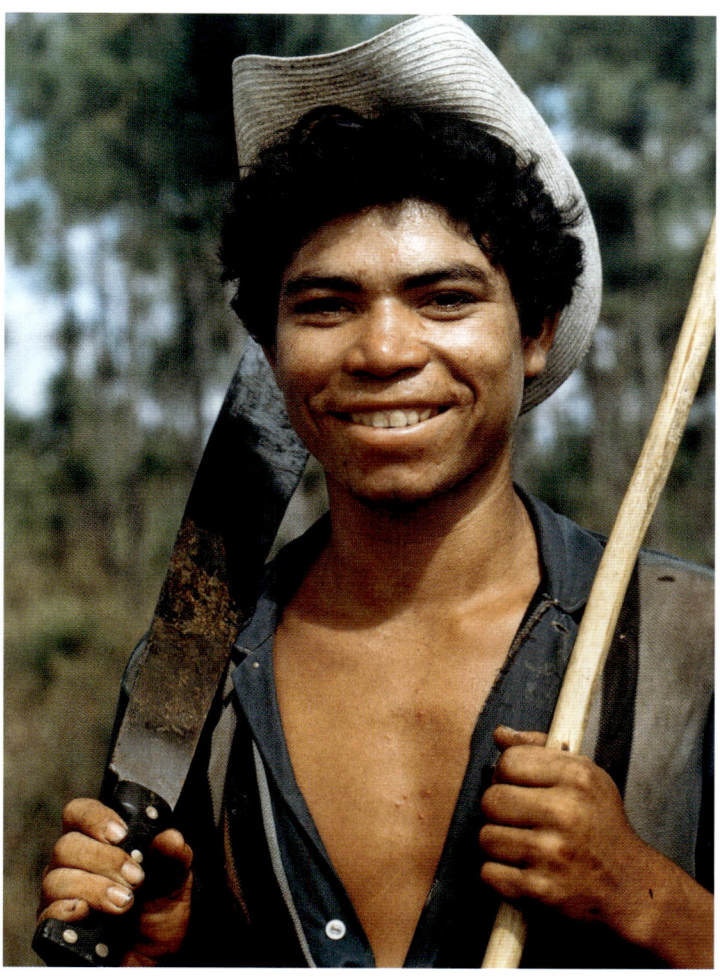

Man from Nicaragua: this image was selected for its close-up focus on the man's face and hands, and in particular the opposition of his thumbs.

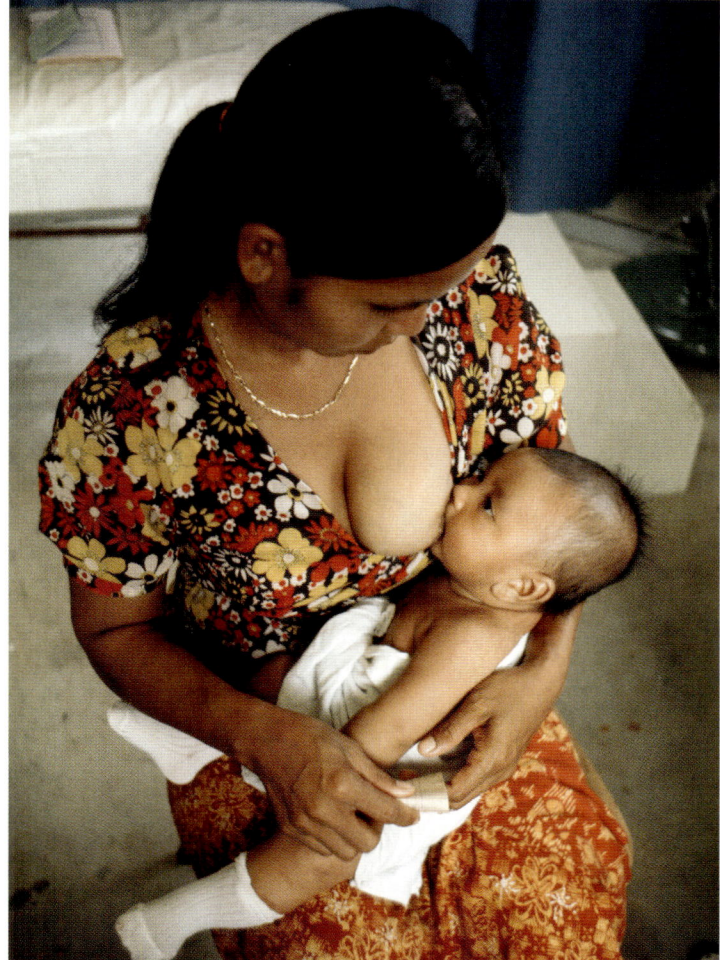

An image of a mother nursing her child. This image was included to show how adults provide food to their offspring.

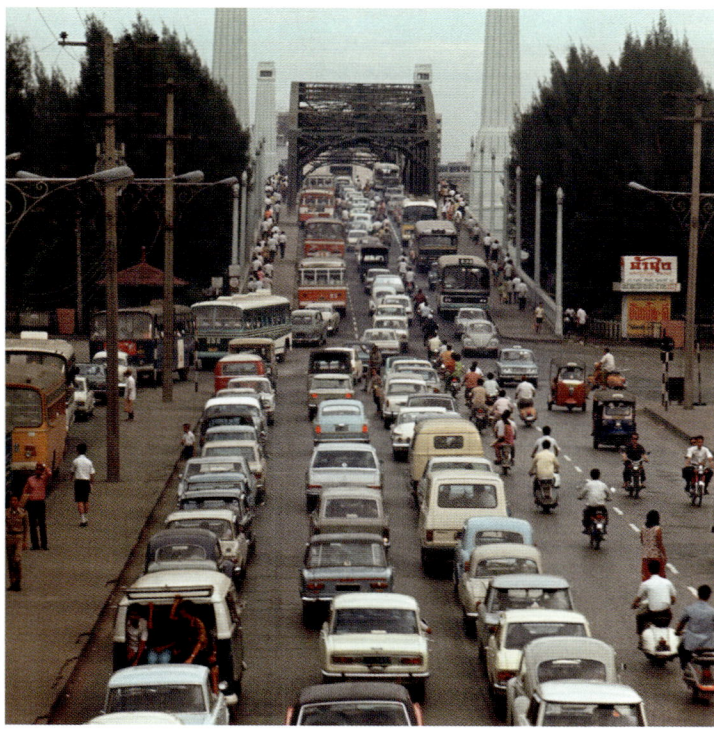

Rush hour traffic, Bangkok, Thailand, 1972: an image to highlight one of the many challenges we face, such as overcrowding in cities, which is just as prevalent now as it was in the 70s.

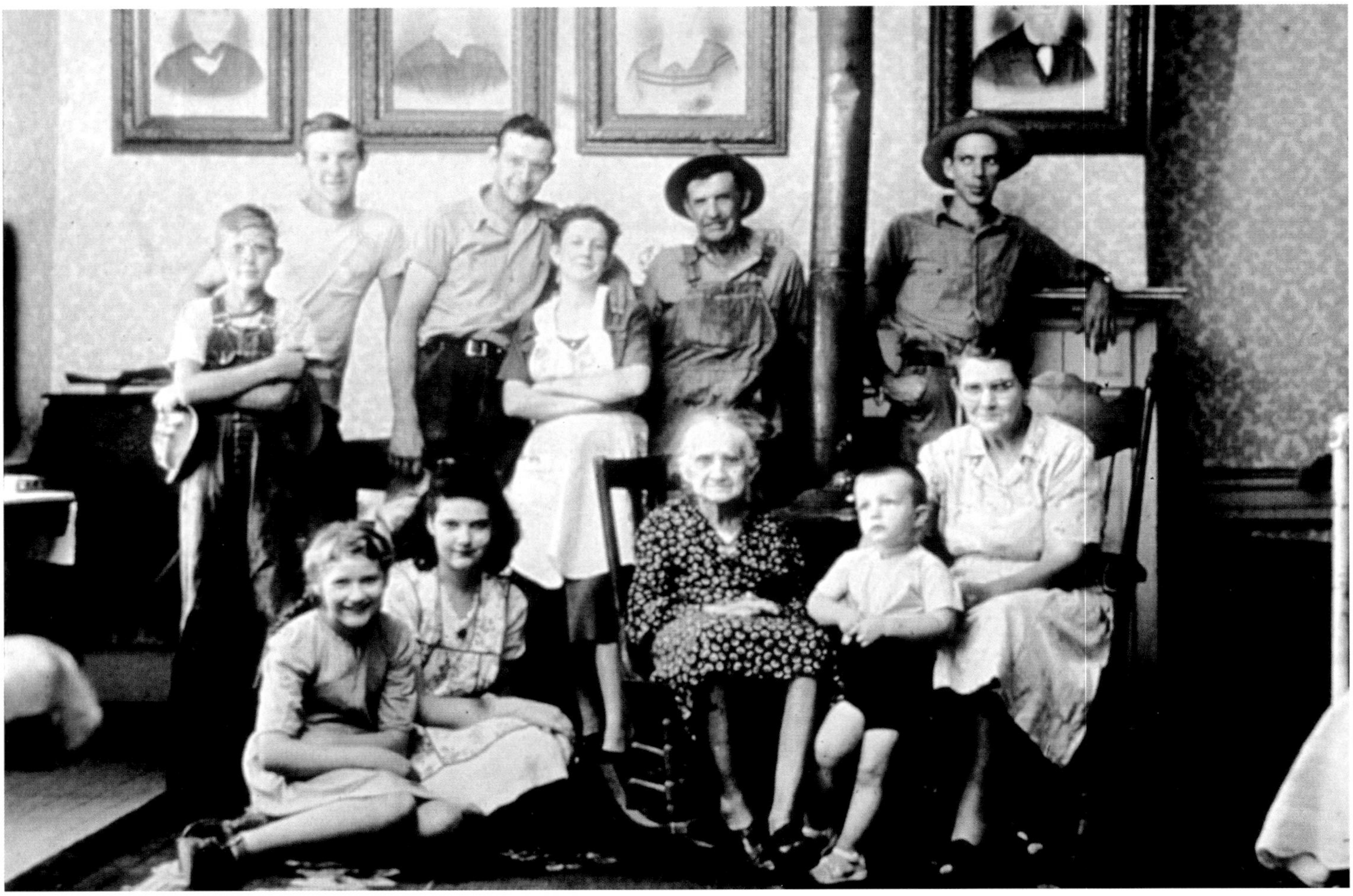

Family portrait: A photograph by Nina Leen of a mid-western American family, showing the five generations of one family together, with the sixth generation of that family displayed in the photographs hung on the wall.

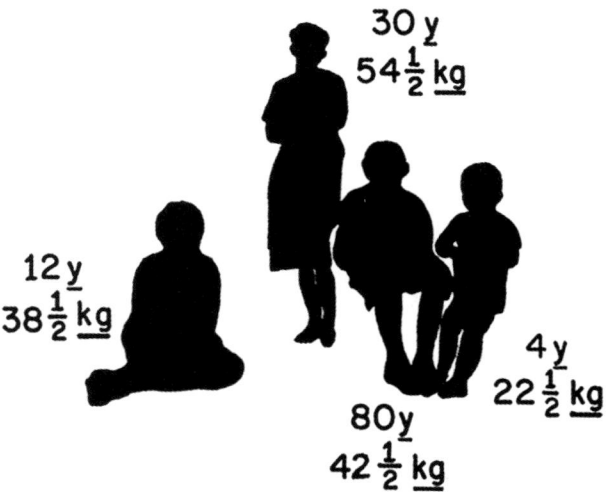

30 y
54½ kg

12 y
38½ kg

4 y
22½ kg

80y
42½ kg

Jon Lomberg used the family portrait above to create a visual guide that gave an indication of the age and weight of certain family members.

was wasted when it came to knowledge sharing. Instead, the ostensibly simple collection of 116 images is a mind-blowing puzzle that has layers upon layers of information hidden inside them, telling the story of our planet and our civilization.

The Music

The music had less of a cerebral sentiment. Not only was the team uncertain to what extent these foreign beings had the capacity of experiencing music, conveying knowledge is not the point of music to begin with. Whether it is bringing joy, delivering sadness or sending shivers down our spine, music is an emotional experience that plays an important part in all our lives. For the Golden Record, the pieces were therefore selected to bring about this emotional feeling to the fullest. Two goals were set in doing so. The first one was that the musical selection would not only represent the culture of those that built and sent this thing in outer space, but also represent the many different cultures around the world. The second one was simpler in that it needed to be a good record. Well, what is that? For the team it was a record that would stand the test of time. Long before music became abundant and out in the open due to the internet, the team spent days and days on end listening to stacks of records from around the world and taking notes. This deep dive resulted in a quirky record that includes pieces of music unlike anything you have heard before.

The selection of the music had the team sometimes hunting up rare records from impossible places. The back of an Asian appliance store in New York became the sourcing place for the Indian and raga musical pieces on the record. Wherever and whatever they searched for, the criterion was that the piece had to add some form of missing culture to the collection. From Bach to Beethoven, and from jazz to folk, it's all on there. The true idiosyncratic character of the record however, was caused by the worldly pieces. Whether it is the Japanese Shakuhachi piece, the Javanese Gamelan, the breathtaking Chinese Ch'in music or the 16-year-old Pygmy girl singing an initiation song in the forest of Africa, bringing them together on one record made it a truly mind-expanding adventure and listening experience.

Due to this critical selection procedure, there was only room for one Western Rock song. Being the absolute peak of Western music at the time, the team naturally turned to The Beatles. All four band members approved and suggested "Here Comes the Sun" for obvious reasons. Even though the records had an unknown audience, far in outer space, the publishers

did not agree and cancelled the contribution. Johnny B. Goode by Chuck Berry in the end became the chosen addition to represent Western Rock, a song that invokes a quintessential feeling of the genre (Berry's nickname was "Father of Rock & Roll" for a reason, of course). Perhaps more importantly, it is a song that speaks to the relentless, unending and hungry journey the Voyager spacecraft would undertake. With a little imagination, the chorus can be seen as a cheerleading anthem for the record and the mission itself: go, go, go Johnny go, go! Passing the test of the team, it also passed the test of the cultural zeitgeist of that time. In the spring of 1978, Steve Martin performed a skit on Saturday Night Live, where he imagined the first positive proof that other intelligent beings inhabit the universe in the form of an incoming message from outer space: "Send More Chuck Berry!"

Message & Voice Recordings

What Steve Martin might have not realized, is that the Golden Record actually also contained direct messages going out to these intelligent beings. Linda Salzman Sagan became responsible for the sounds and voice recordings to be included. Greetings in 55 languages were recorded and assembled as the voices from around the world. The greetings were sort of like tweets avant la lettre. People who spoke Arabic, Portuguese, Spanish, Vietnamese, and so on were asked to come in, say something unscripted from the heart, as short and sweet as they could. The result ranged from "Peace and happiness to everybody" to "Greetings to our friends in the sky, we long to meet you someday." To add a youthful note to the collection, Carl and Linda's son Nick was asked what he would say to aliens if he had the chance. At age six, he rocked up to the studio with an apple juice and a book under his arm, and when the light turned red said, "Hello from the children of Planet Earth."

Overall, a fantastical feat was accomplished in six weeks. A powerful and all-encompassing collection of images, music and greetings that conveyed the spirit of our planet and the people that inhabit it, then and now. But...was it worth it? Making a message also requires taking into account the successful receiving of the tale you are trying to tell. Even if it is a message in a bottle, what are the chances of that bottle being found? In other words, how likely will there be intelligent life actually demanding "More Chuck Berry!?"

THE PROSPECTS

Will it Ever Be Played?

Wondering whether the Golden Record will ever be played leads to a bigger question that the likes of Pink Floyd have asked many times: is there anybody out there? Statistically speaking the numbers almost compel that there are other life forms that have evolved into a state of intelligence and civilization. The bigger you think space is, the more probable it is that others are out there (and space is pretty big!).

Cosmic scales are almost unfathomable. Our Milky Way Galaxy alone already has about 200 billion stars and the universe, as far as we know it, hosts about 200 billion galaxies. What makes the scale especially big is the sheer distances between these stars and galaxies. Only imagining where the most nearest star is already puts this into context. If you took a grain of sand on a table that represents the Sun, Earth would be right next to it with a microscopically small distance in between. Our entire Solar System would fit on a table of about 6 feet. If the next star were also a grain of salt, how far you would need to place it from the table? 6.5 miles (10.5 km).

These sheer distances make it clear that space consists of precisely that, space. Although the universe offers plenty of possibility for there to be other intelligent life that might be able to place the stylus on the record and give it a spin, traveling the distances to do so remains something we cannot really comprehend. Since Voyager does not have the escape velocity to leave our galaxy's gravity, the space in which it will travel is relatively small compared to the complete universe with all its potential.

Besides the factor space, it is perhaps equally important to take into account the factor time. The chances of the Golden Record ever being played are highly dependent on how long our civilization, as well as others, lasts. Imagine it was the other way

around and some other intelligent life form far away, created the Voyager spacecraft with the Golden Record attached. Due to our own advancement in technology, astronomy and space travel, we would have only been able to pick up their signal or presence for the last 50 years. The five billion years before that, those spacecraft would have flown by us without any of us knowing. In fact, maybe they did.

Where the trajectory of the Golden Record will remain a mystery, the fact that it served as a mirror is set in stone indefinitely. Both the making of, as well as the actual contents that were etched into the grooves of those two disks, ultimately offered an unprecedented opportunity to reflect on what we are, what our planet is and what we might have to offer ourselves as well as others. A testament to the human spirit, the hardship we have overcome and the way people have flourished through the ages. A version of humankind. Not "fake" or "inauthentic", but something that is worth living up to.

With this, the Golden Records form a heartbeat of humanity that will never stop beating. They will be there in outer space, giving life to the Voyagers long after humanity has faded away. In a way, these disks have made humanity immortal. In our minds, we are metaphysical beings. We can transcend our experience and see the totality of being, or even of the cosmos. However, in our lived experience, we have always been defined by our finite nature. Human existence is about life and death, about the coming into existence of the world, as well as its demise. The Golden Records offer a new way to approach this philosophical distinction. It puts the metaphysical part of our being, the ability in our thinking to transcend ourselves, into real life. It carries us and our world into a distant future. "There is no end" as Ed Stone simply put it.

This may in the long run be the only evidence that we ever existed. When you know that about something you're working on, you treat it with great respect.

Frank Drake – Golden Record Technical and Pictures Director

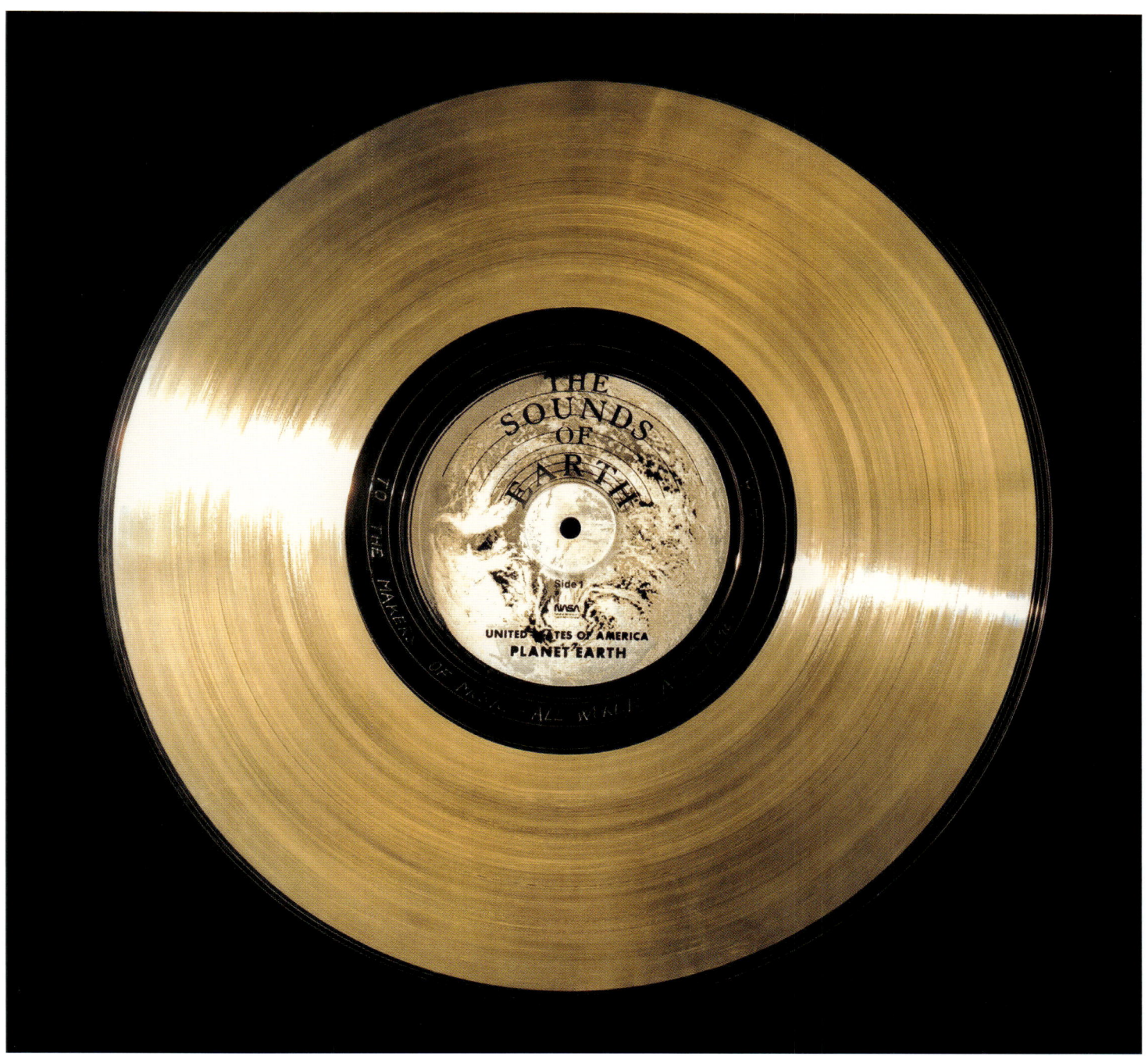

A ROCKY START

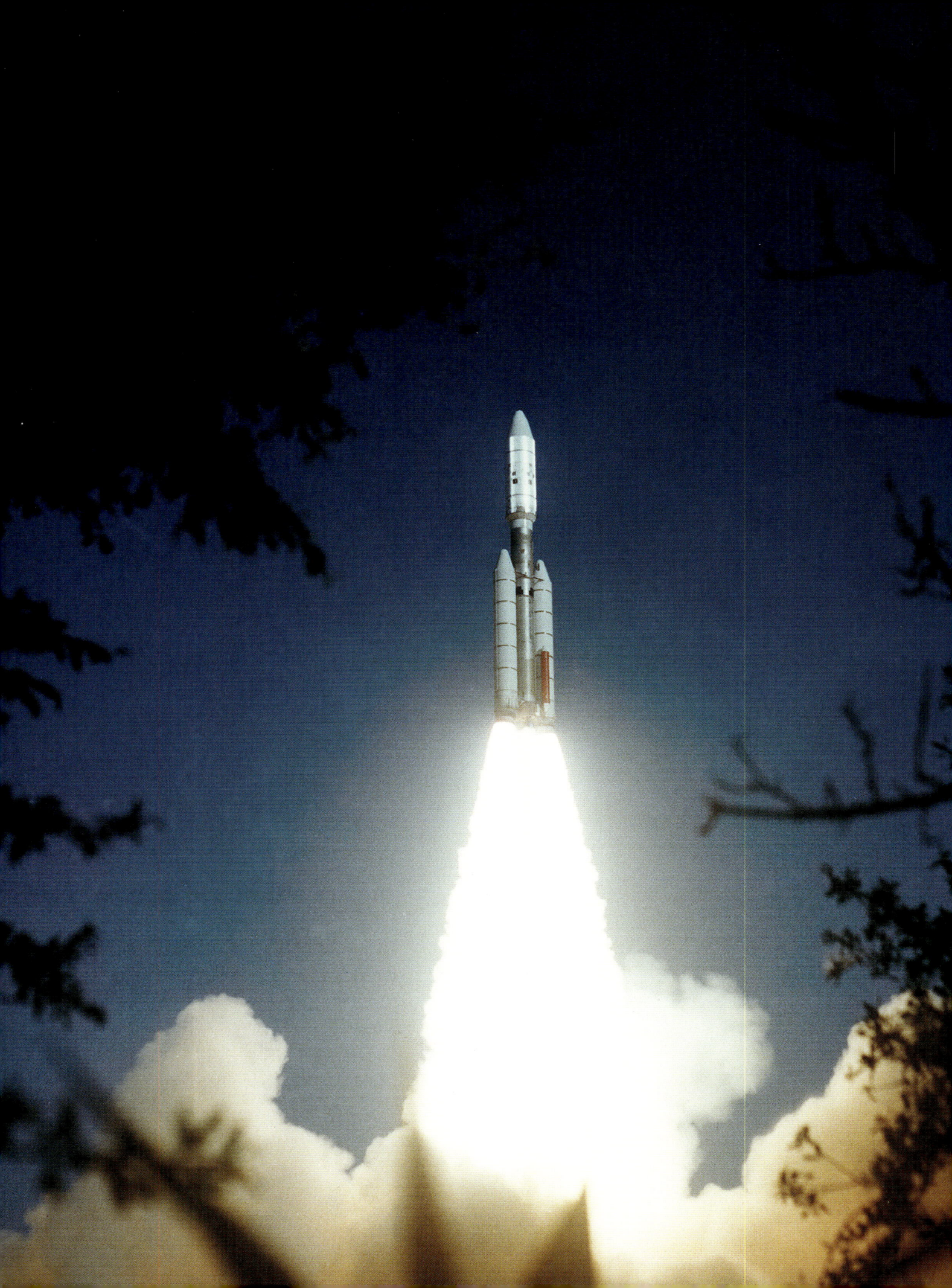

A ROCKY START

Introduction

It was in 1965 that Gary Flandro first discovered the rare alignment of the outer planets in the Solar System. Twelve years later, there were now two Voyager spacecraft built, and mounted onto the top of a Titan rocket, ready to embark on that grand voyage that Gary had first plotted during his stint at JPL all those years ago.

In 1977, Jimmy Carter was just sworn in as the 39th President, and the Commodore PC, the world's first all-in-one computer system was released. A small company called Apple was also incorporated at the beginning of the year. However, in 1977 the genre of space exploration was becoming much more mainstream in popular culture. Images of the first test flight of NASA's Space Shuttle, mounted to the top of a Boeing 747 was broadcast around the world. When the first Star Wars film was released,

filmgoers were lining up outside cinemas for hours to be able to watch it, setting the record for the highest-grossing film of all time. Space exploration was definitely a hot topic.

Both Voyagers launched from the Cape Canaveral launch facility LC-41 at NASA's Kennedy Space Center in Florida. Voyager 2 launched first, commencing on August 20th, at 10:29 local time. Sixteen days later Voyager 1 launched from the same launch facility on September 5th, at 8:56 local time.

Even though Voyager 2 launched first, Voyager 1 would quickly overtake and reach the Jovian System sooner, as Voyager 2 had been launched into a longer, more circular trajectory that would take the spacecraft an extra four months to travel to Jupiter.

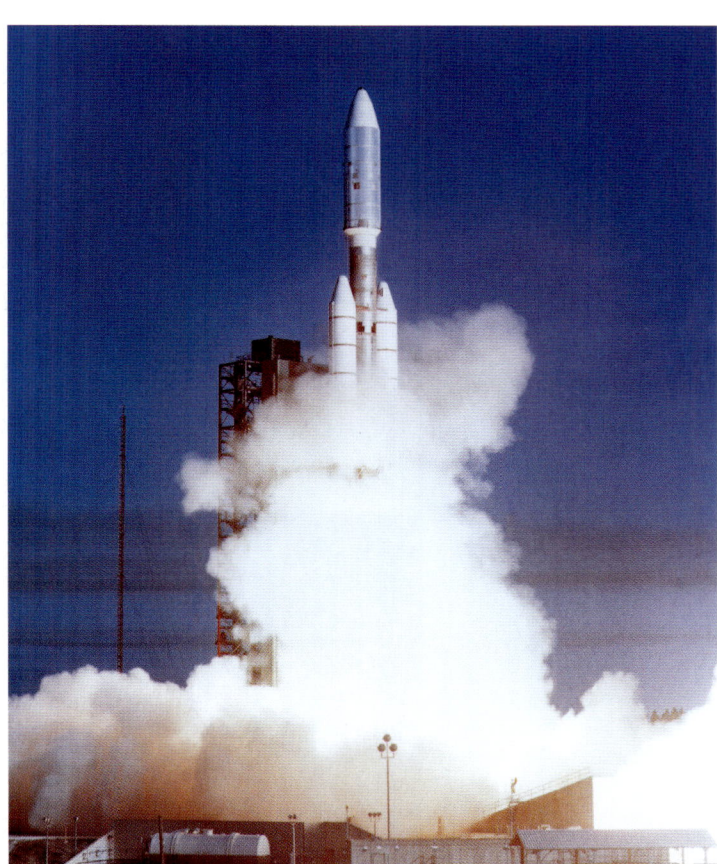

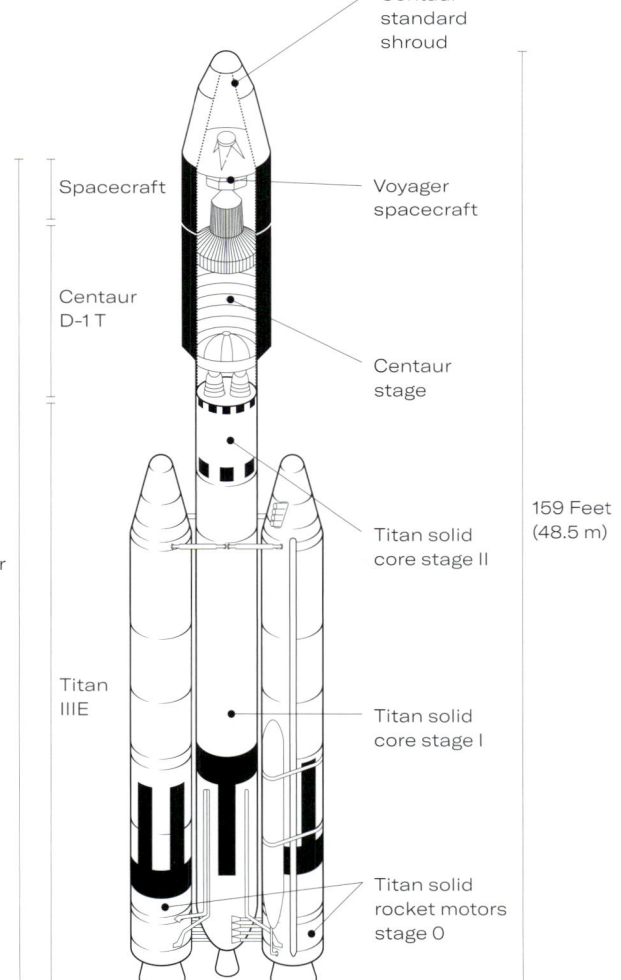

Centaur
standard
shroud

Spacecraft

Centaur
D-1 T

Titan/
Centaur
launch
vehicle

Titan
IIIE

Voyager
spacecraft

Centaur
stage

Titan solid
core stage II

159 Feet
(48.5 m)

Titan solid
core stage I

Titan solid
rocket motors
stage 0

Left: Voyager 2 launching from Cape Canaveral, Florida on August 20th, 1977. Voyager 1 launched later (above) on September 5th, 1977.

Titan-Centaur IIIE in launch configuration.

A Rocky Start

Just by holding this very book in your hands right now, it is clear that both of the Voyager spacecraft successfully achieved the majority of mission goals and objectives that they had to complete. What is a little less known is how close both spacecraft were to not even making it past our own Moon. This certainly would not have been the Grand Voyage that people had envisaged.

The launch of Voyager 2 initially seemed to have gone well. However just minutes after the rocket had reached Earth's orbit, mission control was alerted by the craft's orientation system that there was a suspected gyro failure on board. To add further problems to the mix, it seemed that the spacecraft was sending incomplete radio data transmissions, making it impossible to understand what was causing the gyro failure. Voyager 2 had only been flying for 53 minutes before another issue arose. The science scan platform boom was scheduled to unfold from its stowed launch position, but no data was transmitted back to mission control to confirm that the boom was fully extended. Without the scan platform being in its proper position, the mission objectives for Voyager 2, to gather any scientific data and photographs, would have been severely compromised.

The following day, Voyager 2's communication transmissions were starting to improve, mission control was able to recover some of the lost data that was sent during launch to find out what was causing the unknown pitch and yaw error. It turned out that the violent shaking of the Titan rocket during launch had initiated fail-safe routines on board Voyager that was designed to protect the spacecraft. The Voyager spacecraft has three gyros that make sure that spacecraft flies in a true and steady orientation.

The Attitude and Articulation Control Subsystem (AACS) that controls the gyros showed that one gyro was not working properly after launch. The acquisition of the navigation star Canopus in combination with the Sun lock helped Voyager finally to fully stabilize itself on its three axes for the rest of the mission.

Now that Voyager 2 was flying well, they moved onto solving their next question. What was the status of the scan platform boom? Had it fully extended? Flight controllers suggested that there was an issue with a micro-switch sensor that would confirm if the boom had deployed to within 0.05° of normal. They decided to switch on the plasma science instrument, to take measurements of the known stream of solar wind particles relative to the movement of the spacecraft. This smart thinking allowed mission control to determine that the boom had extended out to within 2° of full deployment position. Further tests, using the wide-angle camera also more accurately, confirmed the boom and platform position. The flight controllers were still concerned that the boom, although extended, was not locked into position. Any movements from the boom would ruin any attempted photograph. On August 29th, they planned a special maneuver that would spin and rotate Voyager 2 as fast as possible. The energy from the sudden jolting movement would determine if the boom was fixed, or it would firmly lock it in place if not.

Voyager 2 automatically aborted the planned sequence when it detected new flight instability. The automatic protocol on board made the spacecraft to roll until it would reacquire sight of the Sun and Canopus reference star. In the end, the flight controllers felt comfortable that the boom arm was sufficiently fixed, also expecting the arm to stiffen further as it would venture into the colder reaches of space and decided to not run any further maneuvers. Voyager 2 could now carry on with its mission. While the first of eight trajectory correction maneuvers was executed two months later than originally scheduled, the probe was now on its way to Jupiter.

Fast forward to September 5th. It was now Voyager 1's turn to embark on the longest journey in human history. The Titan IIIE thrusters were ignited at 8:56 local time, propelling Voyager 1 high up into Earth's atmosphere. The Titan-Centaur was the most powerful rocket used since the Apollo program. However, the Titan stage of this rocket did not live up to its billing.

The booster is not accelerating fast enough, due to an incorrect mixture of fuel and oxidizer in the rocket. Struggling to find the power required to escape Earth's gravity, the Centaur booster engines drink an extra 1,200 pounds (545 kg) of fuel to climb to its initial orbit. The ground team at the Launch Center in Florida have 45 minutes to calculate whether the Centaur booster has enough fuel left for its second firing to give Voyager 1 its final push onto Jupiter. The rapid calculations suggest that the booster will make it. The Voyager team back home waits with bated breath. Less than an hour after reaching its initial orbit, the booster fires for the final time. It makes it. With a mere 3.5 seconds of fuel left in the tanks, Voyager 1 is now joining Voyager 2 towards Jupiter. Albeit, just by the skin of its teeth.

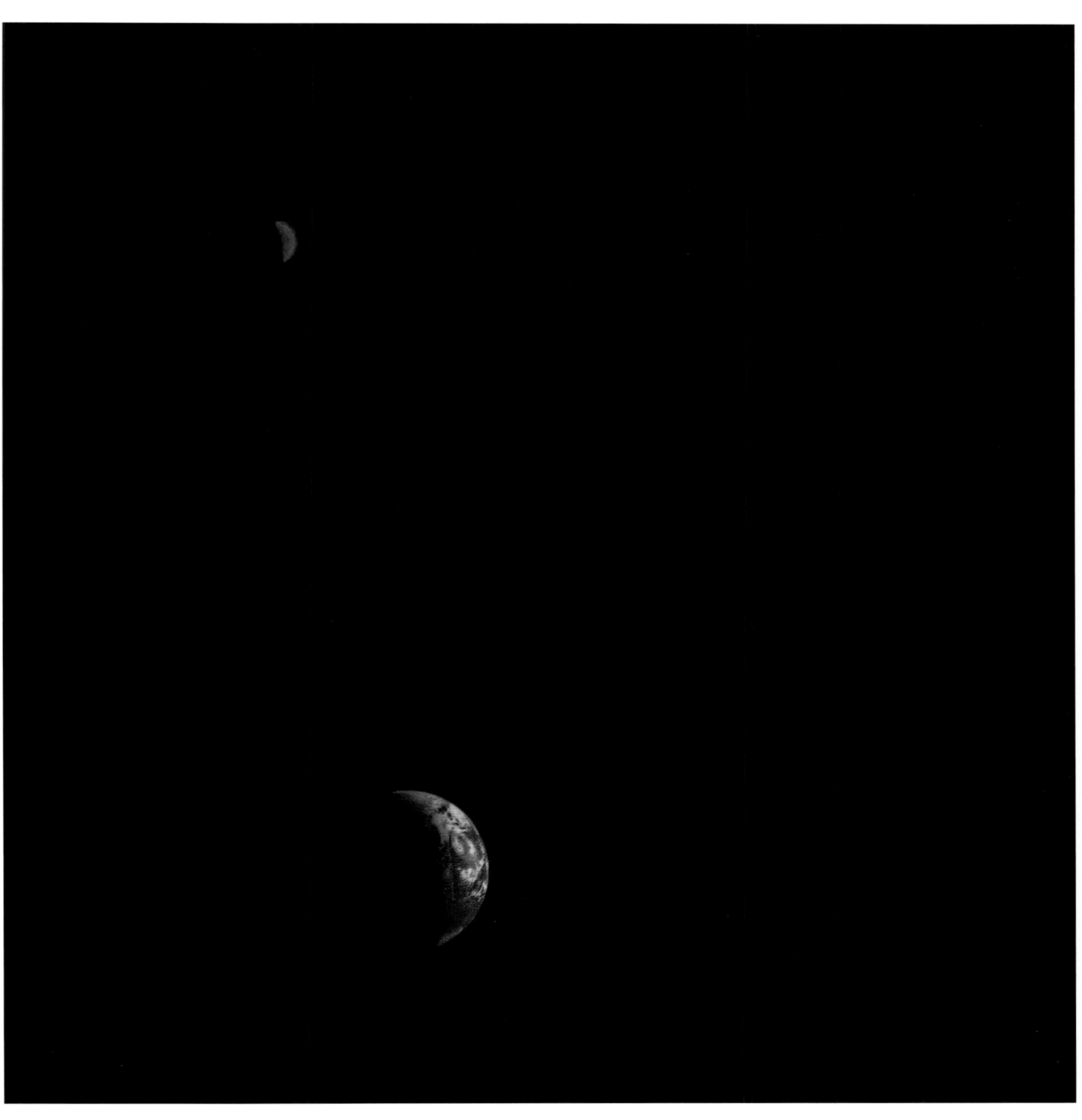

The first ever photograph of the Earth and Moon within one
single frame. Taken by Voyager 1 on September 18th, 1977, from
a distance of 7.25 million miles (11.66 million km) from Earth.

The Launch   Earth & The Moon   Voyager 2   October 12th, 1977

Voyager 2 also captured The Earth and
The Moon in one frame on October 12th, 1977.

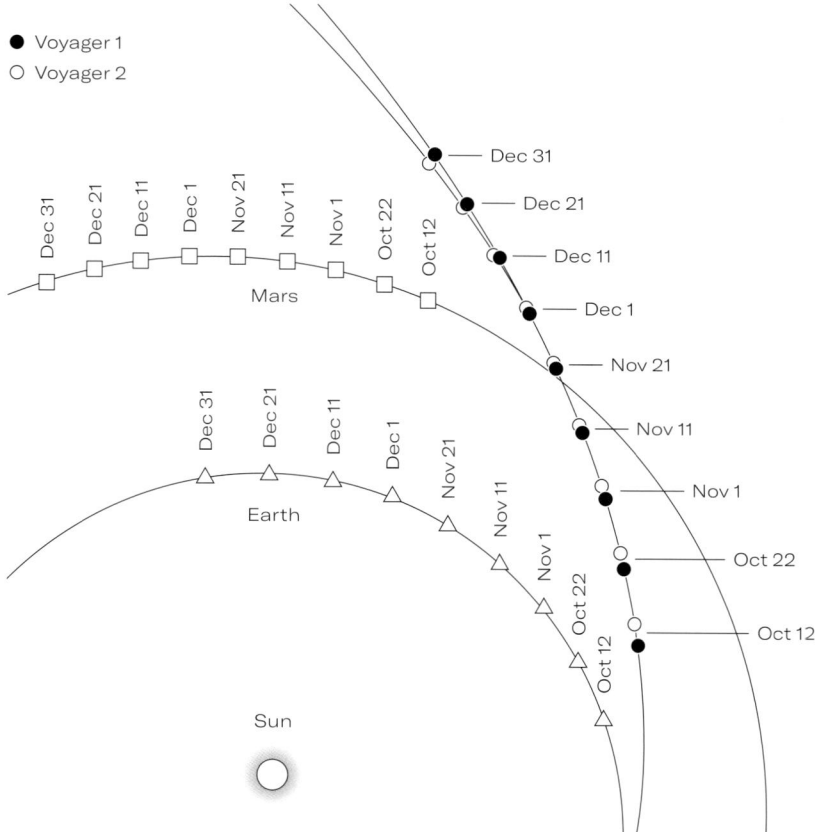

● Voyager 1
○ Voyager 2

Dec 31
Dec 21
Dec 11
Dec 1
Nov 21
Nov 11
Nov 1
Oct 22
Oct 12

Mars

Dec 31
Dec 21
Dec 11
Dec 1
Nov 21
Nov 11
Nov 1
Oct 22
Oct 12

Earth

Dec 31 — Dec 31
Dec 21 — Dec 21
Dec 11 — Dec 11
Dec 1 — Dec 1
Nov 21 — Nov 21
Nov 11 — Nov 11
Nov 1 — Nov 1
Oct 22 — Oct 22
Oct 12 — Oct 12

Sun

The early trajectory and timeline of both Voyager spacecraft as it leaves Earth's orbit, past Mars and Voyager 1 overtaking Voyager 2.

First part of the Journey

Getting a rocket to break through Earth's gravitational forces is one thing, powering a spacecraft to Jupiter under its own steam is another entirely. To reach Jupiter, the Voyager probes required as much energy for the two-year journey as it needed to get the rocket off the ground and into Earth's orbit. Of course the spacecraft could not benefit from any planetary slingshot assists during their first leg of the journey.

Over the following 20 months, both spacecraft would start to significantly slow down due to the loss of energy as it traveled further away from Earth's influence. Slowing down from around 23 miles per second (36 km/s) to 6 miles per second (10 km/s). Both Voyagers would travel (relatively) slowly through the asteroid belt, and would only start to accelerate again once each craft approached the gravitational pull of the Jovian system. From this point on, the Voyager craft could rely on the planets it would visit to push its way through the Solar System.

Even though Voyager 1 was launched after Voyager 2, it had a shorter trajectory that was optimized for a close approach to Saturn's largest and most interesting moon, Titan. Both spacecraft entered into the asteroid belt on December 10th, 1977, with Voyager 1 overtaking the other nine days later.

If you construct a mental image of the Voyager probes traveling through the asteroid belt, it is probably one of a spacecraft maneuvering around millions of small, dark, floating rocks. The reality

is quite the opposite. Traveling through the vast emptiness of the edges of the inner Solar System was fairly quiet, the density of the asteroid belt being so low that no maneuvering was necessary to avoid collisions. The harsh conditions of interplanetary space could cause some serious damage to the craft's sensitive equipment. The ground crews would periodically test all the systems aboard to make sure everything was in good working condition for when the spacecraft would start to make their approaches to Jupiter.

One major issue did arise during Voyager 2's journey through the asteroid belt. In April 1978, Voyager 2 started having trouble receiving control commands transmitted from Earth. This caused Voyager to autonomously fix the issue, which made the spacecraft switch from its primary to back up radio receiver. Unfortunately, a new error developed in the backup receiver. A failed capacitor in the receiver meant that only signals sent at a very precise frequency could be utilized. With only a narrow frequency being detected, the communications team back home had to calculate and adjust the frequency, as the signal became distorted and thus undetectable, as the speed at which Voyager 2 receded changed due to Earth's orbital motion and, later, due to the acceleration of the spacecraft. Voyager 1 exited the asteroid belt on September 8th, 1978, and started its distant observations of Jupiter on January 6th, 1979. Voyager 2 exited the asteroid belt later on October 21st, 1978, and started its distant observations of the Jovian System, 6 months later on April 25th, 1979.

KING OF THE GODS

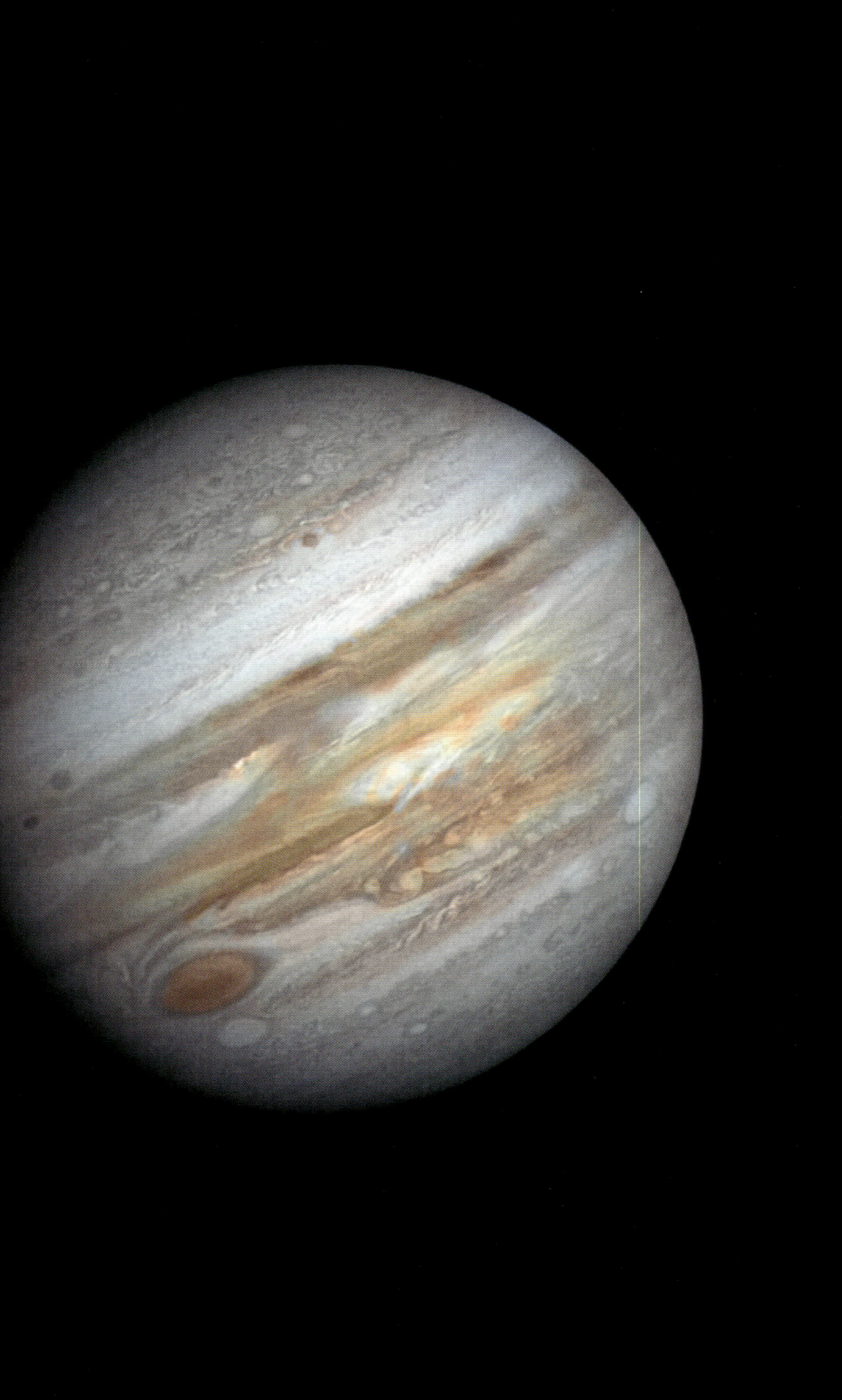

THE FEROCIOUS FIRST

A journey to Jupiter from Earth will let you pass Mars first, and then take you right through the asteroid belt, and eventually drop you off at the biggest marble of them all. The fifth planet from the Sun indeed is the largest planet by a long stretch. Its scale is probably best put into context with a simple comparison. Take all other planets, moons, comets, and asteroids in the Solar System and mash them together. Collectively, all these objects would still not have as much combined mass as that of Jupiter.

This giant is mainly made up of gaseous and liquid matter, primarily composed of hydrogen and a quarter of its mass being helium. Research has shown that there is no well-defined solid surface, but it might have a rocky core of heavier elements. The planet has a diameter of 88,846 miles (142,984 km). That means you would need 1,321 single Earths, to fit into one Jupiter. It is also not afraid to show itself. The size and (relative) closeness to Earth makes Jupiter one of the brightest objects in our sky—the fourth brightest after the Sun, Moon, and Venus. So bright in fact, that sometimes you can see Jupiter shining in the daytime when the Sun is low in the sky.

Its high visibility has made the presence of this planet known since ancient times. It was the Babylonians that first spotted and documented it on stone tablets that date between 350 and 50 BCE. They referred to the planet as Marduk, named after the primary god in Babylonian religion. A little while later the Ancient Greeks had other plans for it and called the planet Dias, more commonly known as Zeus. The Romans coined the name of the planet that we know it by today. In Latin, it was called Iovis stella (the star of Jupiter) after the King of all the Roman Gods, he who ruled the light and the sky.

Throughout time, the influence of this King of the Gods was thought to be felt by people. Over the centuries, the presence of Jupiter at certain times of the year was believed to have an altering effect on people's mood and personality, as documented in the horoscopes of the Middle Ages. Astrologers coined the term "jovial" accordingly, referring to the uplifting and happy astrological influence that the Jovian System would have on people.

Left: Jupiter captured by Voyager 2 on July 6th, 1979 at 19:27:11.

THE STARRY MESSENGER

Actual knowledge concerning Jupiter followed a different path. That started with Aristarchus of Samos, who got it pretty much correct back in around 240 BCE. The Greek astronomer and mathematician was the first person recorded who proposed the idea of Heliocentrism, the idea that the Sun is the center of our Solar System and that the Earth and all the other planets revolve around it. Quite the announcement considering the Ancient Greeks believed in mythology.

It would take about 1900 years for a big next leap in our understanding. During the 16th Century, a remarkable period in the history of astronomical enlightenment, our knowledge of the Solar System and science vastly expanded. This period perhaps can only be rivaled with the series of new discoveries that were to be revealed by the Voyager mission.

Polish-born Renaissance astronomer Nicolaus Copernicus was the first to predict and calculate a heliocentric model of the known Solar System in his book, "On the Revolutions of the Heavenly Spheres." His hypothesis was later improved by the work of Johannes Kepler, whose 1609 work "Laws of Planetary Motion" described how the planets orbit around the Sun.

A year later, a man from Pisa wrote a letter. In this letter, dated January 7th, 1610, he wrote about the three bright stars that he noticed next to Jupiter when looking through the telescope that he famously built. This man from Pisa was called Galileo Galilei. The Italian astronomer, physicist and engineer continued to observe these three stars over the next week, seeing all three stars move to different positions around Jupiter. In certain cases, some of the stars disappear from sight and then reappeared a few days later. On January 13th, he recorded an extra star even further from the planet, taking the tally up to four. As he continued to observe, it became apparent that these four bright stars were in fact four large moons.

Emboldened by this discovery he published "Starry Messenger" two months later. A publication that had a momentous effect on our understanding of the Solar System. In it, Galileo wrote about his discovery of these four moons—Io, Europa, Ganymede and Callisto, collectively naming them the Medicean Stars, in honor of the four royal Medici brothers. The name did not stick, and this group of moons are now referred to as the Galilean Moons, after their original discoverer.

In this publication, Galileo claims that these four moons orbit around Jupiter, similar to how our Moon orbits around Earth. He elaborated further on this discovery, claiming that seeing these satellites orbit around another planet is proof that the Earth was

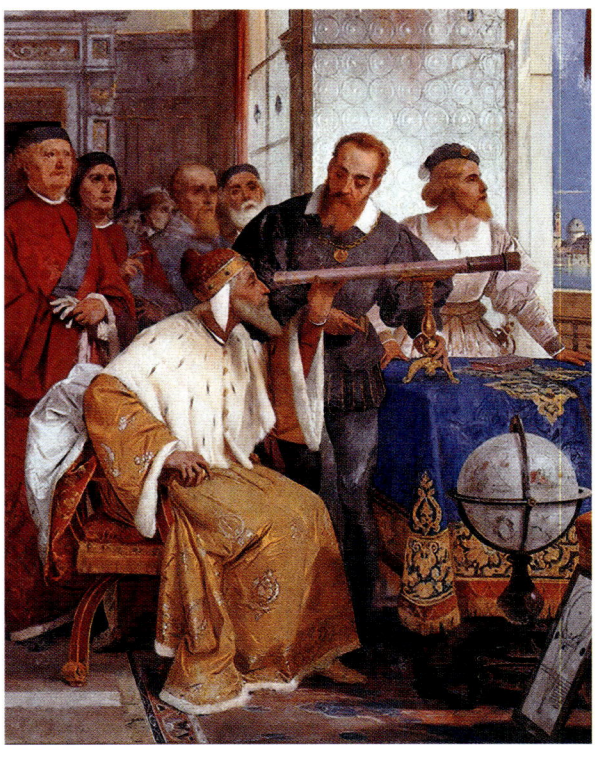

Galileo demonstrating his telescope to the Doge of Venice, part of a fresco by Giuseppe Bertini.

not the center of the universe, and that the Earth and the other four planets known to us at the time orbited around the Sun. Naturally, the Roman Catholic Church was not too keen on the statement. After refusing to renounce his scientific claim, they eventually placed him under house arrest for the remainder of his life.

This golden age of astronomical discovery continued. A few months after Galileo's death in 1642, Isaac Newton was born on Christmas Day. Years later, the proverbial apple finally dropped in front of Newton. Inspired by the previous work and theories from Johannes Kepler and Galileo Galilei, he announced his Law of Universal Gravitation, which would provide the foundation for much of our understanding of how objects in our Solar System behave.

In 1665, English astronomer Robert Hooke and Italian astronomer Giovanni Cassini both (separately) documented Jupiter's "Giant Red Spot" for the first time. Over the next three centuries, telescopes became more powerful and refined, allowing for better and clearer sightings of Jupiter and its storming swelling bands of color. However, it was not until the 1970s that everyone on Earth would be able to view truly mesmerizing close-up images from the king of the Solar System.

JOURNEY TO JUPITER

Imaging of the still-distant Jupiter commenced January 6th, 1979. Already these far distant images were of higher resolution and detail than any photograph taken from Earth. The photography and data collection became more extensive as the spacecraft got nearer to their targets.

On approach, the images of Jupiter were assembled into time-lapse sequences, so for the first time scientists could truly appreciate the dynamics of the belt and zones, and the highly dynamic nature of the storms. Voyager 1 closed in on Jupiter and its moons on January 6th, 1979, beginning its encounter phase as it sped through the Jovian System at 21,748 mph (35,000 km/h). It reached its closest approach on March 5th, passing just 217,479 miles (350,000 km) above Jupiter's cloud tops, capturing thousands of images over the course of its flyby. Voyager 1 completed its Jupiter mission in early April.

Voyager 2 started its approach a little later that month, taking a different route through the Jovian System. During its July 9th closest approach, it would pass Jupiter from 354,181 miles (570,000 km) away, 136,701 miles (220,000 km) farther away than the pass of Voyager 1. The different course was not just planned to get a closer look at Europa, but also had a more ambitious motivation. Through this route, Voyager 2 could make use of the right amount of force from the gravitational pull it would receive when passing by Jupiter, to slingshot itself to Saturn in a way that it would allow it to continue its journey on to Uranus and Neptune.

In total, both spacecraft collected more than 33,000 pictures of the Jovian System, expanding our knowledge of the Solar System with its data, while at the same time stunning the world with some of the most intriguing and fascinating space images ever seen.

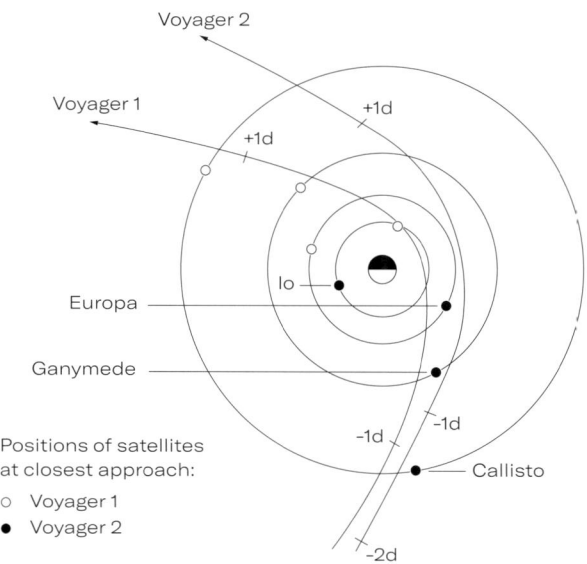

The positions of satellites at the closest approaches from Voyager 1 and 2.

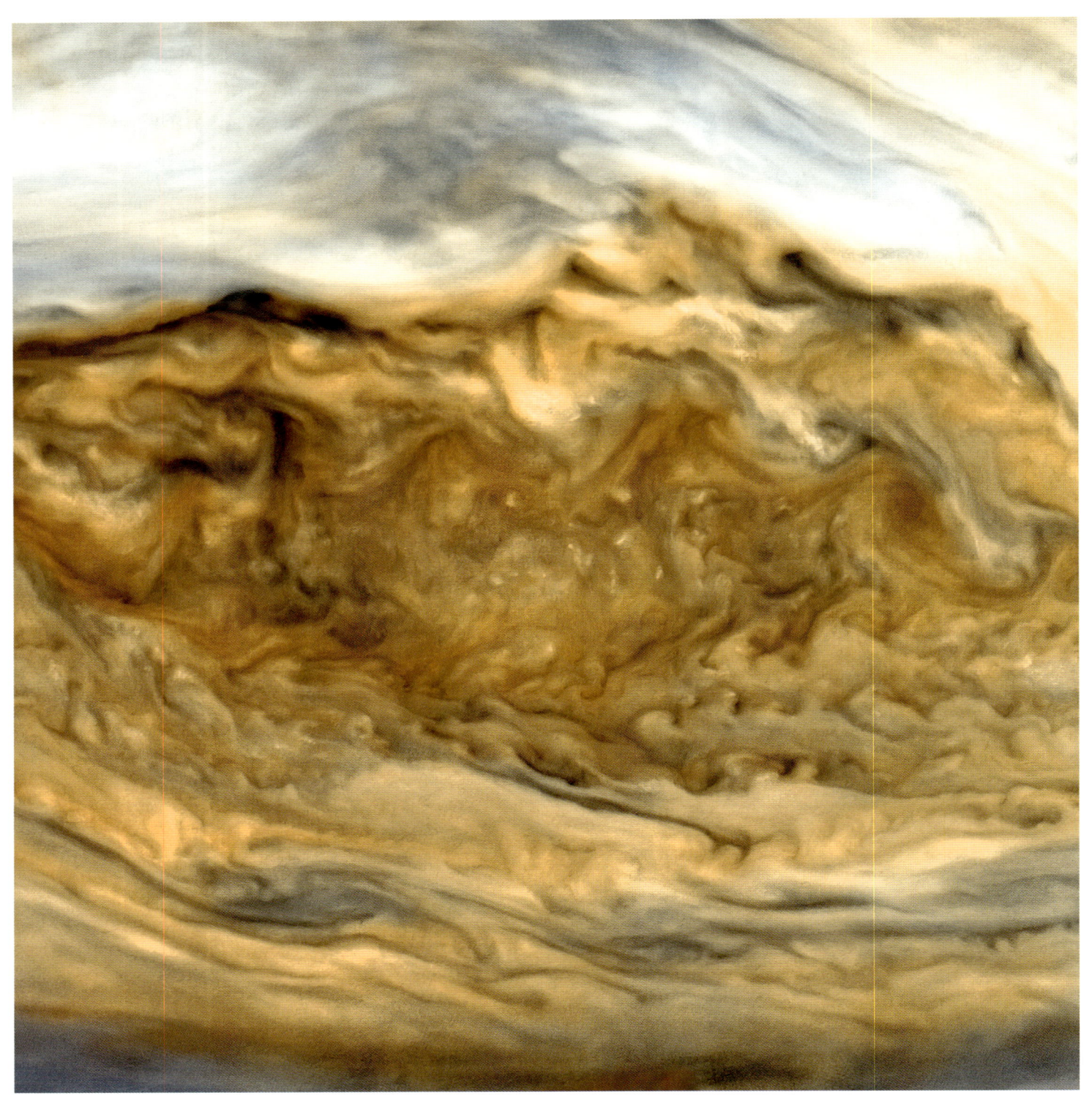

A close-up of the Great Red Spot, showing its discrete clouds.

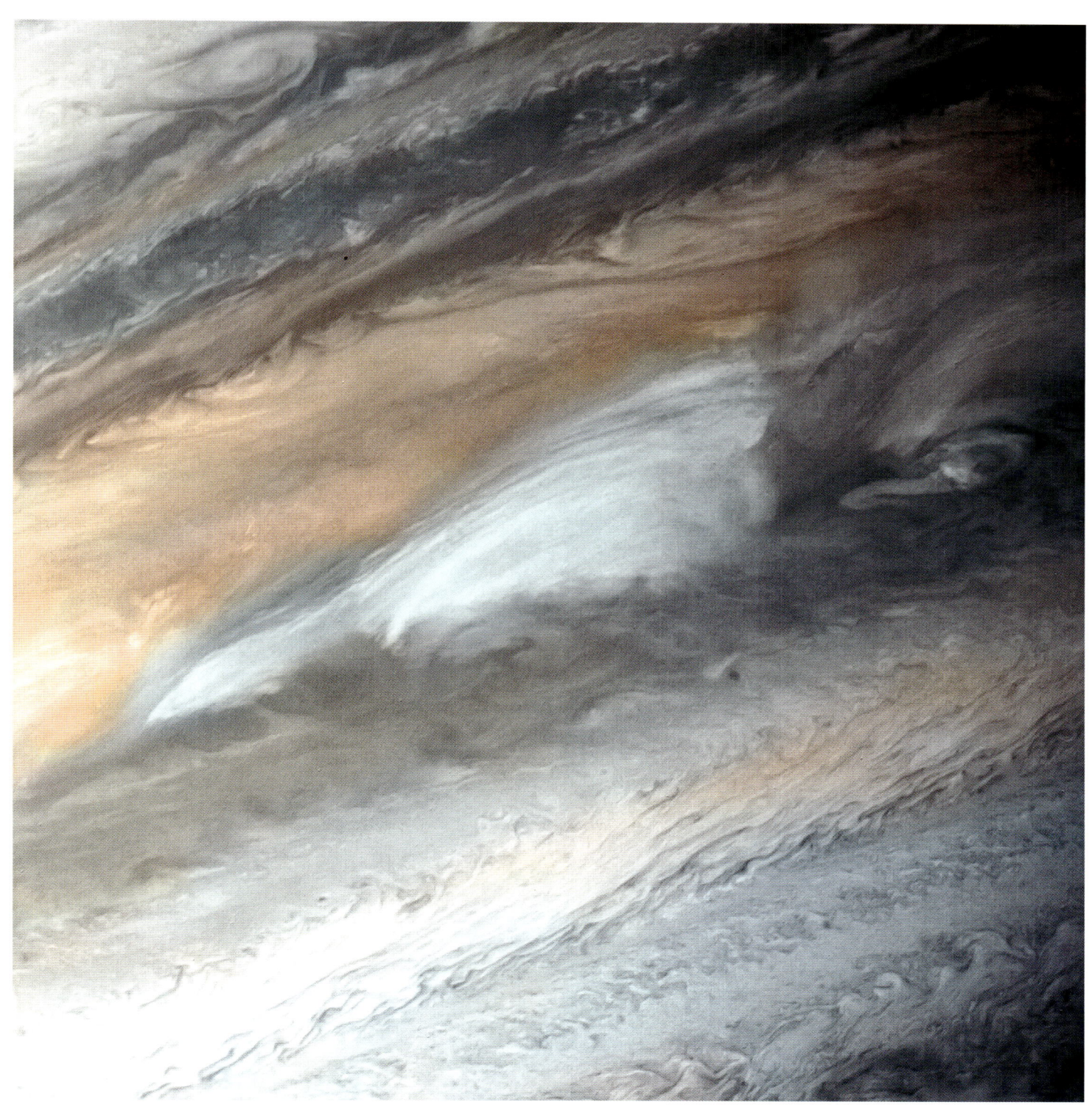

OUT OF SYNC

The harsh radiation conditions around Jupiter were well known to the Voyager team. The earlier Pioneer probes had detected and relayed this information back to Earth during their missions. This discovery was an important warning to all future missions. The Voyager engineers used this knowledge to make sure that the intense radiation conditions would not wreak havoc on the onboard electronics and computer systems. They installed an extra 50 pounds (23 kg) of Tantalum to shield the sensitive systems. Tantalum is a dense "transition metal" that is strongly resistant to corrosion and very well adapted to blocking high-energy solar protons and cosmic ray radiation. Unfortunately, even with this extra thick metal shielding, Voyager 1 was not able to fully cope with the levels of radiation with which Jupiter bombarded it.

As Voyager 1 left the Jovian System, the craft sent back a series of smeared and blurred imagery. It turned out that radiation had damaged and altered the code that was controlling the computer's clock system. The alteration in code meant that the clock in the control scan platform became slightly discordant with the other clock that controlled the imaging camera's exposure time. This meant that the scan platform started to move (commanded by one computer) before the shutter was closed (commanded by the other computer). This led to smeared pictures and caused a lot of useful data and imagery to be lost from Voyager 1's visit to Jupiter. Fortunately, the imaging teams had enough time to code a workaround for the upcoming Voyager 2 flyby that would happen a few months later. By simply planning their sequences to wait two seconds after the shutter closed to start the slew to the next image location, they cleverly avoided any potential future problems with radiation further damaging the computer processors.

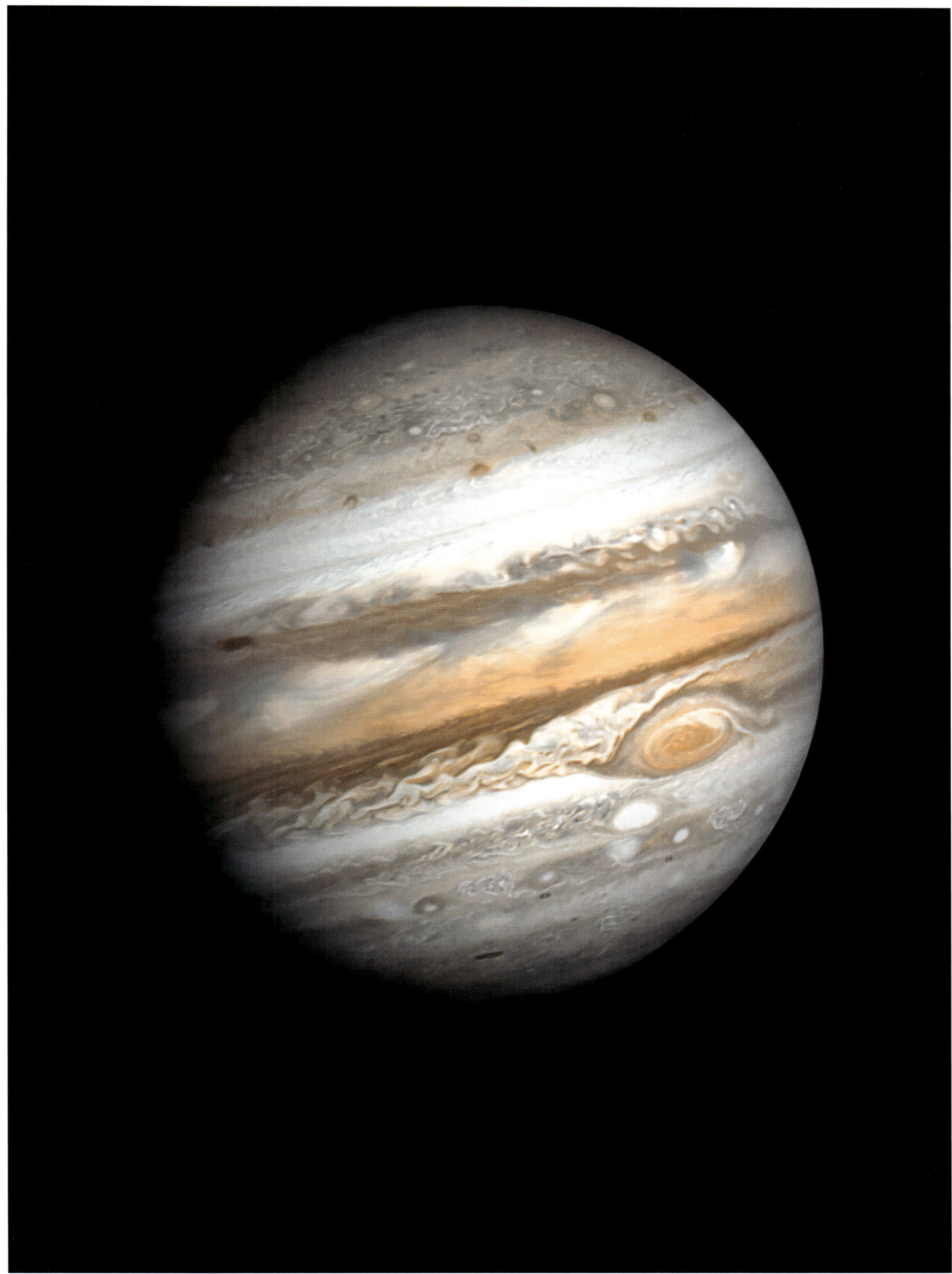

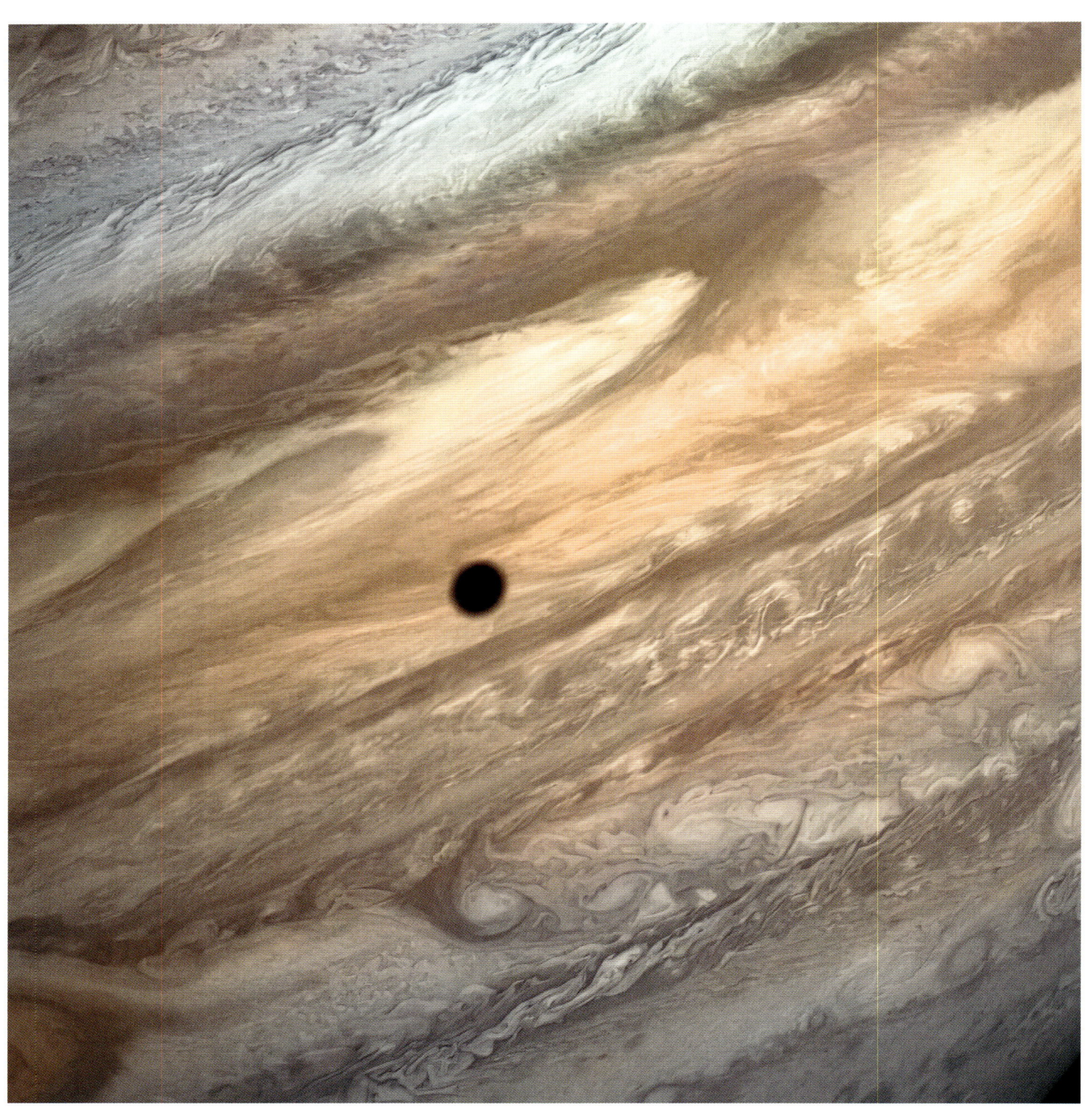

Jupiter's cloud tops with the circular shadow of Io in the middle.

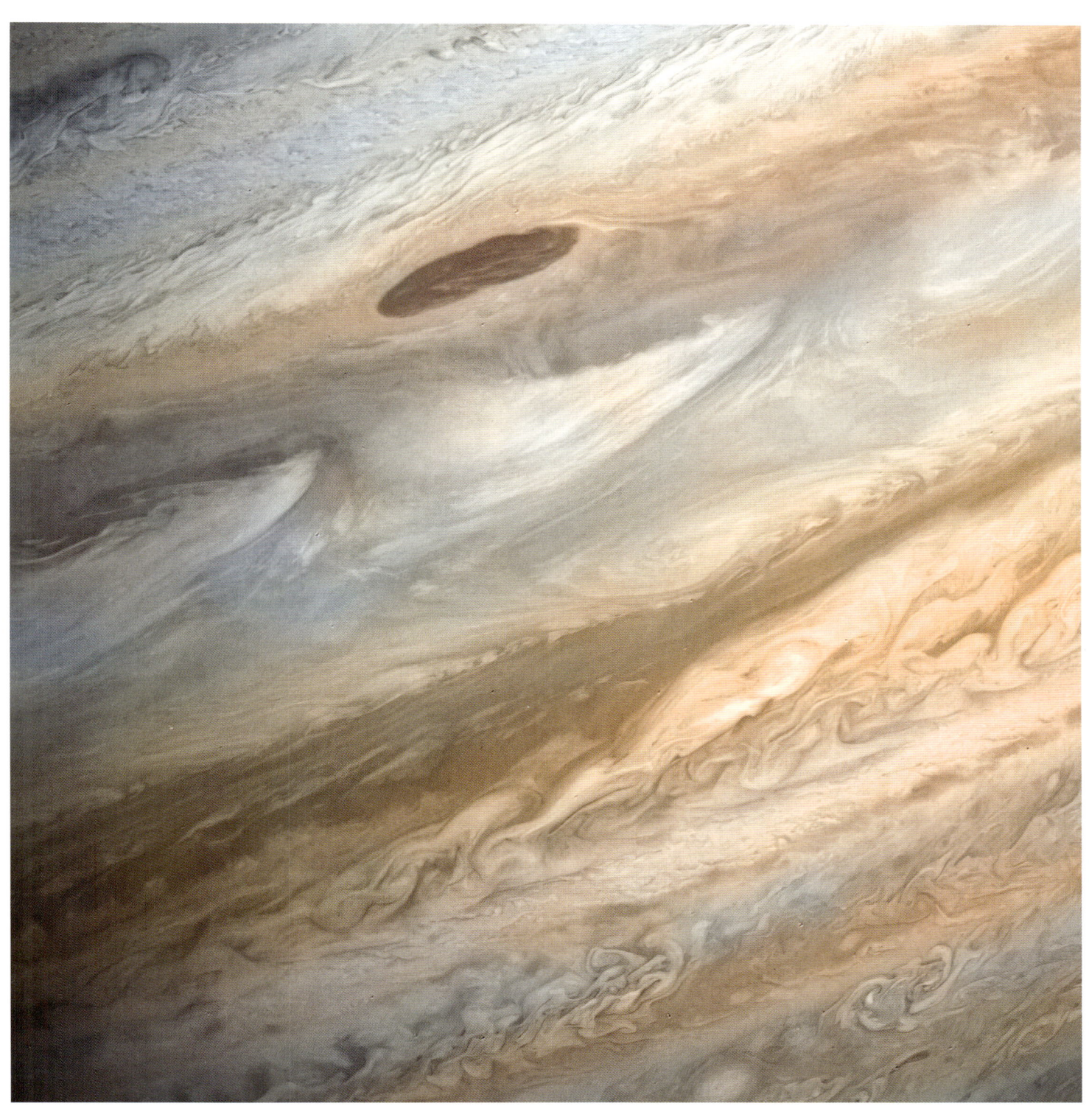

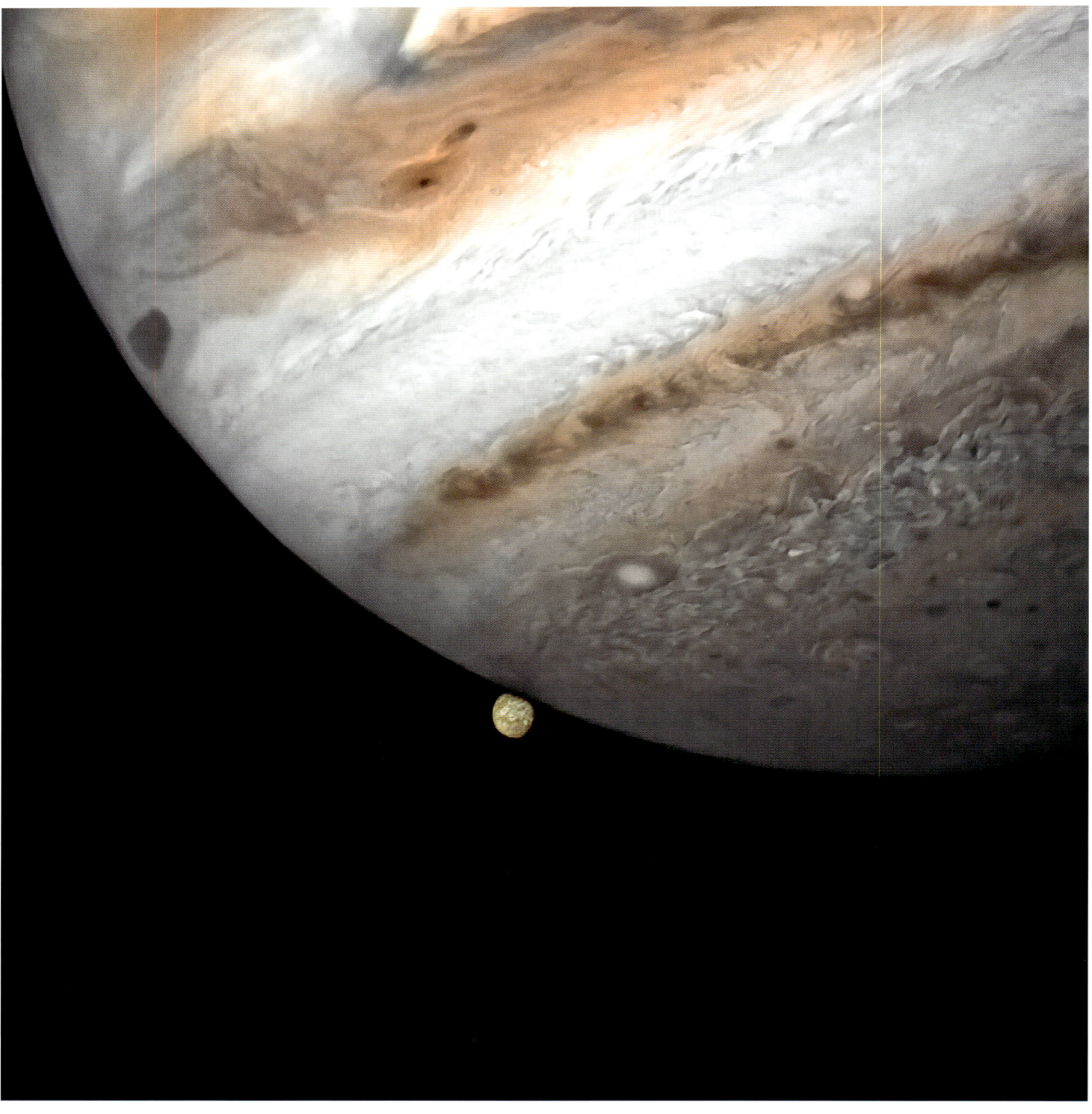

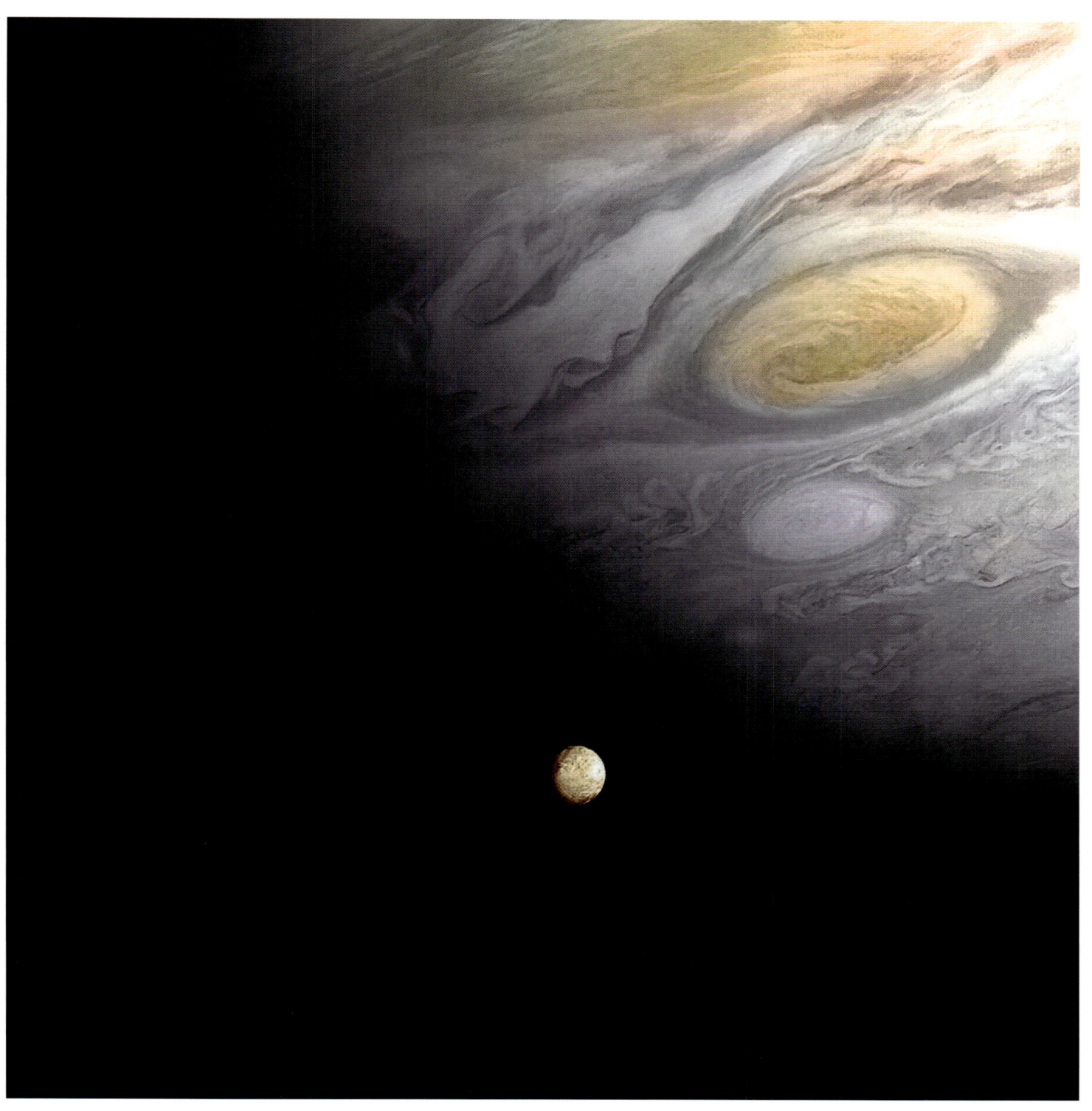

We roar along the rust belts, the great
red spot, the polar vortex, the caress
of solar flares, ruffle the molten methane
and ammonia oceans of me, the storm-
riven non-surface of me and mine,
that which you call skin, a threadbare term
to describe where I stop and others begin.

Alan Sternann Rousselot, Dawn of the Algorithm

A STORMY ATMOSPHERE

Jupiter has always been known for its ferocious character. The swirling bands and spots on the planet have been observed from Earth for many centuries. However, it was not until 1973 that we got really close-up views of Jupiter. As more detailed pictures of its cloud-tops became available, it was becoming apparent just how turbulent the weather is on Jupiter.

Swirling cyclones fight against each other, pushing their 200 mph (320 km/h) winds around the planet. The atmosphere of Jupiter is the largest in the Solar System. Its upper atmosphere is composed of about 90% hydrogen and 10% helium. Clouds constantly cover the surface and are formed from ammonia and ammonium hydrosulfide. These toxic clouds and lightning storms push up against each other to form massive and complicated storm systems, spinning around in a continuous and uniform motion to form some of the most beautiful and stunning patterns seen in nature.

One of these storm systems is the Great Red Spot. This storm measures around 10,160 miles (16,350 km) wide, 1.3 times the diameter of the Earth. The storm cyclone rotates in a constant counterclockwise direction, taking around six days to complete one full circle.

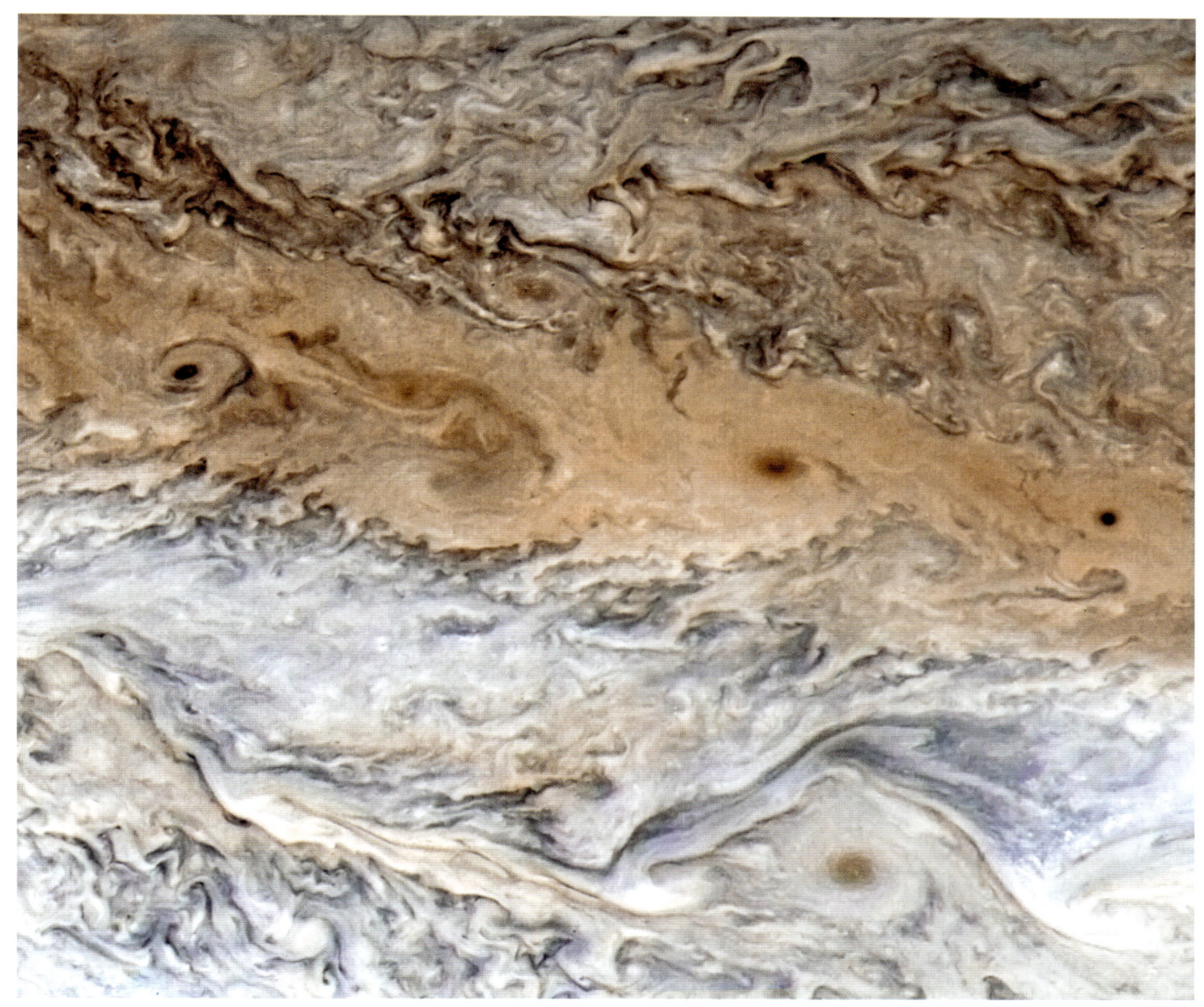

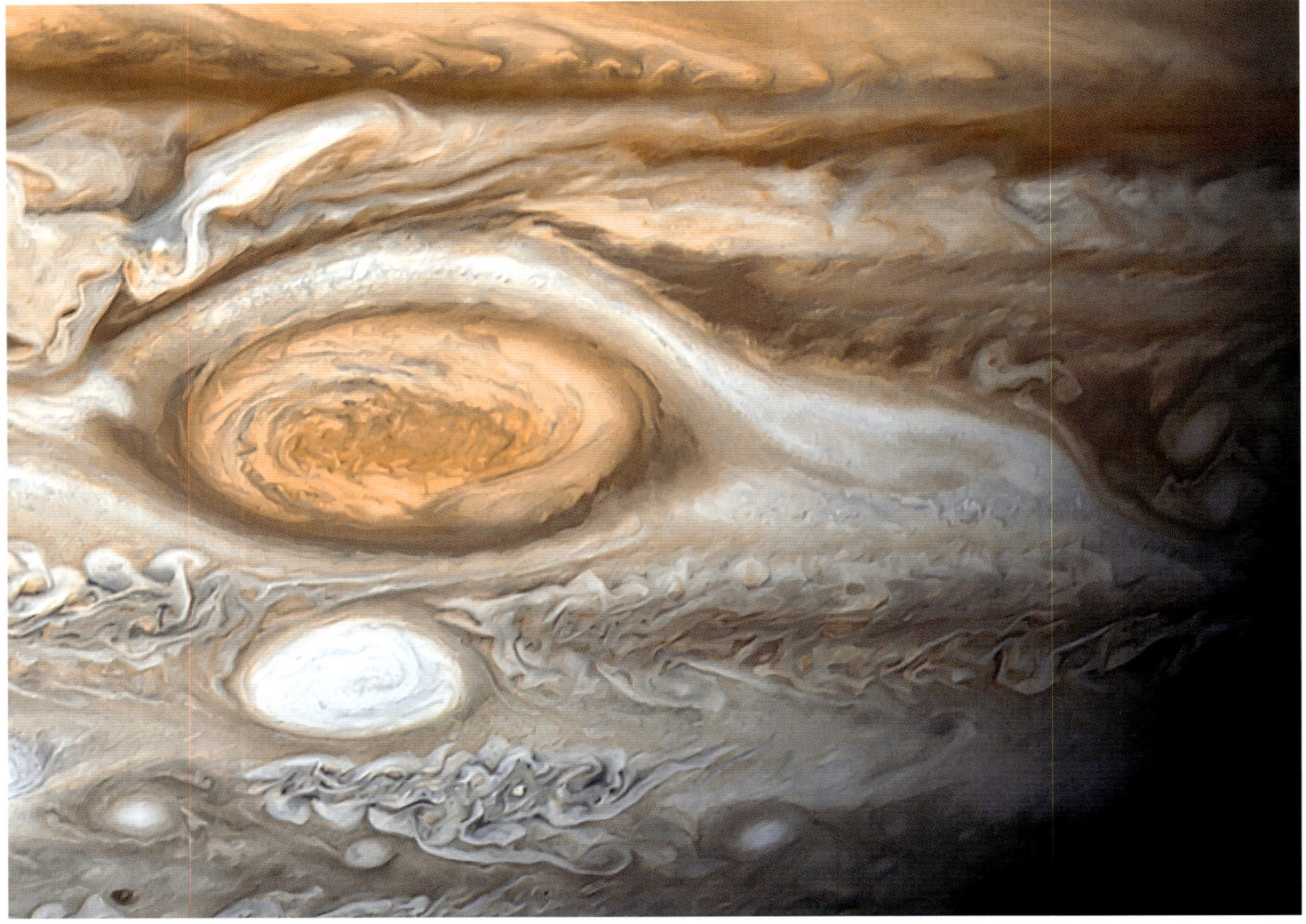

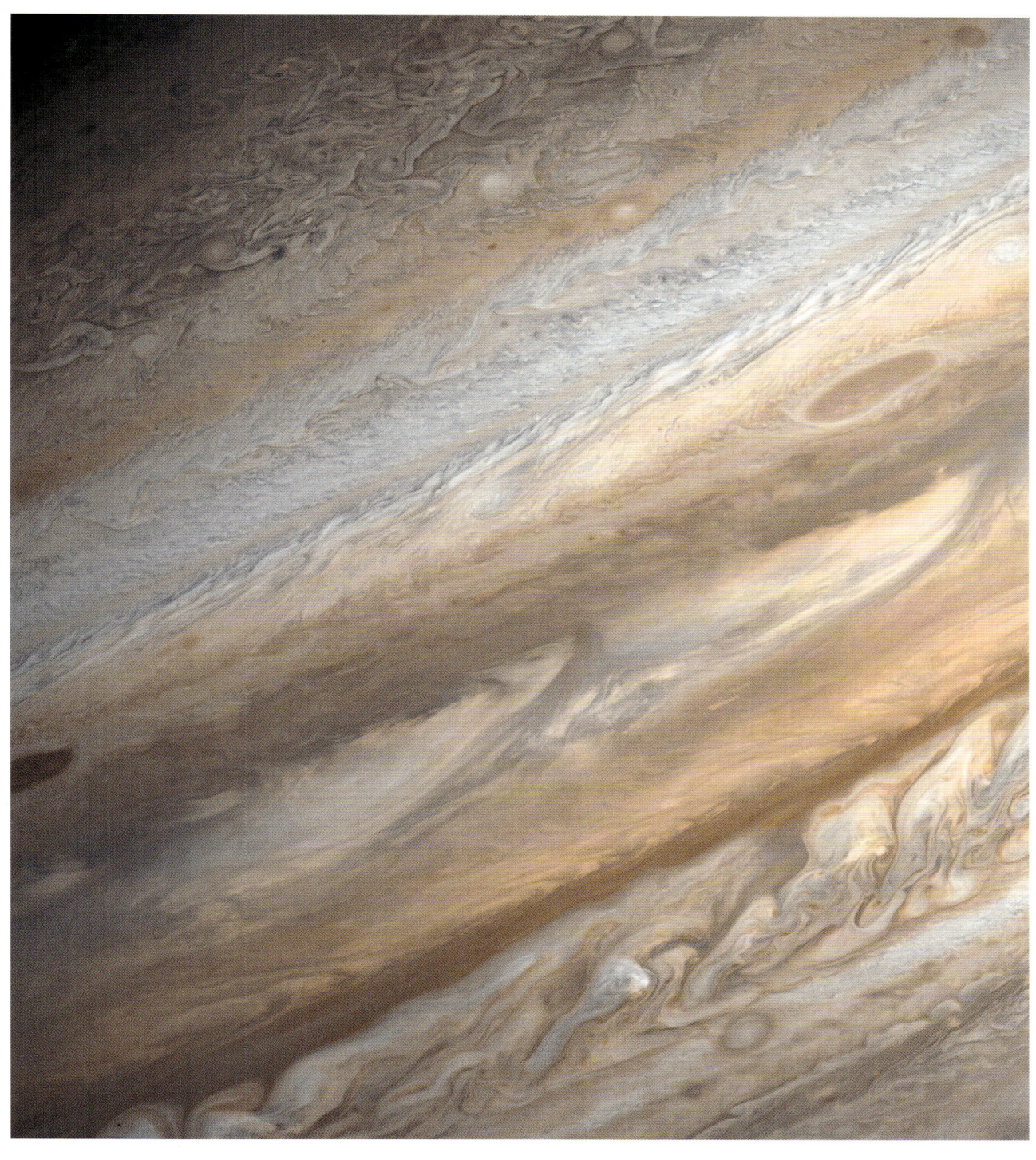

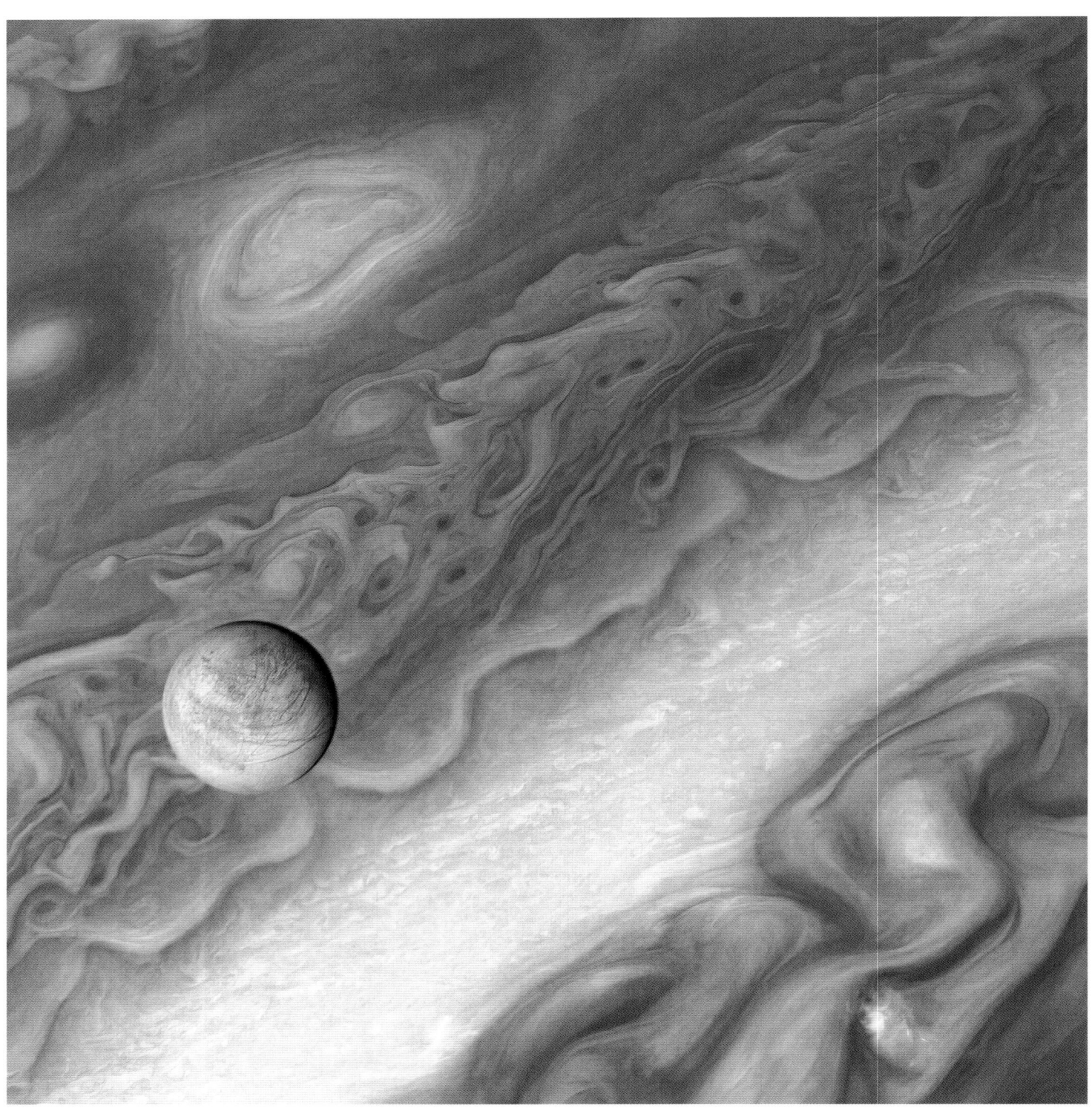

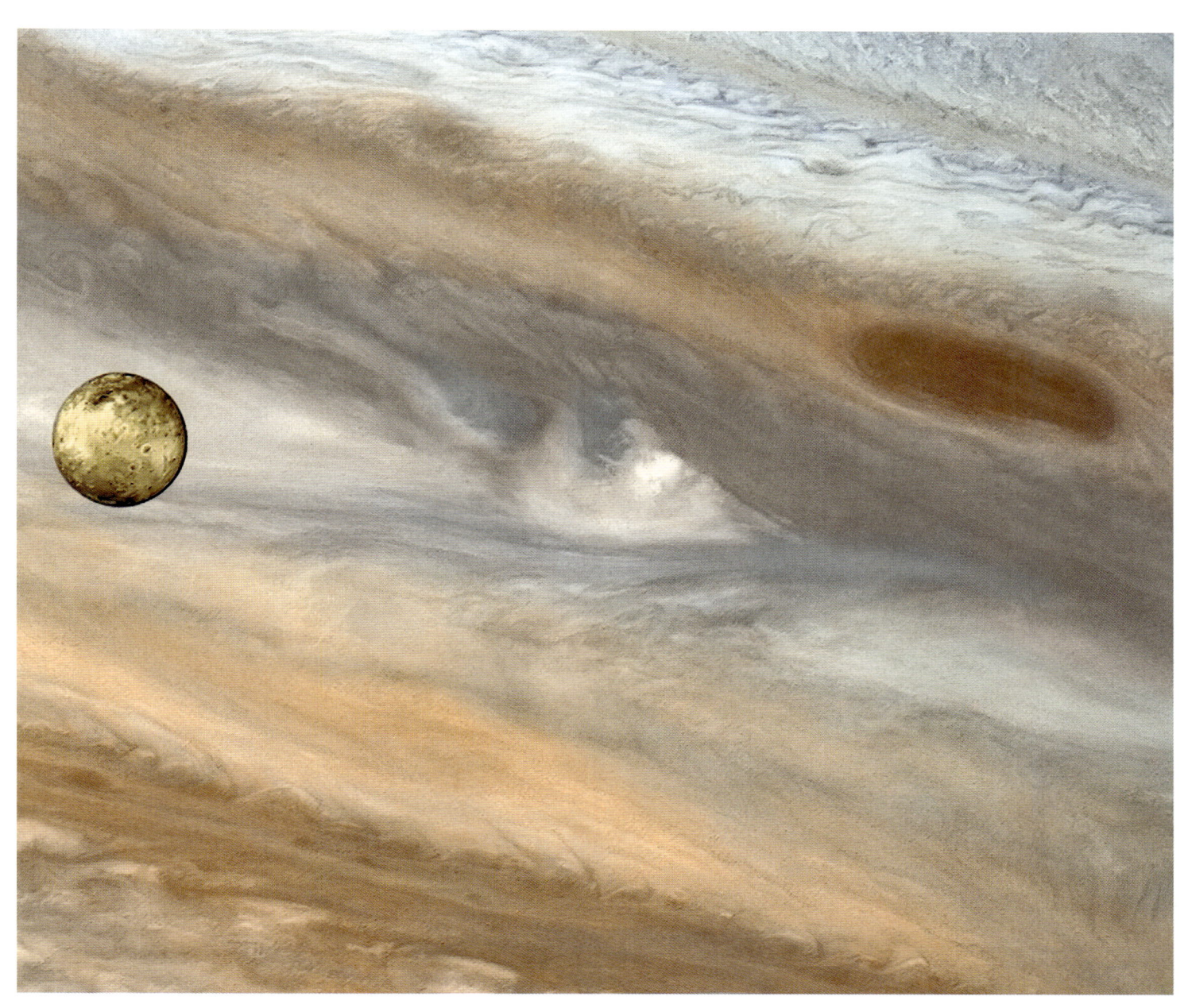

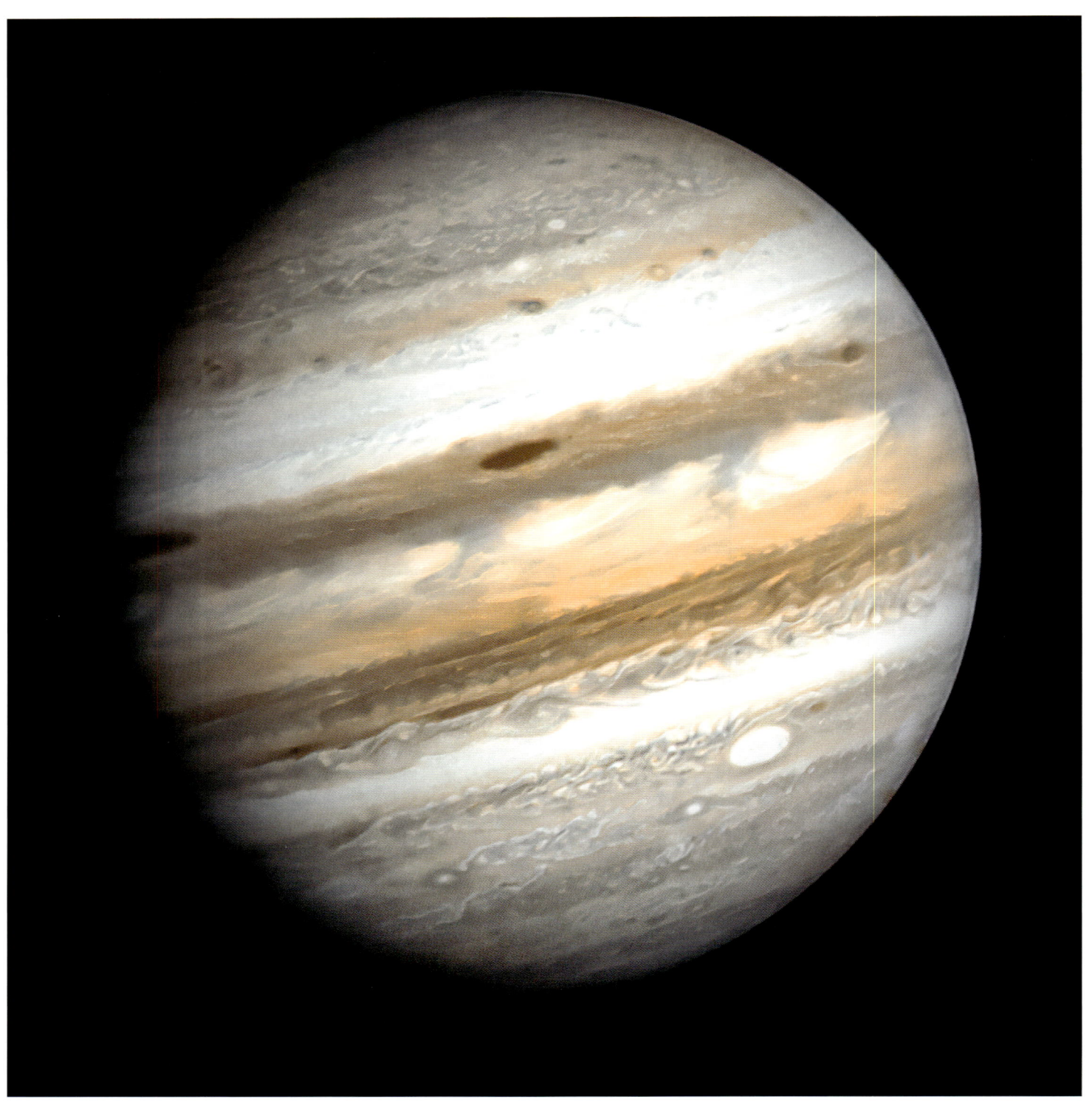

Right: Voyager 2, June 30th, 1979, 13:53:34.

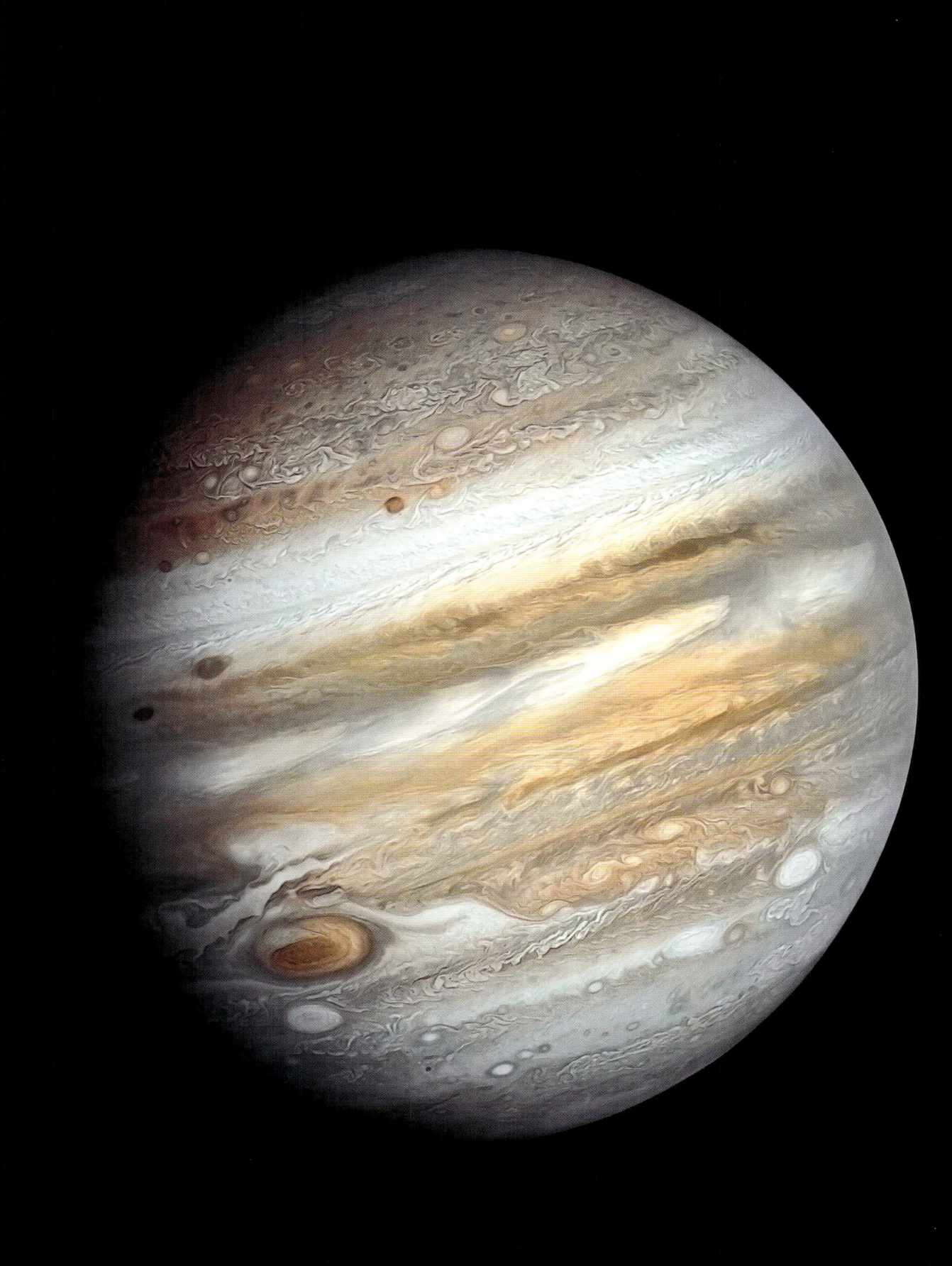

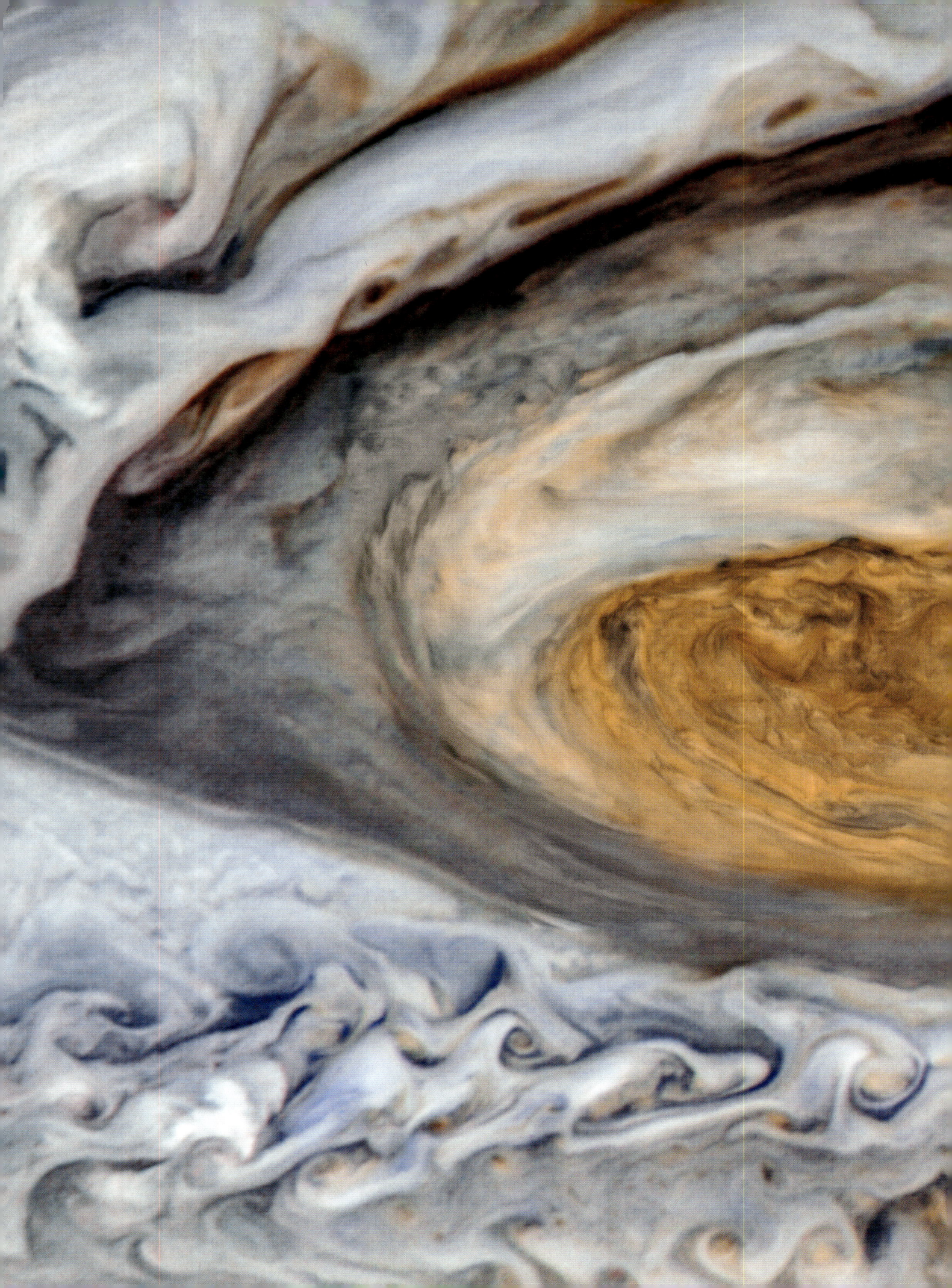

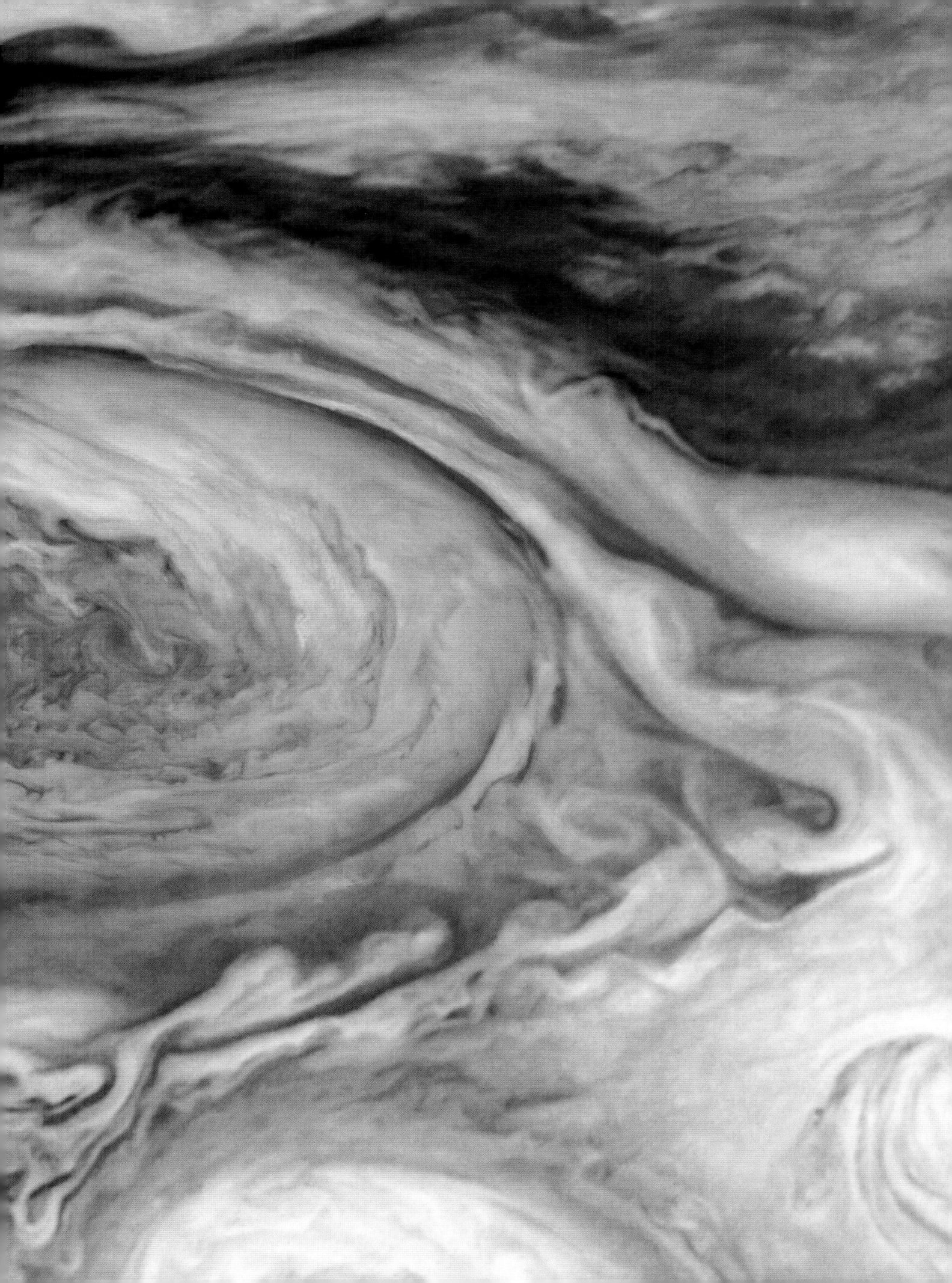

The first mission to Mars did not expect to find craters and river valleys, and yet they did. The first mission to Jupiter didn't expect to find ocean worlds and volcano worlds, but they did.

Alan Stern

THE GALILEAN MOONS

The Jovian System could almost be considered an independent Solar System in its own right. The giant planet is like another sun, with four moons, all big enough to individually be considered planet-sized. In this "Solar System", Jupiter is known to have 79 natural satellites. 63 of the moons measure less than 6 miles (10 km) across, and have only been known to science since the close flybys of the Pioneer and Voyager missions. The four major moons that orbit Jupiter are the same ones Galileo saw through his telescope. The Galilean moons were the first objects found to orbit another body that was not the Sun or the Earth. The impact this discovery had on our understanding of the Solar System, and the physics that keeps it all together, was profound. An impact that has only been matched by Voyager 1 and Voyager 2 visiting these giant moons four centuries later.

Previous page: Voyager 1, March 4th, 1979, 07:08:35.

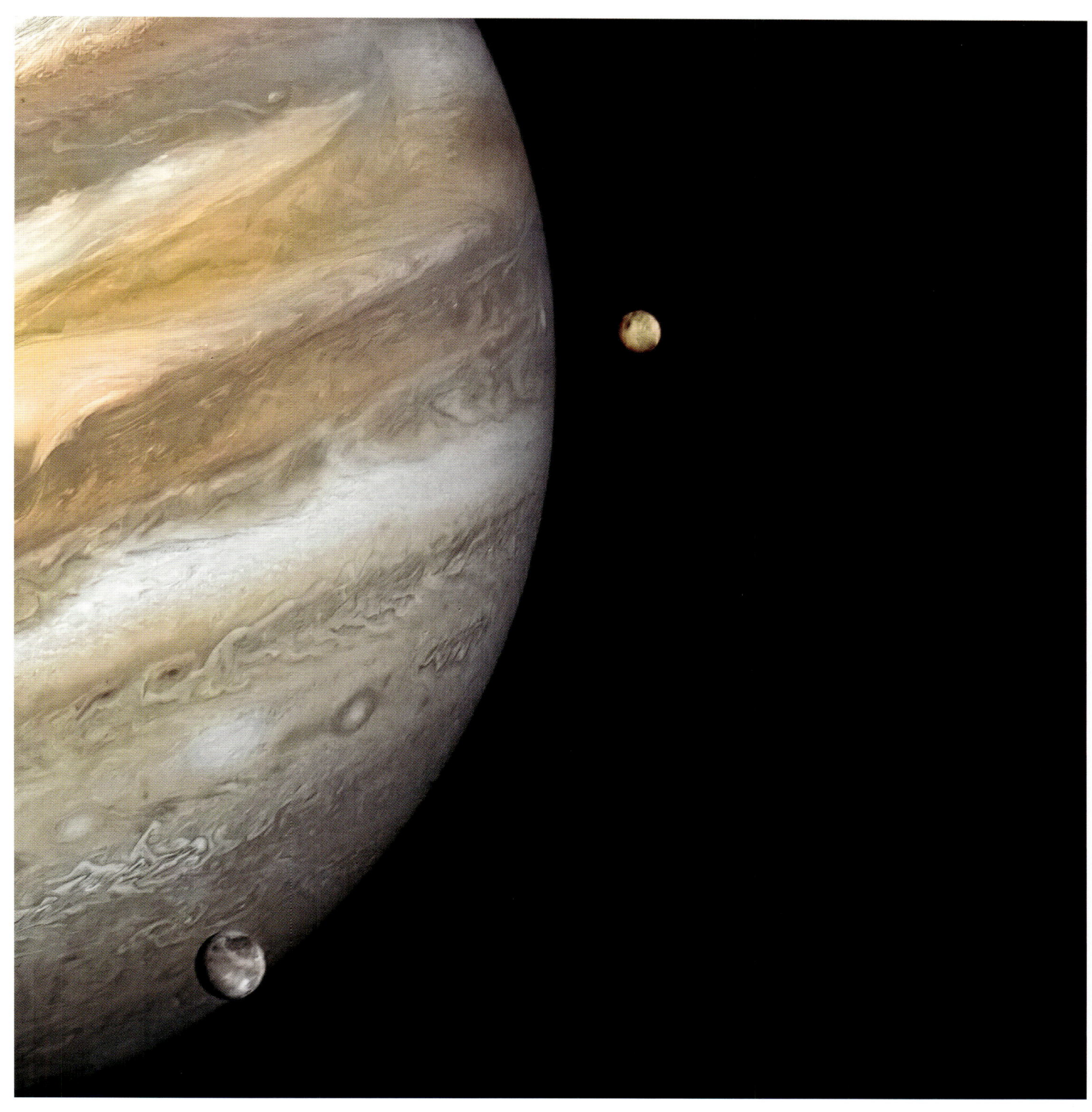

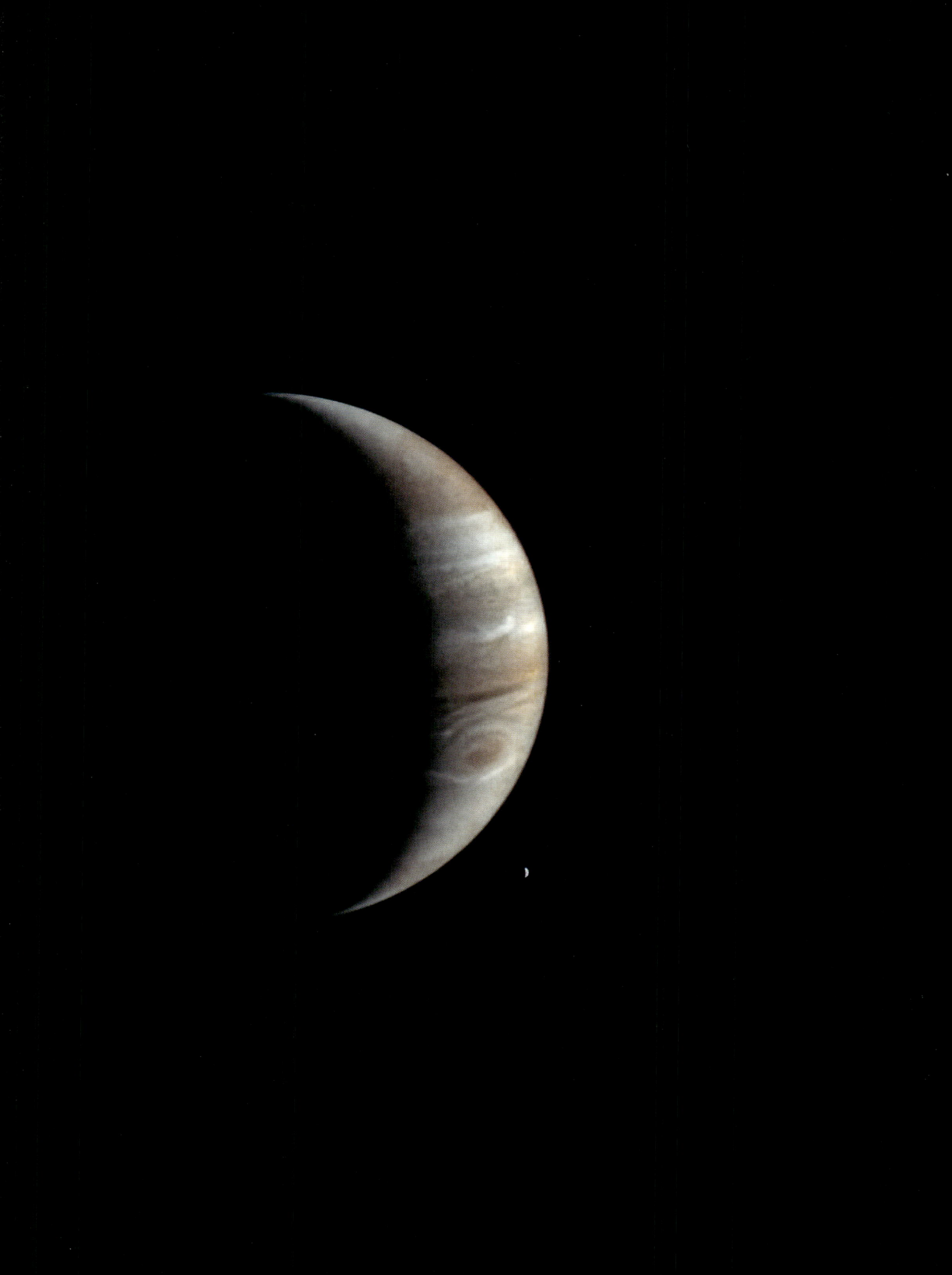

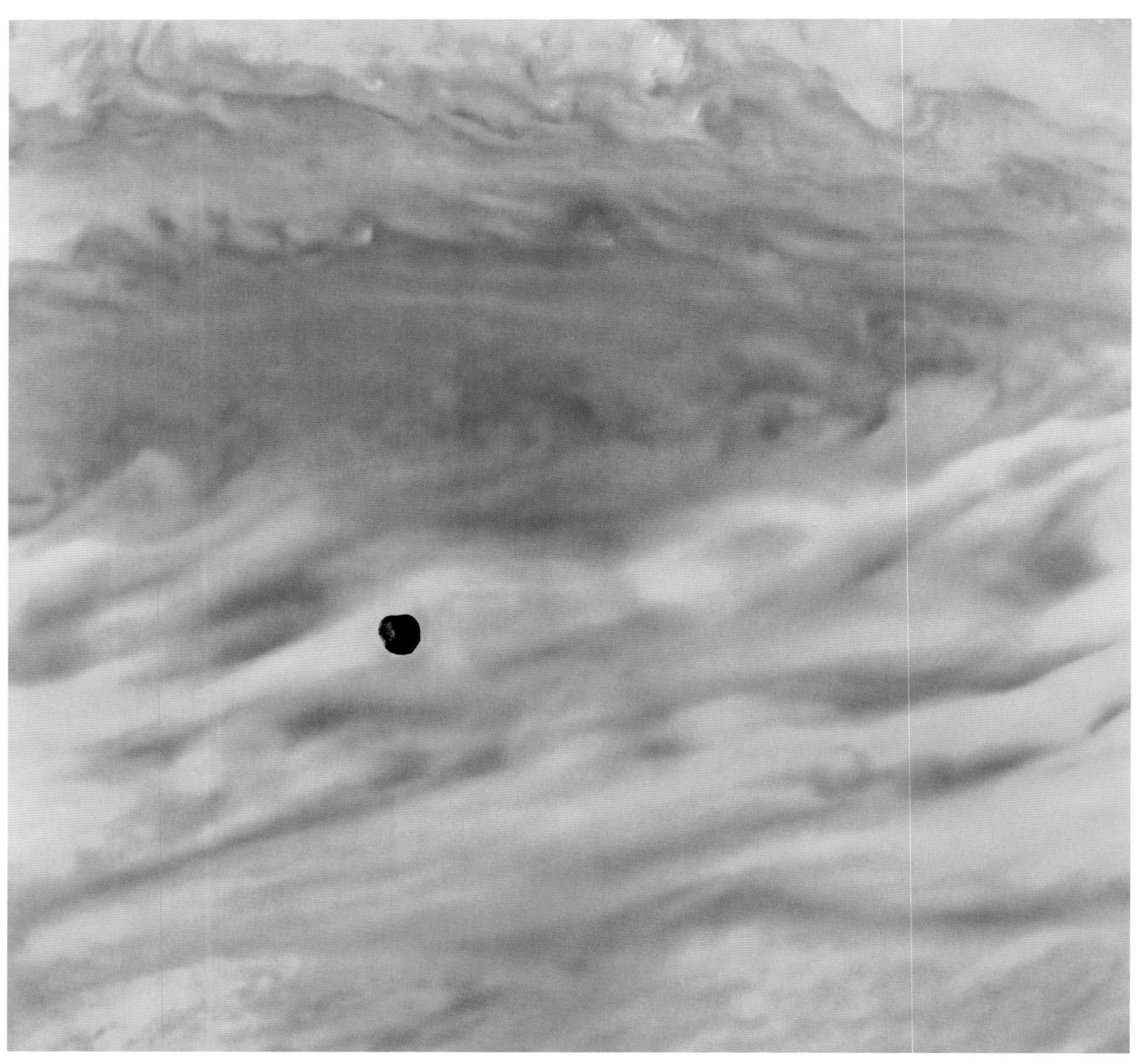

Previous page: Voyager 1, March 13th, 1979, 19:30:20.

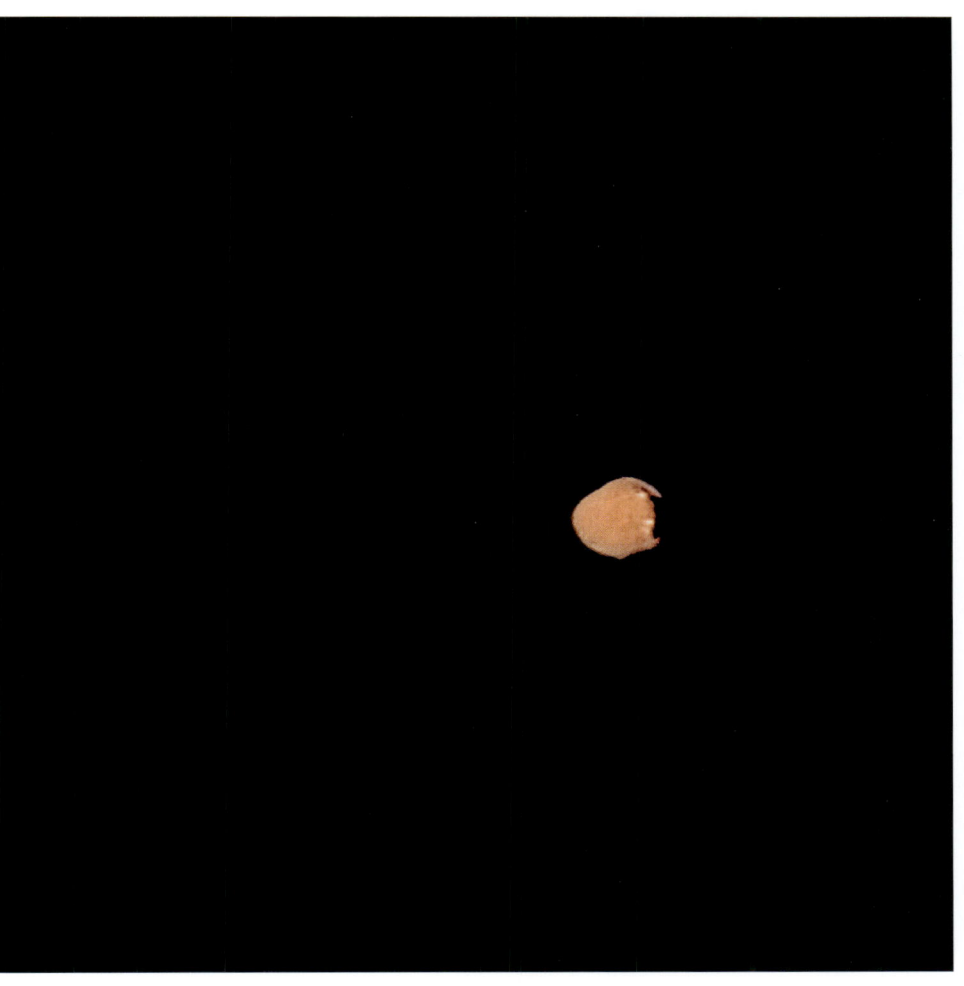

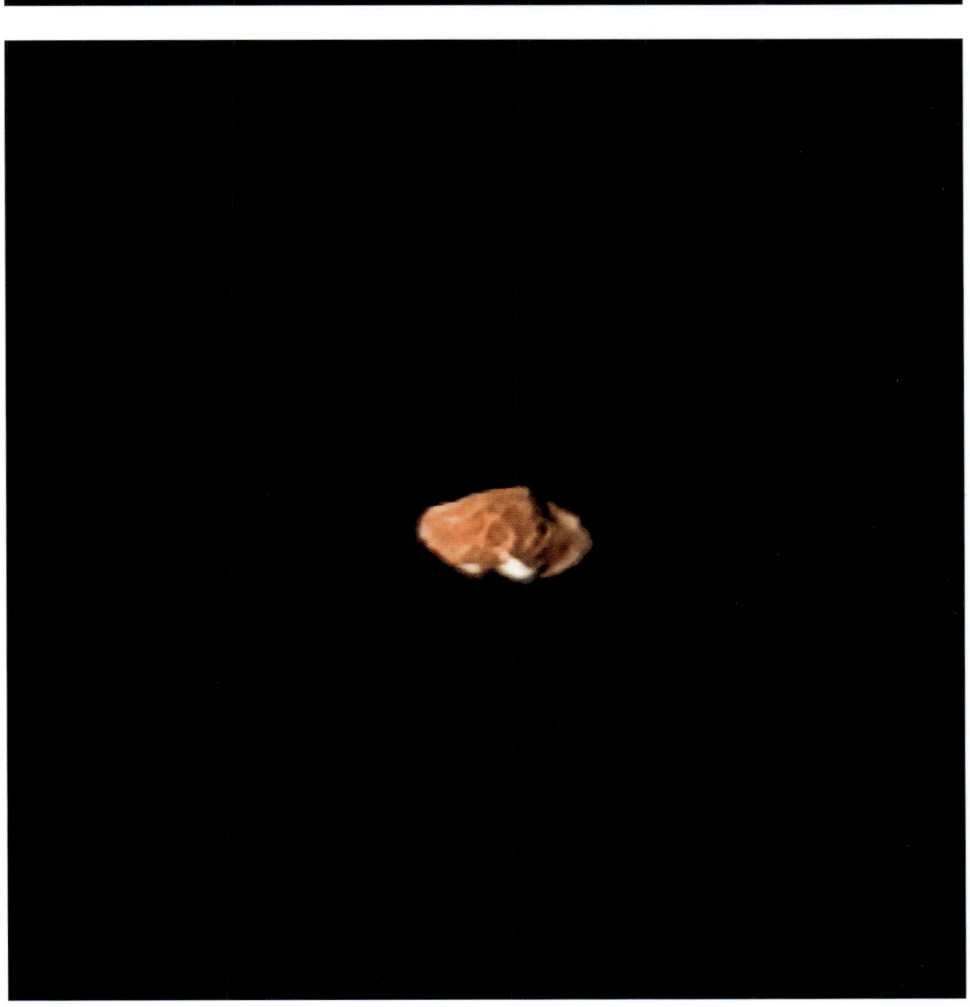

ORBITAL RESONANCE

The orbits of the Galilean moons have led to remarkable and enlightening discoveries over the centuries. In particular, scientific highlights include Newton's formulation of his Law of Universal Gravitation during the 1670s, published in 1687, and the later work on orbital resonance by French physicist Pierre-Simon Laplace.

Their work led to further scientific understanding of what happens when two or more orbiting bodies pass by each other, causing an exact, systematic and regular gravitational pull on one another. It was their study of the Galilean moons, working out the mathematical ratio behind the orbits of Io, Europa, and Ganymede, that would start to explain what was to be discovered on these moons by the Voyagers.

The "Laplace Resonance" is a calculated orbit ratio from three of the four Galilean moons. The ratio showed how each of the three worlds would align with one another in a ratio of 1:2:4. Each time one of the moons passes near one another, their gravitational fields would move the moons slightly closer or away, pulling them off from what would be in perfectly circular orbit. It is similar to how our Moon pulls the tides in and out from the shoreline.

The Laplace Resonance ratio worked out that for every complete orbit Ganymede made around Jupiter, Europa would pass Ganymede twice, and Io would pass by four times. It's at these passing points where each of the moons would meet those huge and powerful forces known as "tidal kneading" and "tidal heating." The results of this pulling have a dramatic effect on the composition and structure of each of these moons.

Right: Io and the day-night boundary of Jupiter.

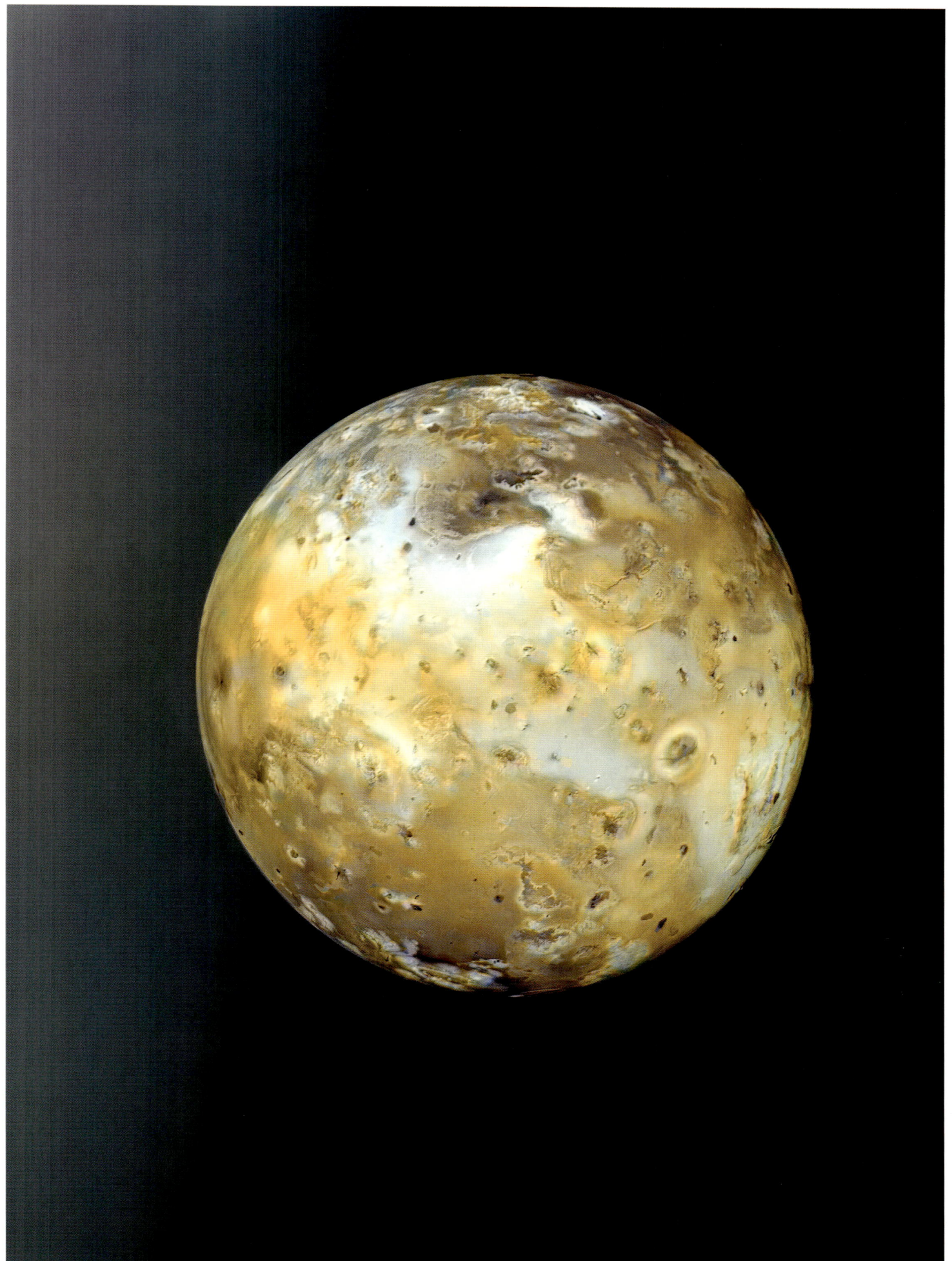

IO

The first close observation of Io was from Voyager 1 on March 5th, 1979, passing close to the South Pole at around a distance of 12,800 miles (20,600 km). The detailed images that returned to Earth showed that Io was distinctly different from the other Galilean satellites. The surface was orange-yellow in color and pockmarked with dark spots and rings dotted around, all of which were originally mistaken to be old crater marks. It was not until three days after the first encounter with Io that one of the most significant discoveries in our Solar System was found by chance.

The Voyager spacecraft took a series of images that had no scientific research purpose. These images would help the navigation controllers to more precisely triangulate the spacecraft's position, heading and location. A process called "optical navigation." Navigation engineers would process these images to enhance the contrast in the photograph, making the faint stars visible enough in order to triangulate the craft's position.

Linda Morabito, one of the navigation engineers, was given the task of calculating Voyager 1's location as it was leaving the Jovian System. One of the images she was processing caught her attention. A strange cloud-like mark appeared to be coming out from Io's surface. It was quickly confirmed that there was no other moon or object behind Io at the time that could have created the illusion of a cloud adjacent to Io. This left only one quite spectacular hypothesis on the table. This cloud crescent and glow was evidence of volcanic activity, expelling ejecta 186 miles (300 km) up into Io's almost non-existent atmosphere.

This one image was the first evidence of active extraterrestrial volcanism. Other planetary researchers further confirmed the hypothesis. Moreover, their research not only showed that Io was not just merely volcanically active, it was hyperactive and harboring the most intense and numerous volcanic eruptions known to occur in our Solar System. What appeared to be craters turned out to be volcanoes. These volcanoes erupt so frequently that the surface of Io is constantly changing. The hyperactive volcanic activity is caused by tidal kneading and heating from the forces exerted on the moon from its orbiting neighbors. It was noted that the surface of Io had notably changed in appearance in between the two Voyager spacecraft flybys of the moon.

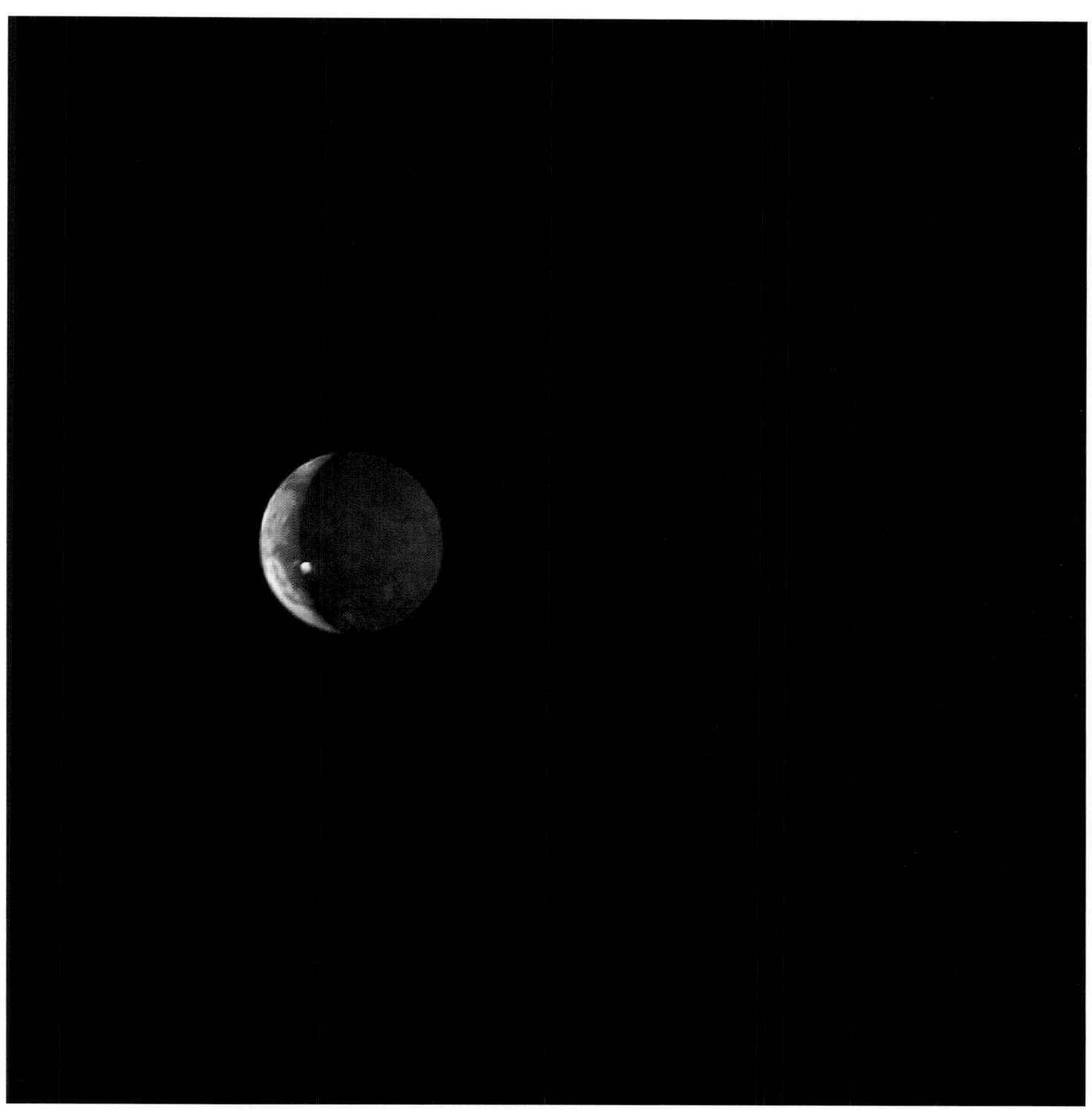

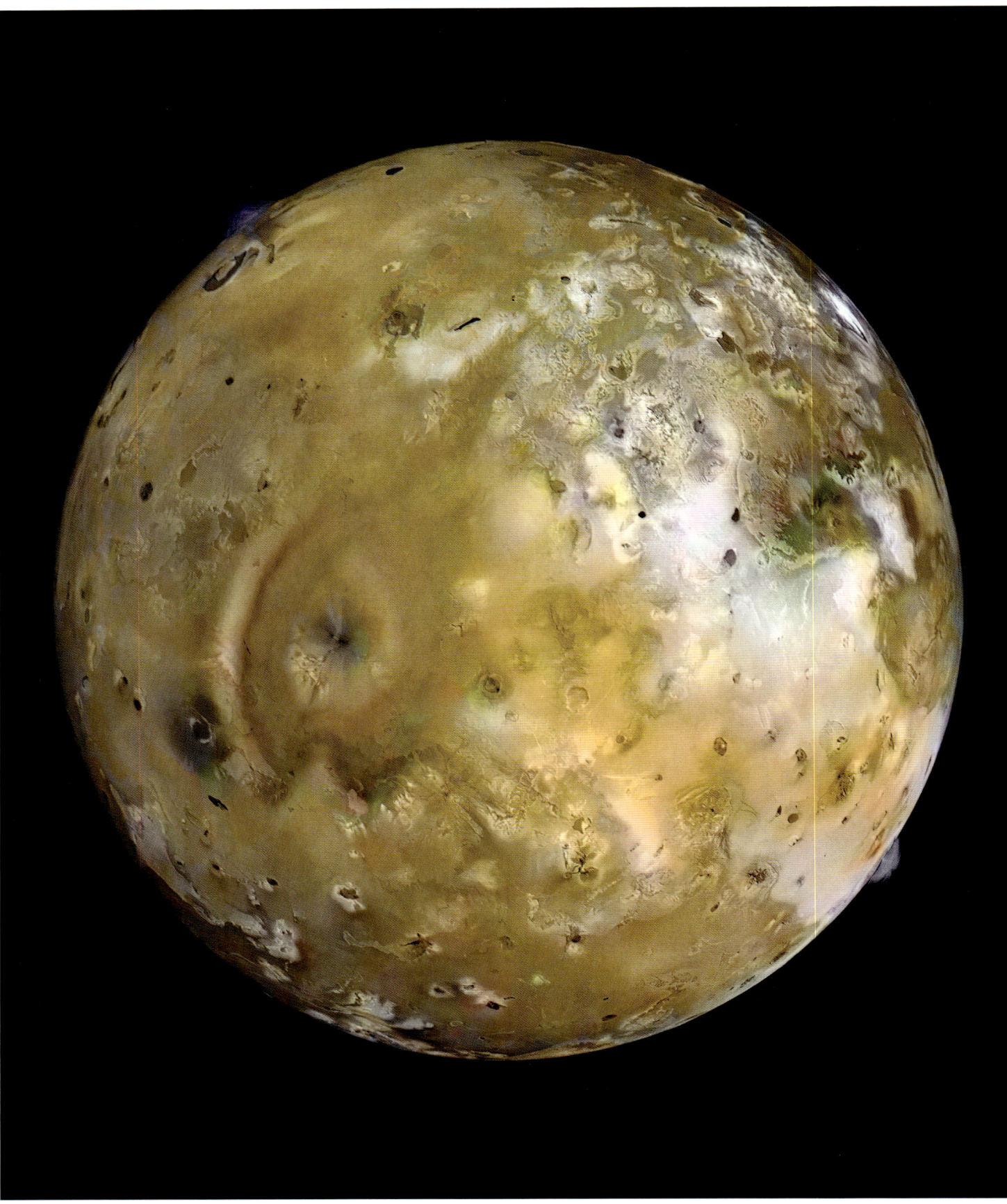

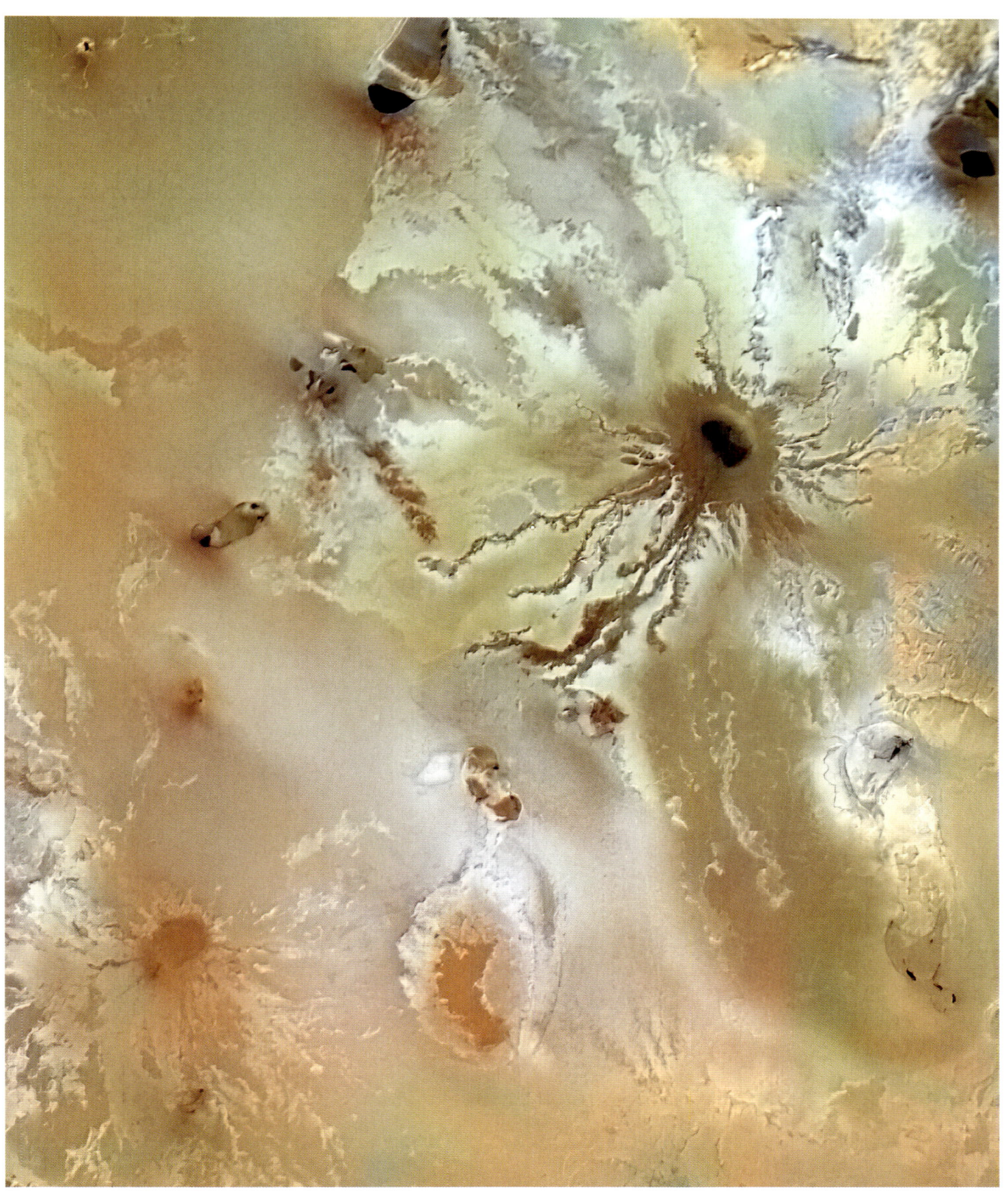

A clear view on the lava flows of Ra Patera.

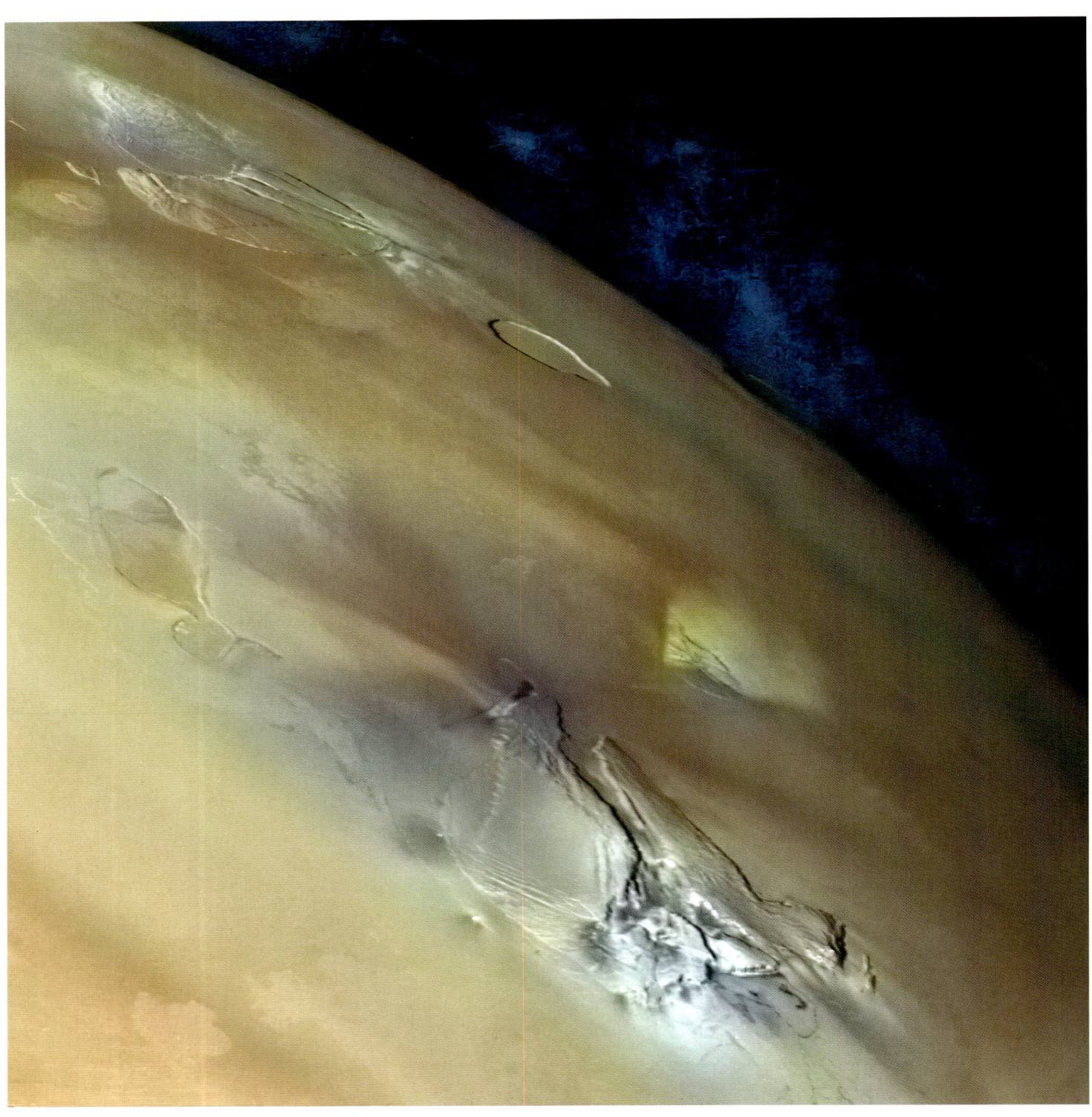

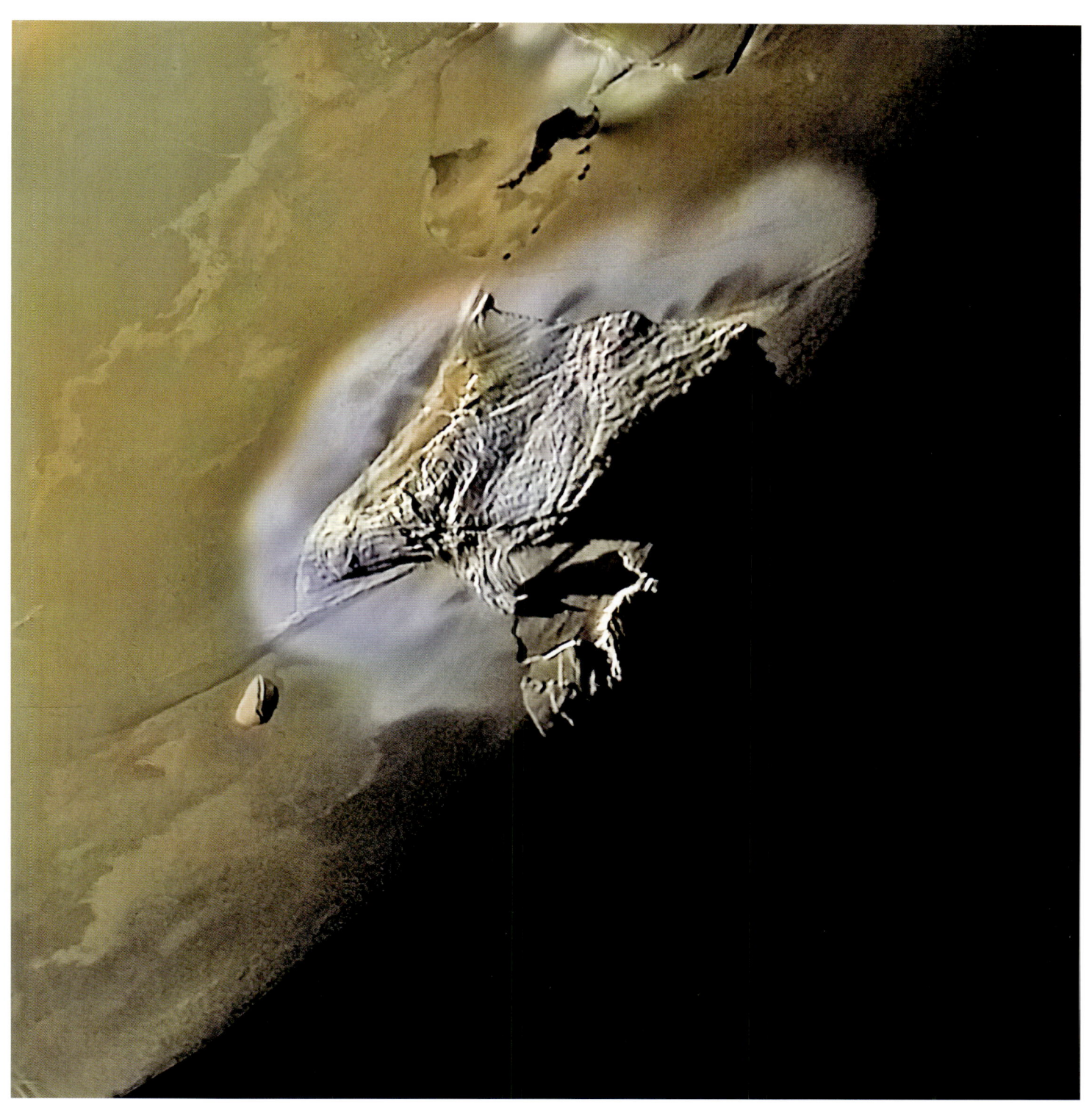

Haemus Mons, a mountain with a staggering height of 6.2 miles (10 km).

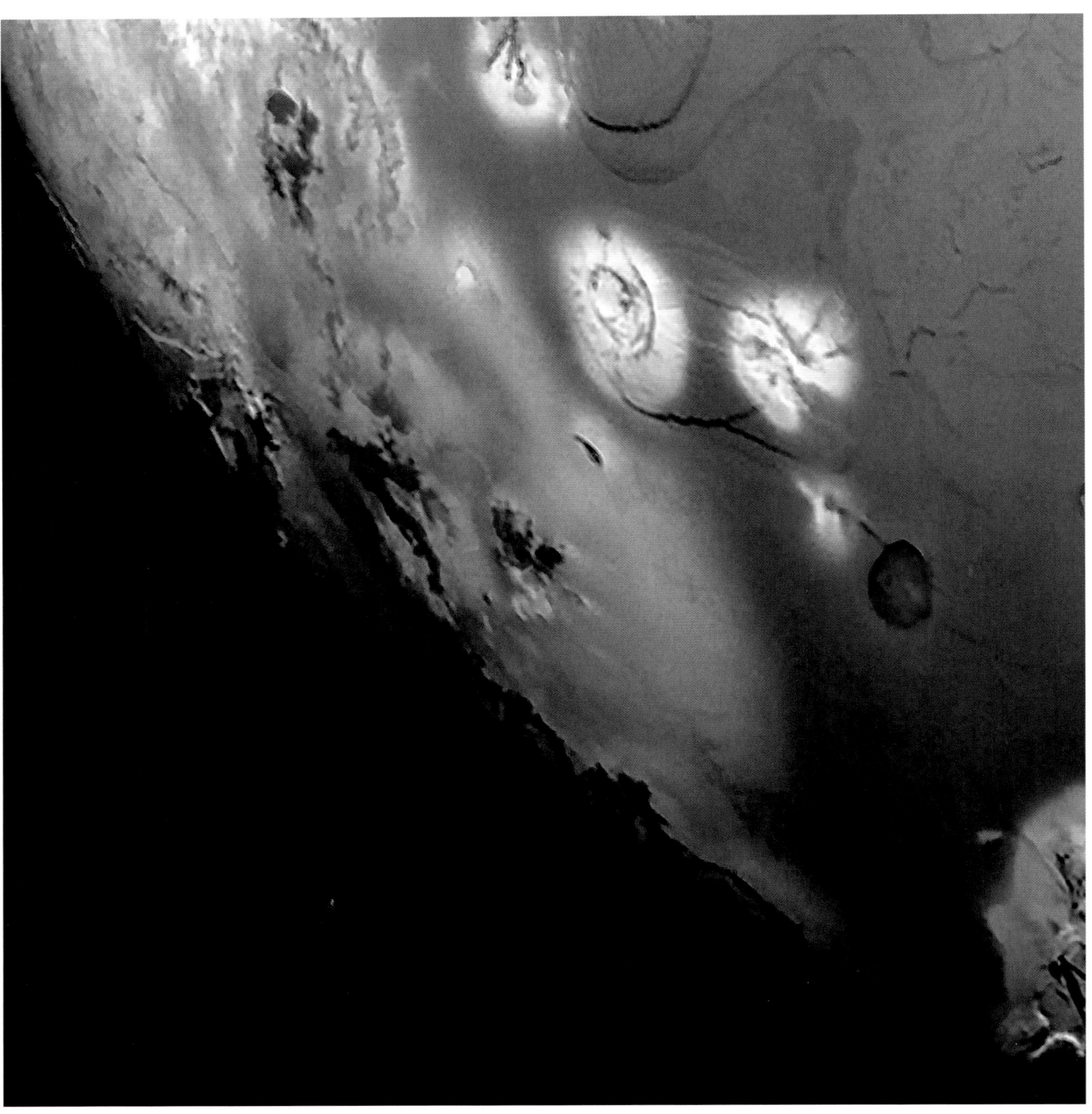

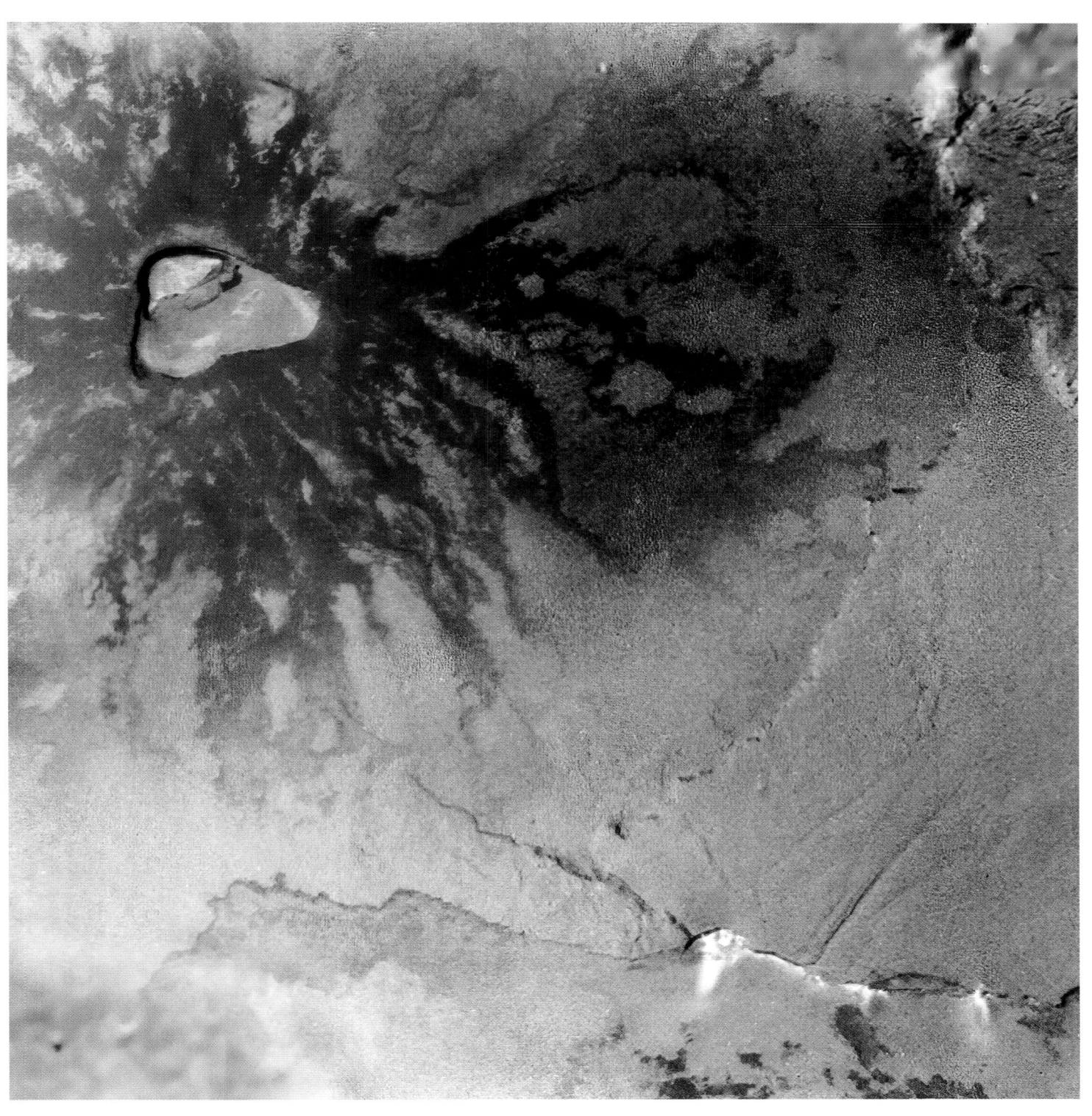

The lava flows of Maasaw Patera.

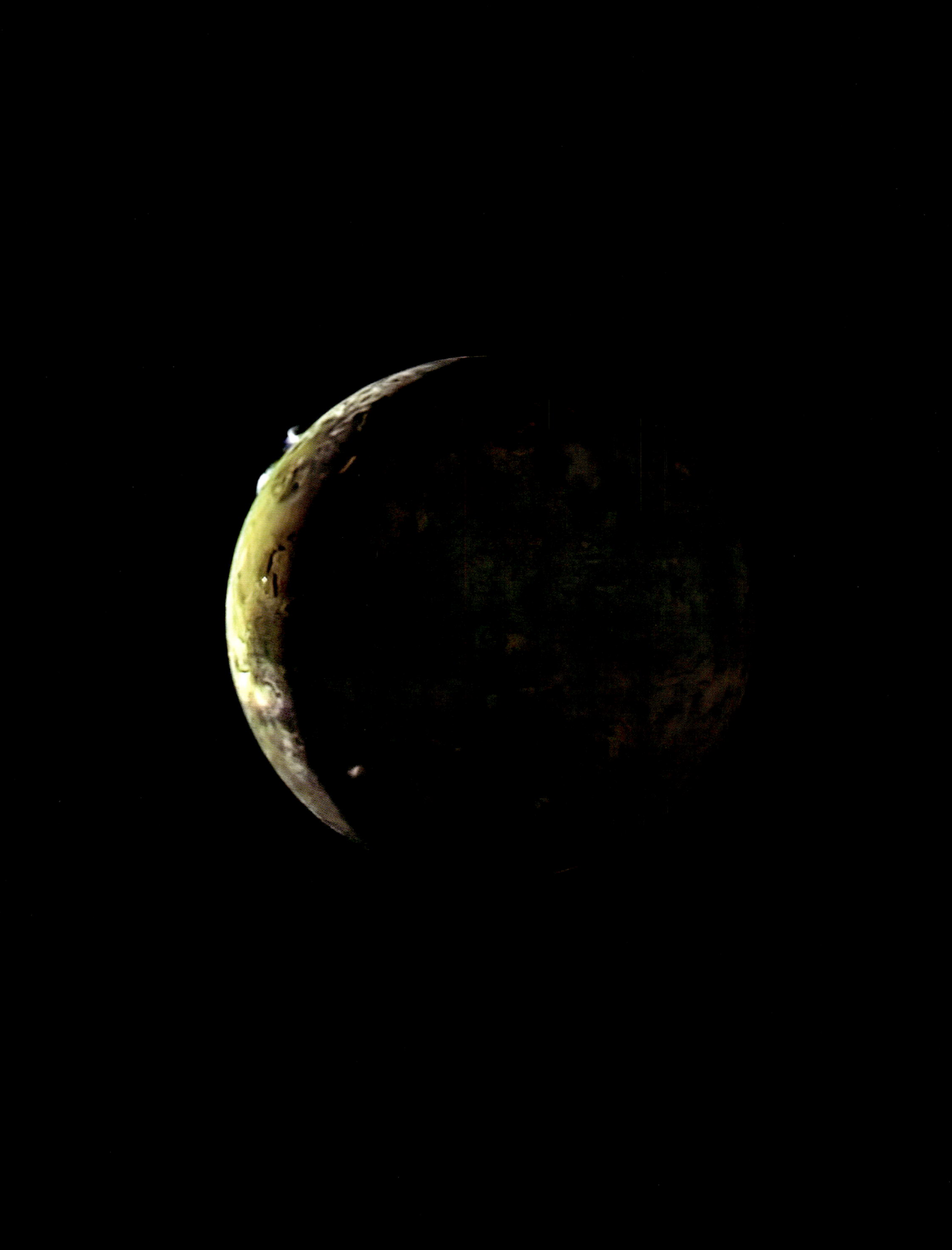

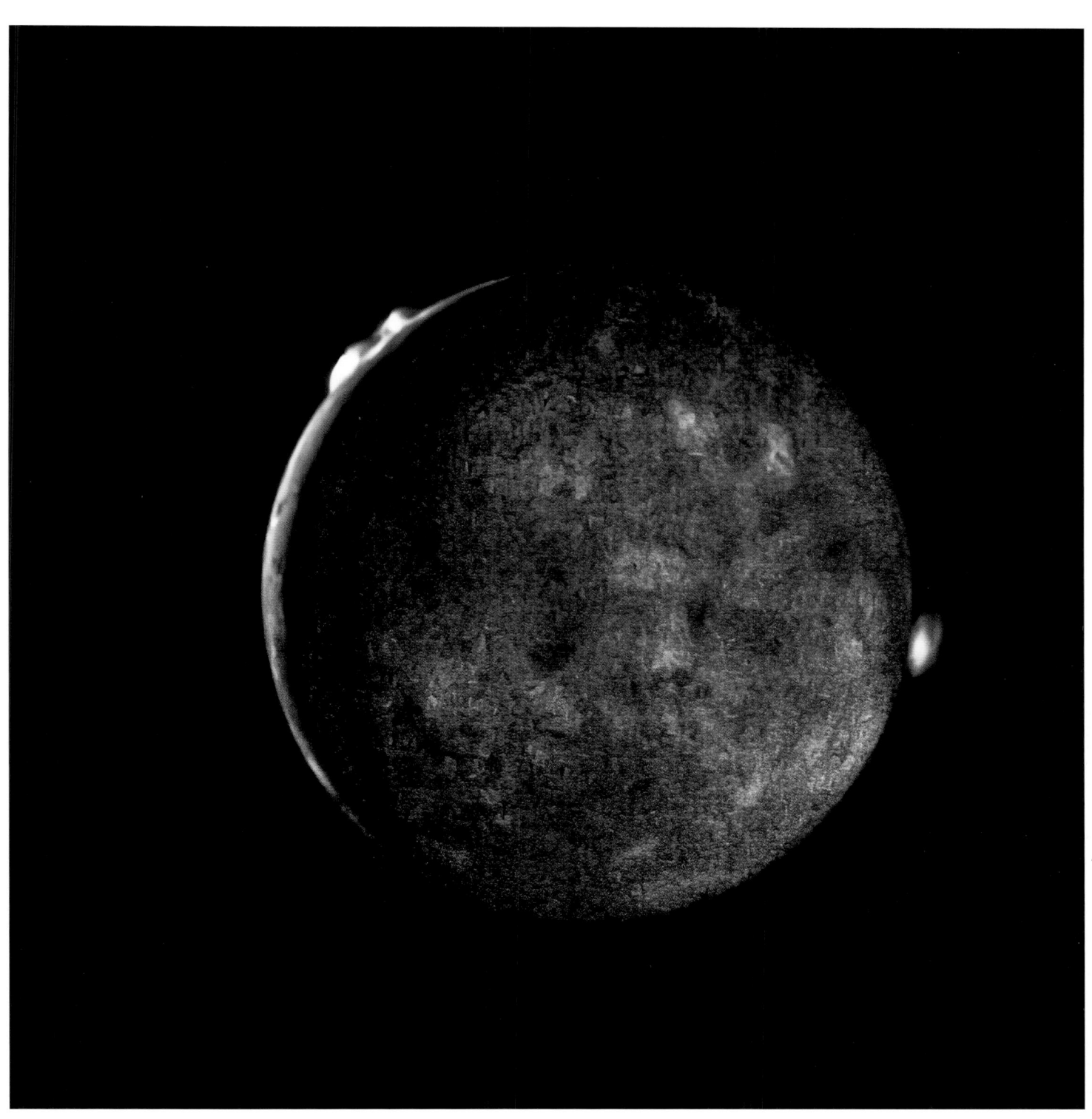

Left: Voyager 2, July 10th, 1979, 04:30:21.

EUROPA

We had to wait until Voyager 2 made its flyby of Europa to get a closer glimpse of this moon. The earlier Pioneer missions only sent back poor imagery and the trajectory of Voyager 1 meant its cameras could only get a distant glimpse of the moon as its route was to take it to the three other Galilean moons. It was on July 9th, 1979, when Voyager 2 passed by Europa at its closest point. The data and images sent back showed in great detail an exciting and an unalike new world with a highly unusually flat surface.

The images of Europa were intriguing to the science community back home. A smooth blue-white ball of ice, unusually flat apart from a crisscross of black lines spread over its surface. What was even more puzzling was what lay beneath the flat, thick sheet of ice that completely wraps around the moon.

In size, Europa is similar to the Moon that orbits our Earth. However, the relief and elevation on Europa is very different from anywhere else in the Solar System. For example, on our Moon you will find an elevation difference between the valleys and mountains of around 3 to 5 miles (5 to 8 km). Europa is very different, its highest and lowest elevation points being a few hundred meters high or deep.

The lack of impact craters from asteroids and comets provided evidence that the flat surface was young and resurfaced constantly. Close-up photos and data collected from the Voyager 2 instruments suggested that this smooth icy surface was similar in appearance to sea ice. A layer of frozen water is floating on top of a saline water ocean, somewhat similar to the ice shelves we have on Earth. It has been calculated that the ocean on Europa is the largest in the Solar System. From further data analysis, it is predicted that Europa has a solid metal core.

The photographs and data provided evidence of internal heating caused by the tidal forces put on the moon by its neighbors. Tidal forces from orbital resonance, where the objects under this force are being pushed and pulled, creates heat from the resulting friction. A lot of heat actually. At this scale, it is capable of turning rock into lava and ice into water. The visible cracks in the ice showed areas of reddish-brown material deposited along the edges, remnants of possible volcanic lava flows that oozed out from between the cracks, very similar to that of underwater volcanic eruptions that occur between the tectonic plates at the bottom of Earth's oceans. The similarities between our warm and saline oceans and those of Europa have led scientists to suspect that it would be a key candidate for being a habitable environment that could support life.

Unfortunately, the Voyager 2 flyby of Europa was fleeting and left the science teams back home hungry with only limited data. Further research of the moon has since taken place, adding to the speculation that there is possibility of life on Europa. However, this hypothesis will remain unanswered until we send a probe to land on the icy surface and drill down through the thick ice to sample the water that lies below.

Right: Europa, Voyager 2, July 9th, 1979, 14:14:22.

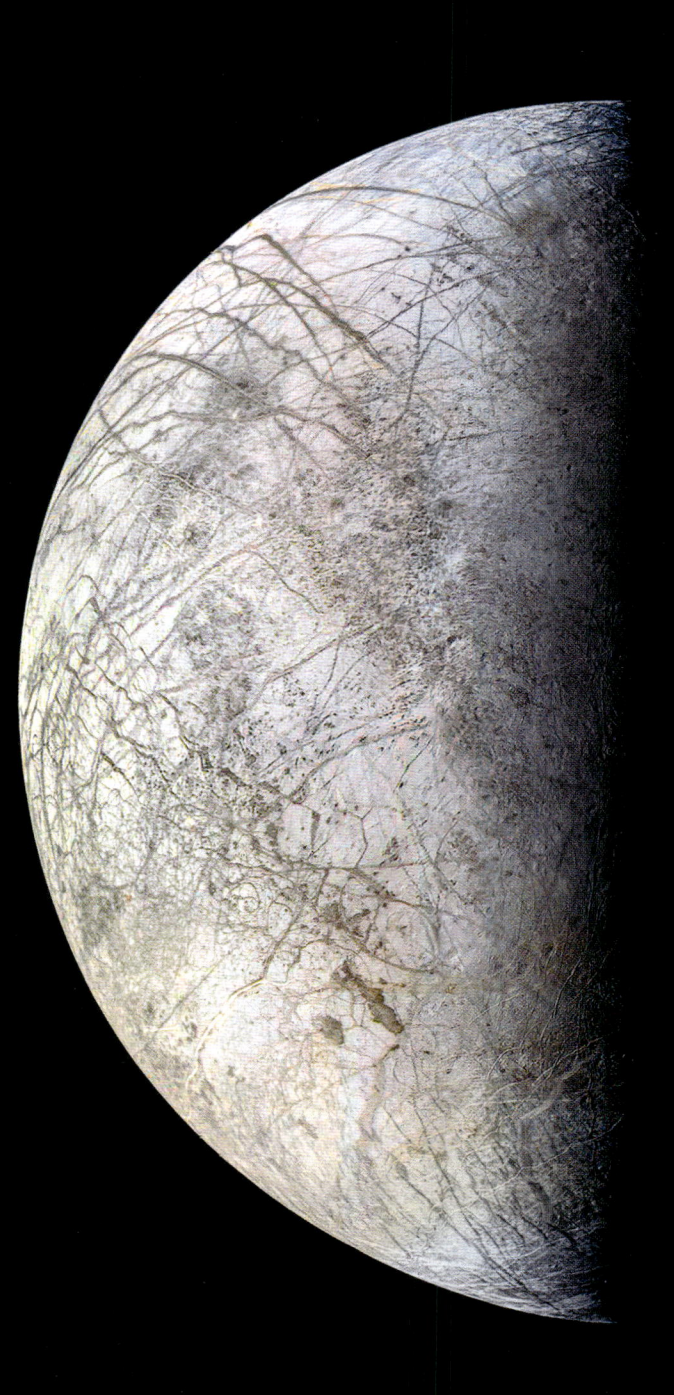

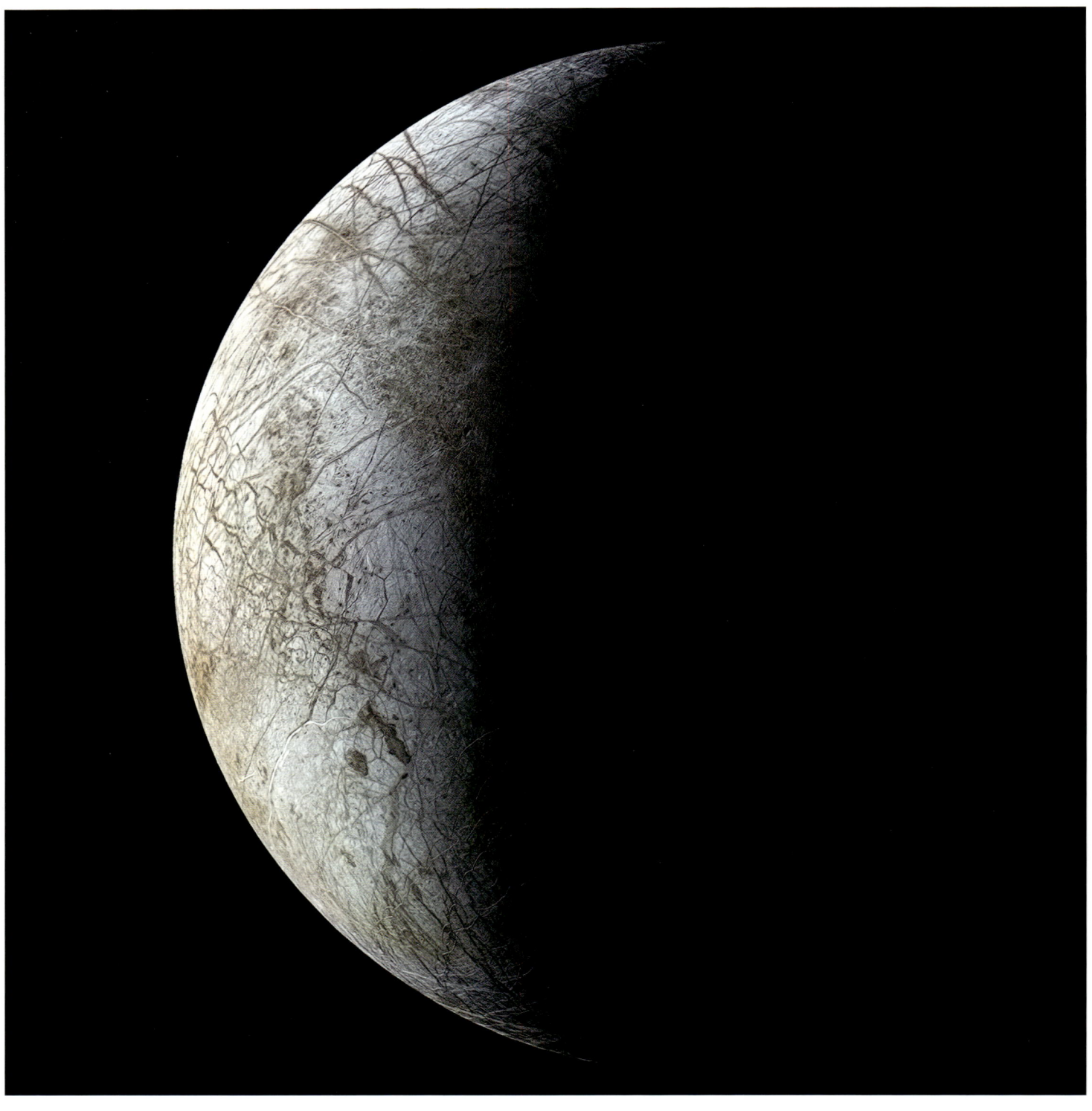

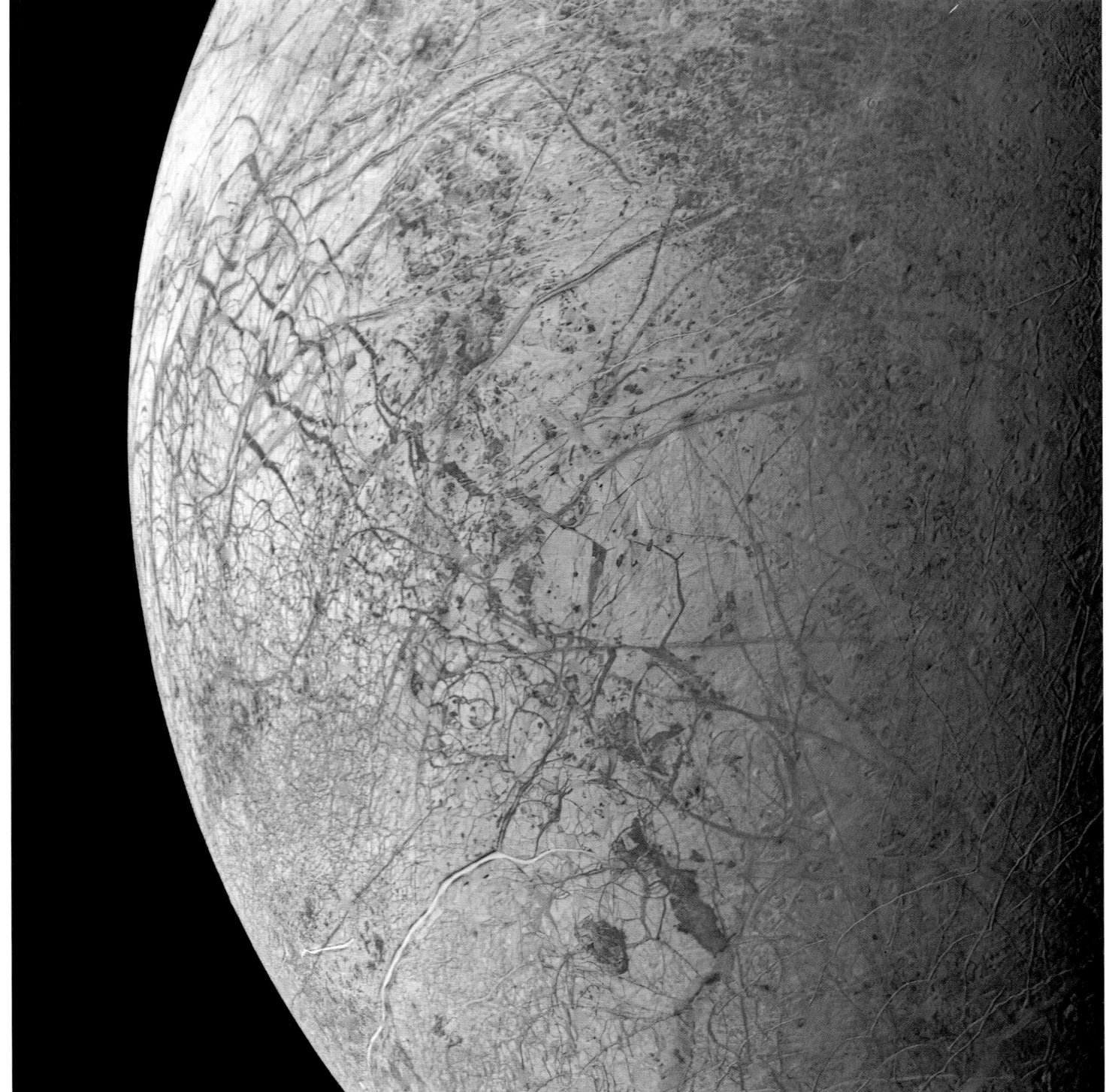

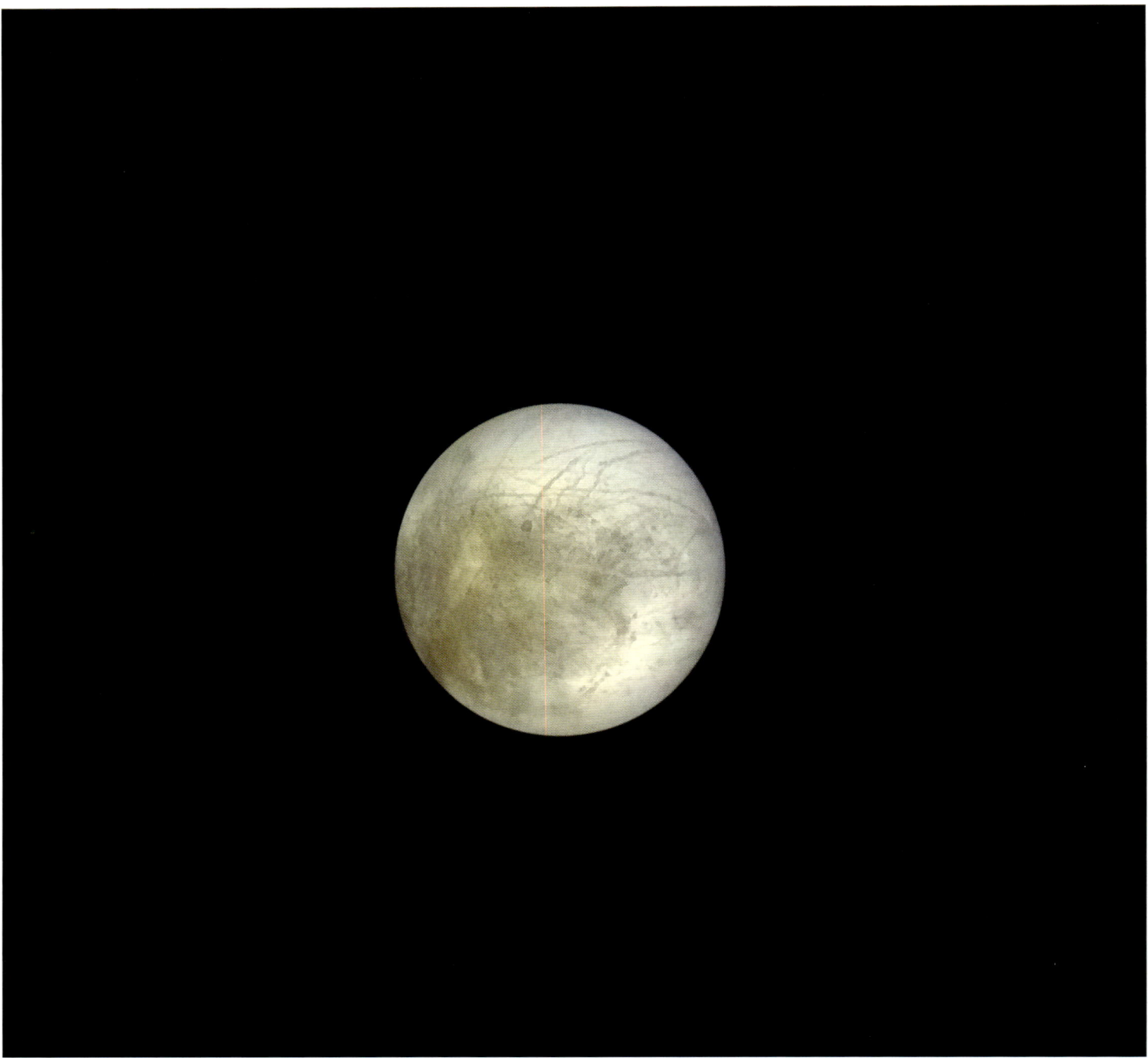

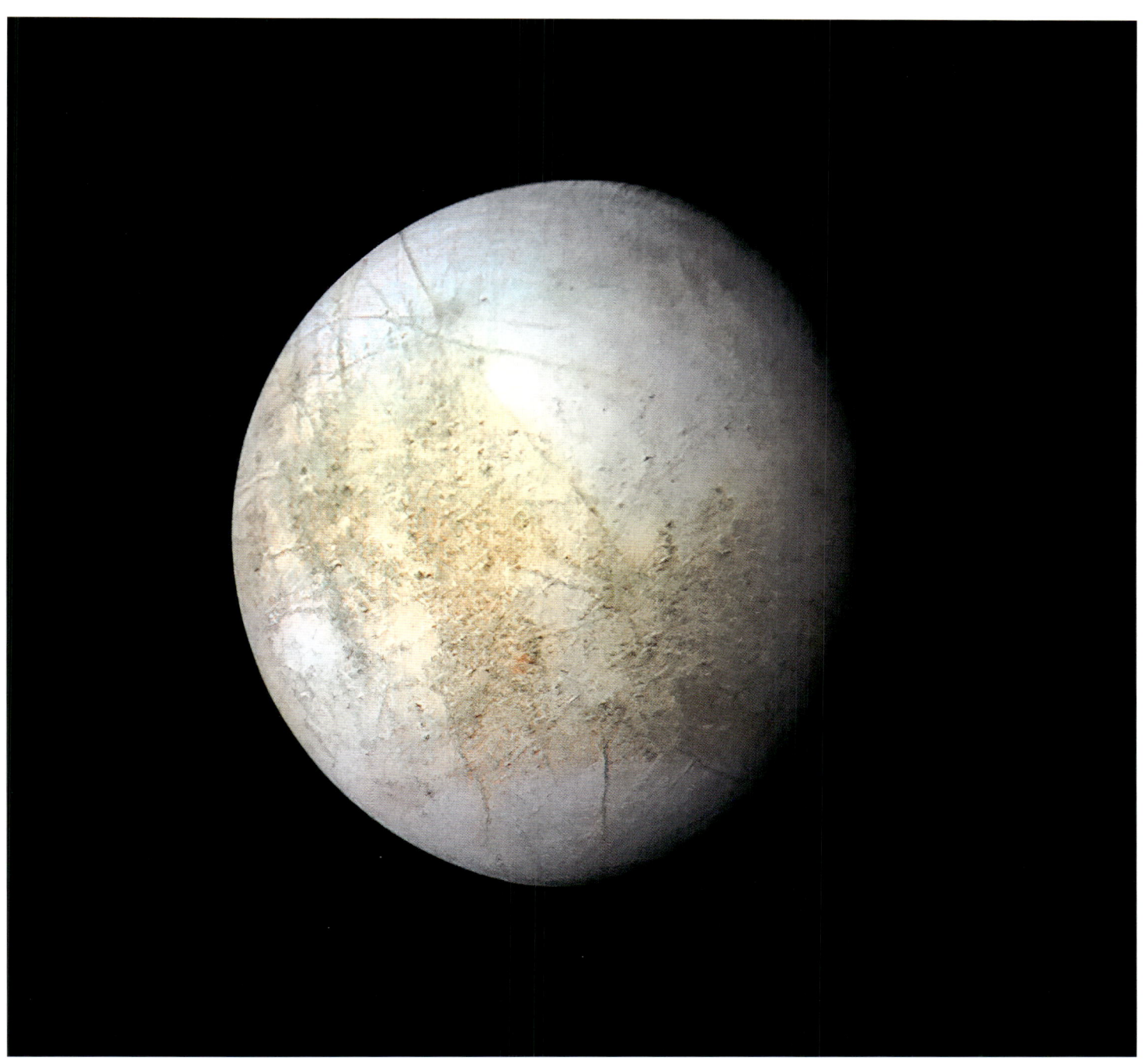

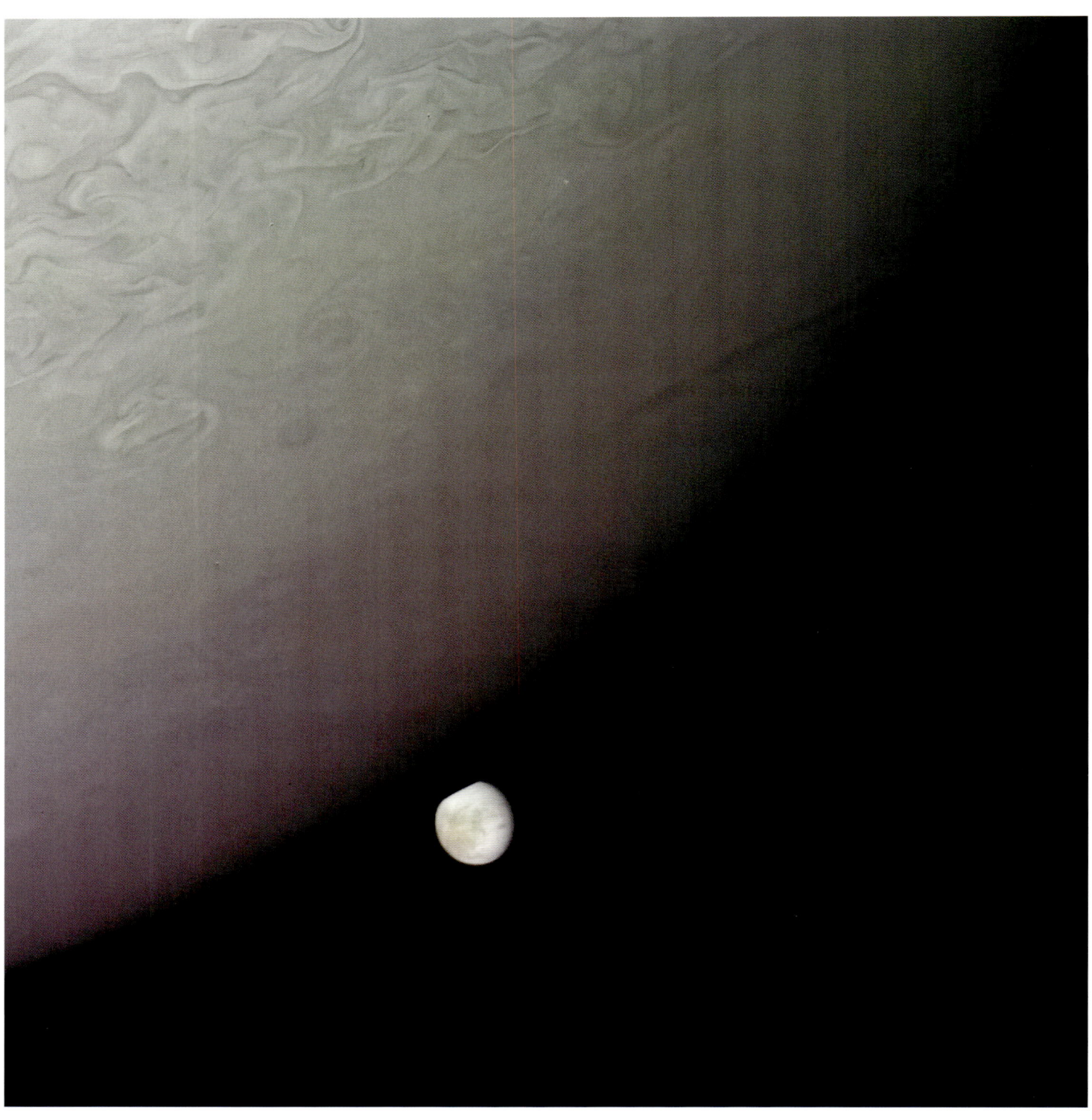

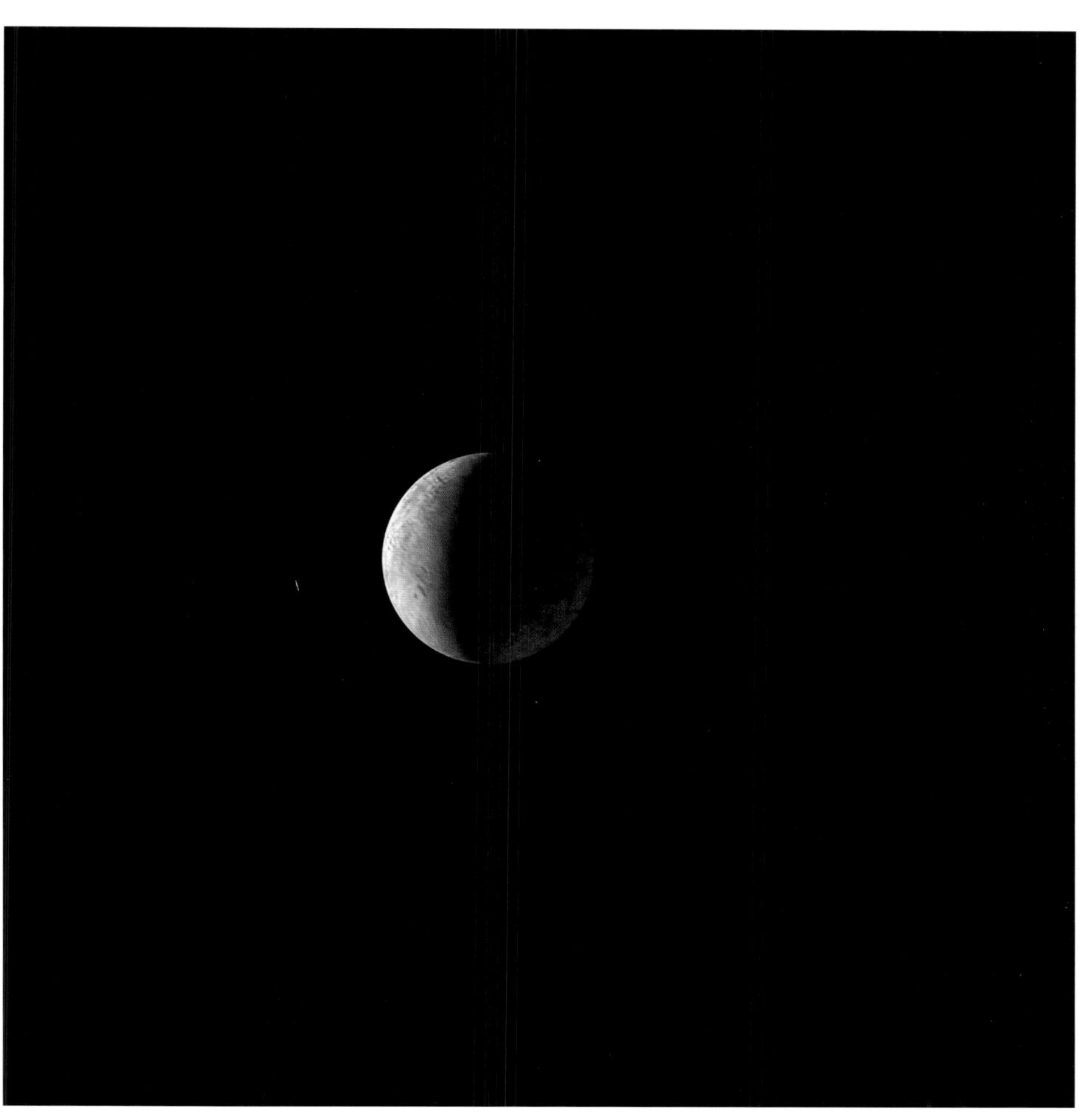

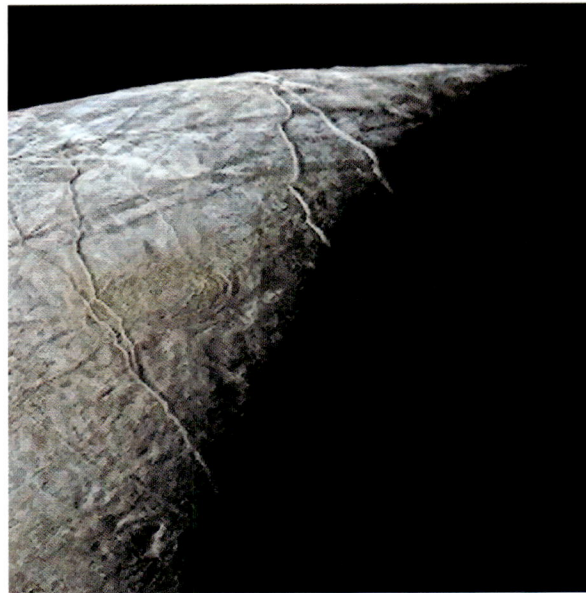

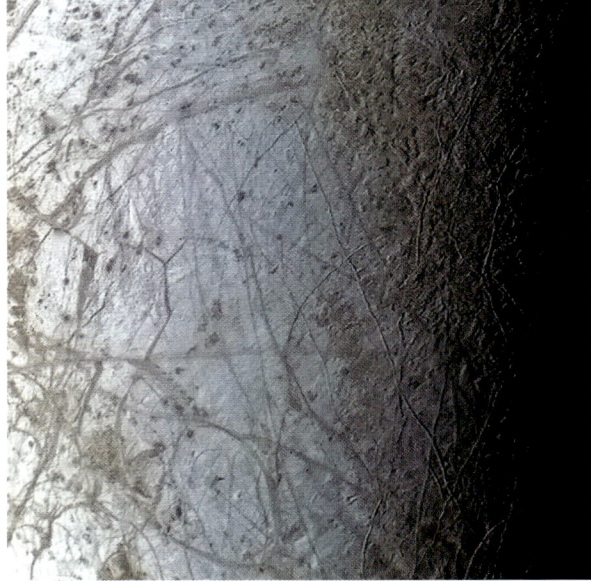

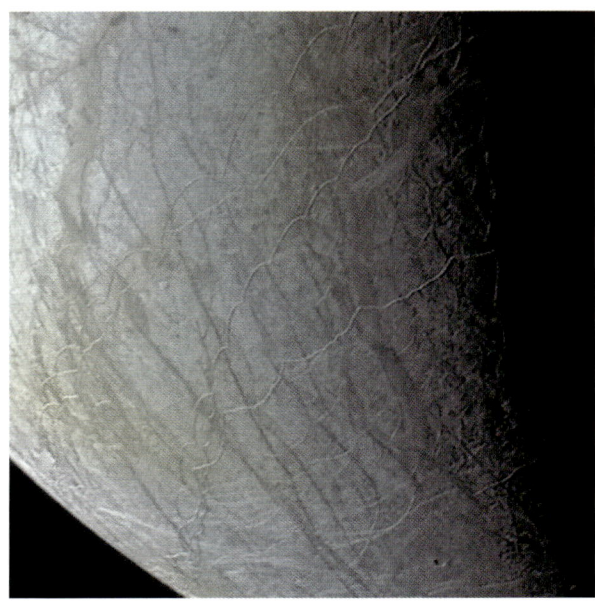

Top: a circular feature on the surface shows
the impact basin called Tyre Macula.

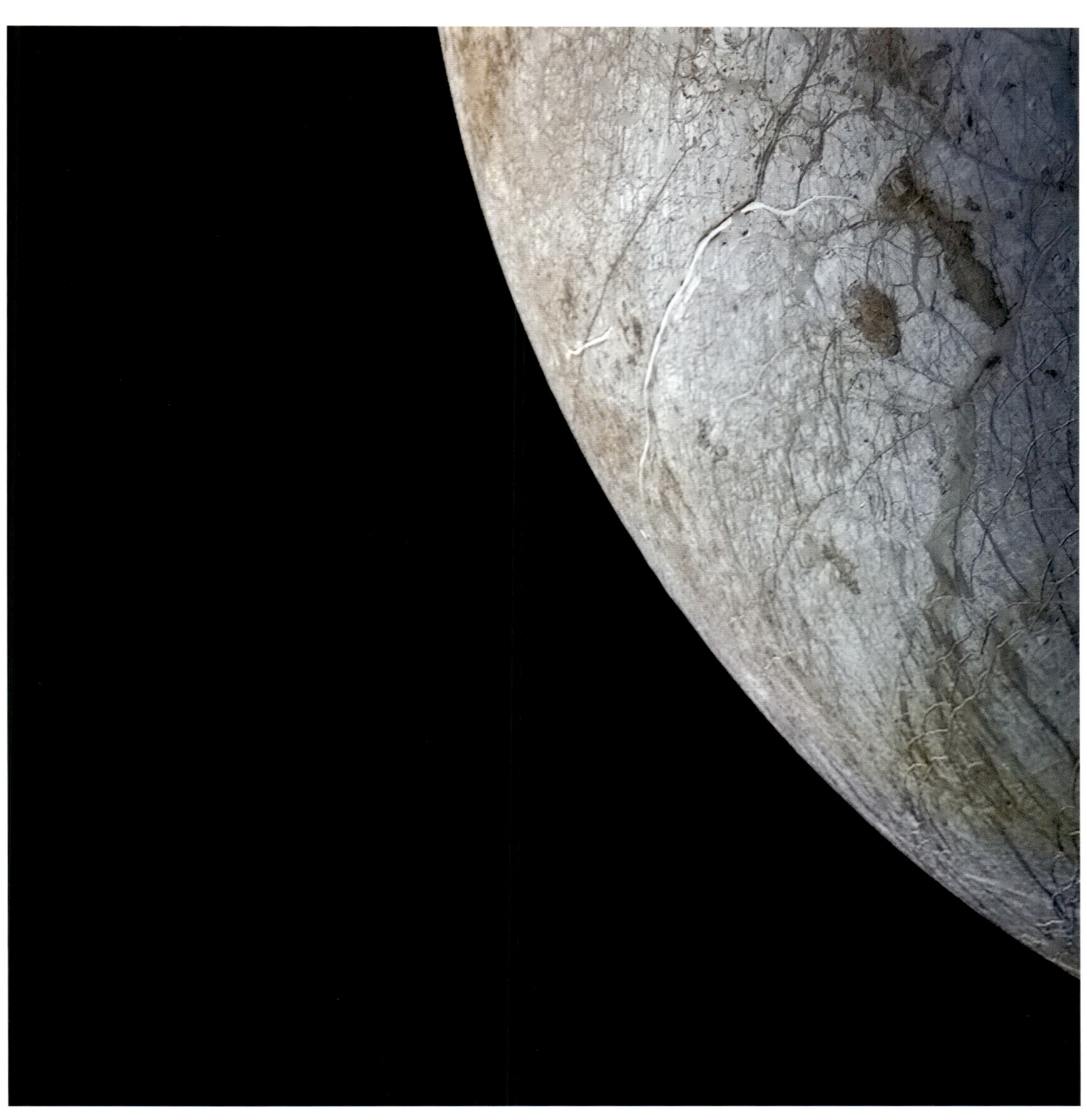

July 9th, 1979, 16:32:45   Voyager 2   Europa   Jupiter

GANYMEDE

It would be almost unfair to describe Ganymede as a moon. At 3,273 miles (5,268 km) in diameter, it is the ninth largest object in the Solar System, larger than the planet Mercury and almost as big as Mars. It is considered the only moon to have its own magnetic field. When the Voyager probes passed by Ganymede, the returning imagery revealed that the moon in places was as old as Jupiter, around 4.5 billion years old.

The surface showed two contrasting types of terrain: ancient and cratered or with massive grooves and flows. The latter suggested that Ganymede's icy crust is, in places, less ancient and is being altered and resurfaced due to tectonic activity caused by the same tidal kneading forces as from the other Galilean moons.

The composition of Ganymede is a combination of rock and ice. The Voyager data suggested it contained a solid iron core and lacked a detectable atmosphere.

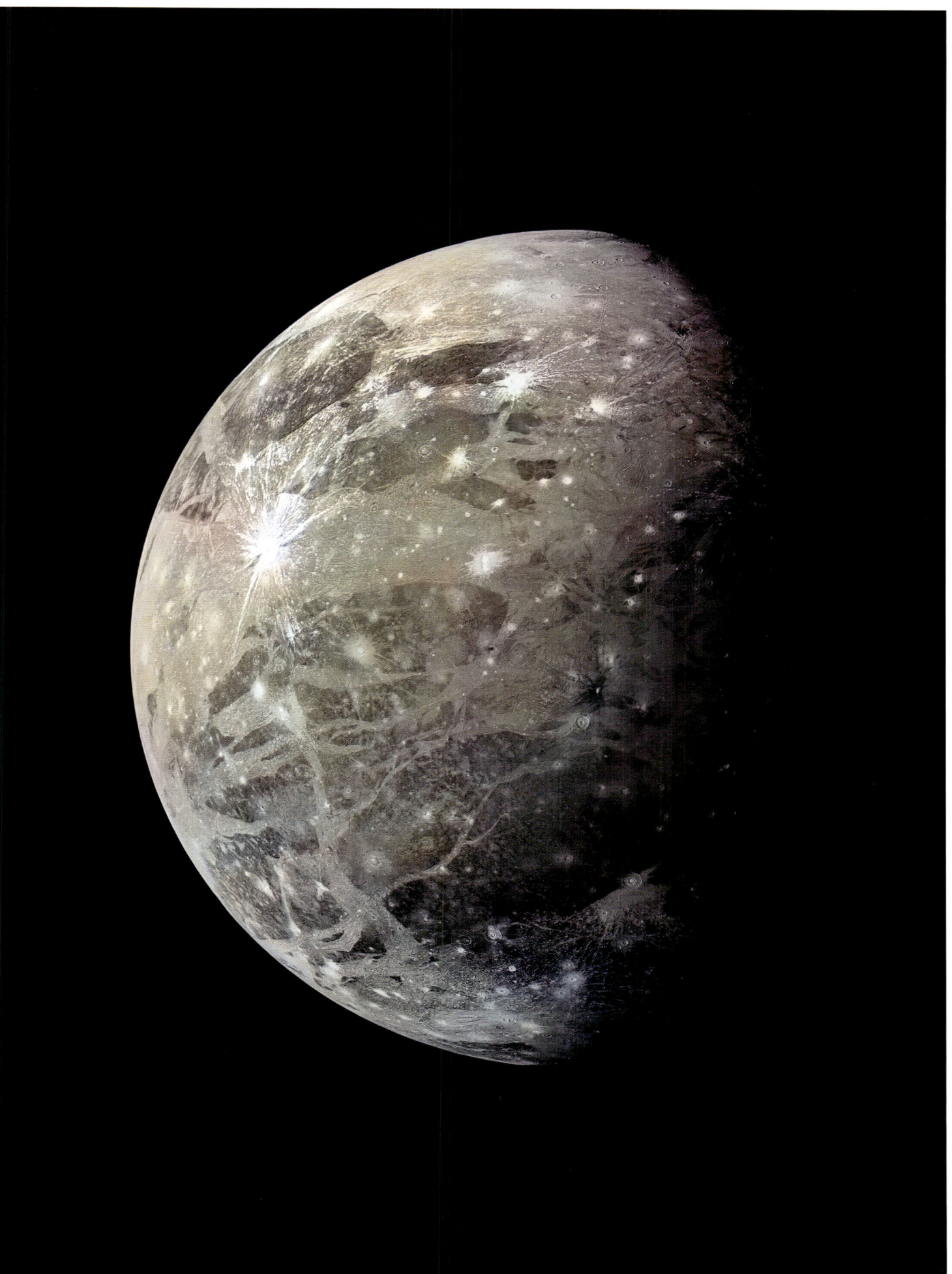

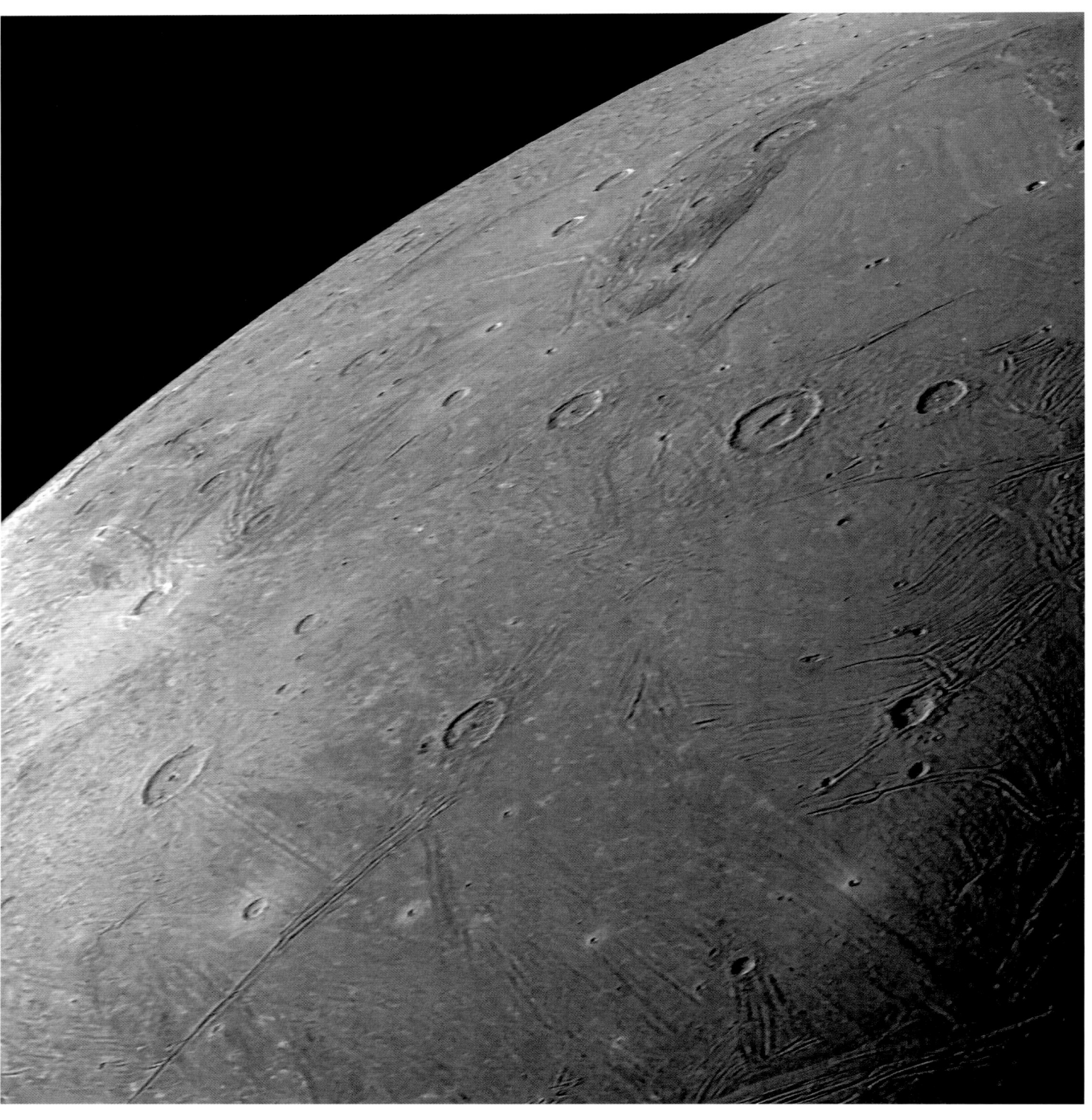

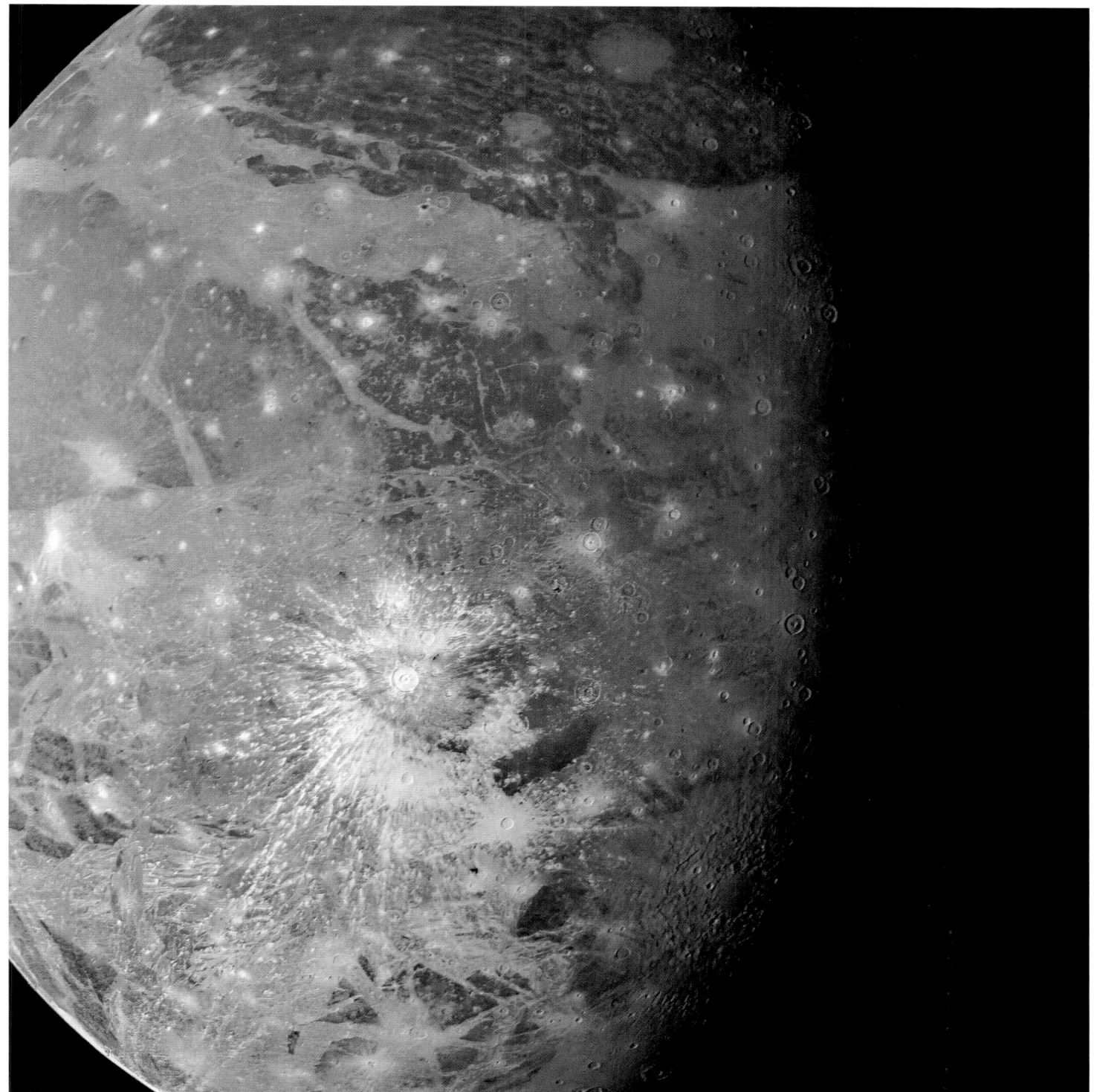

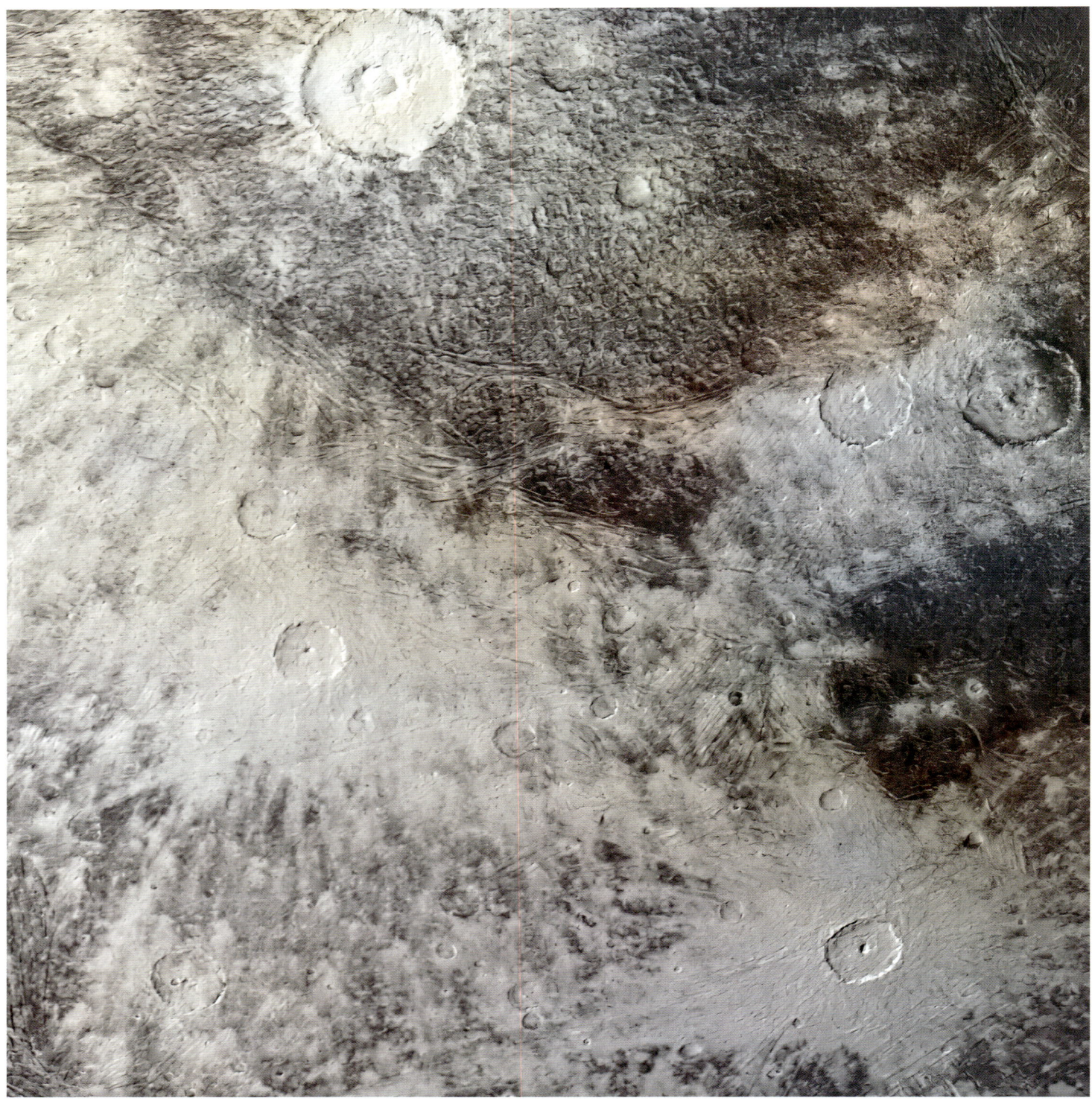

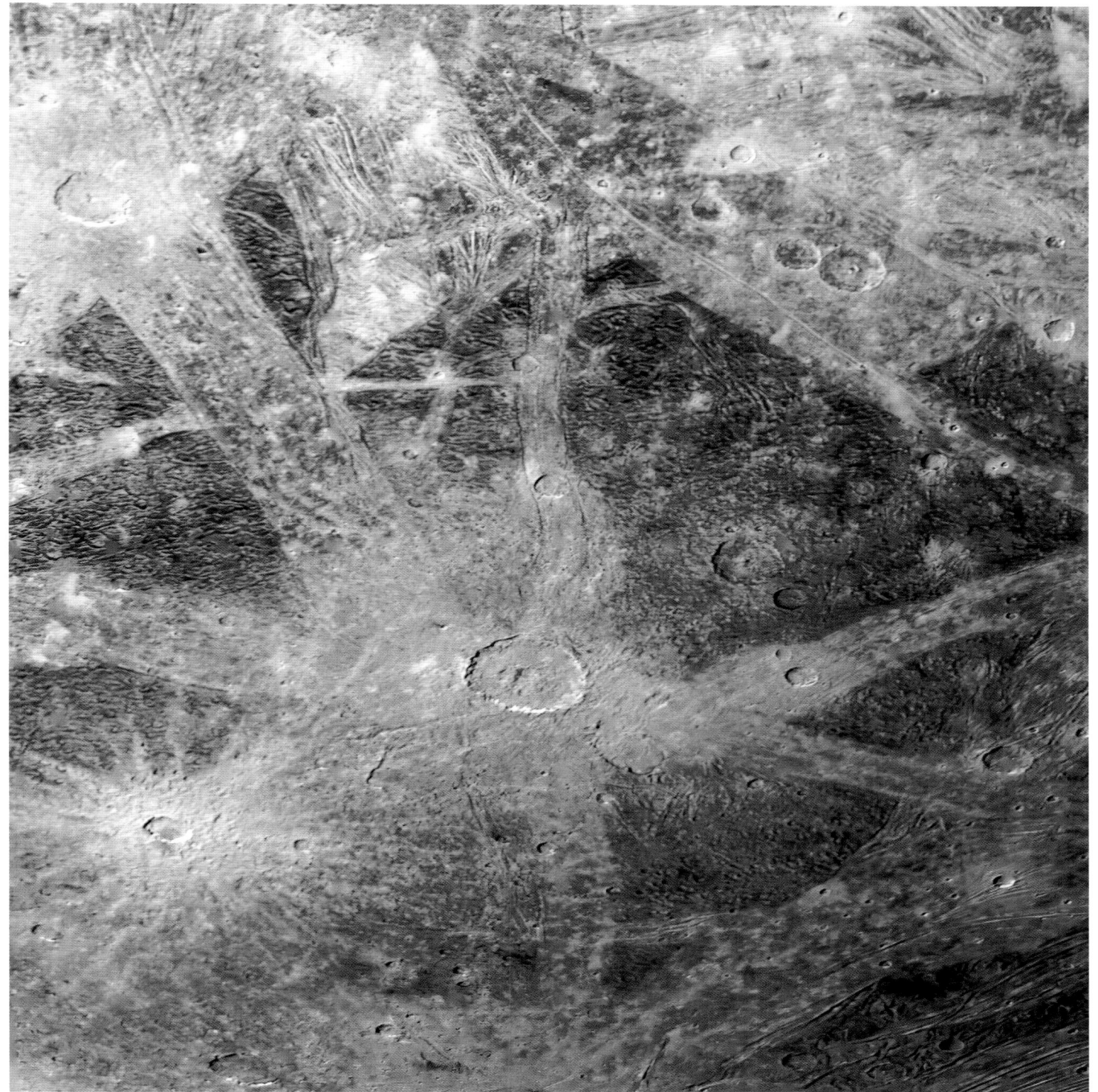

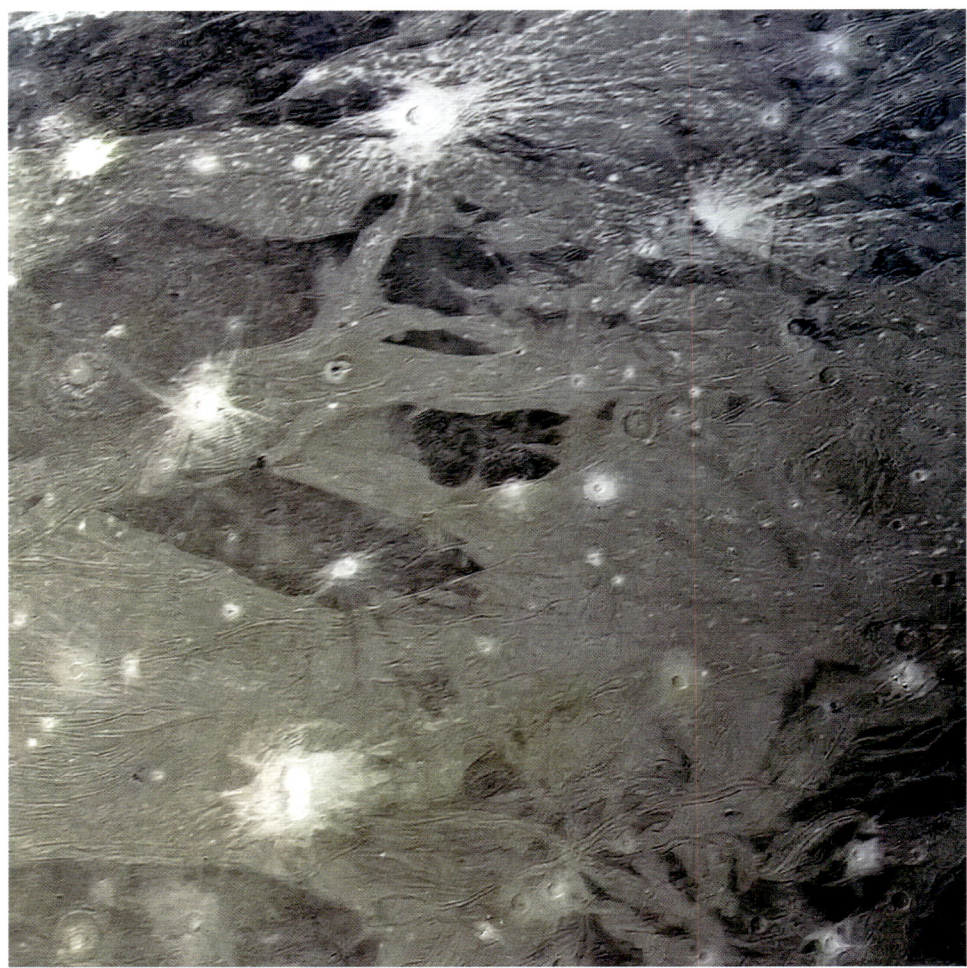

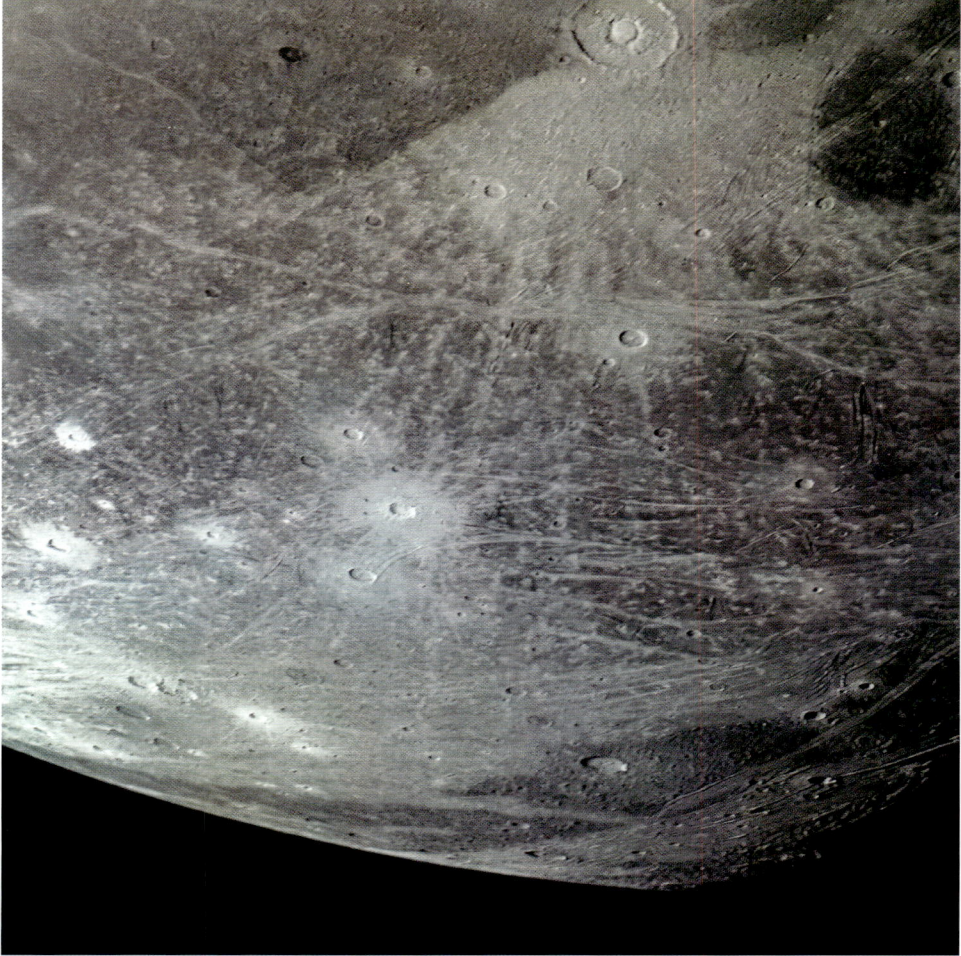

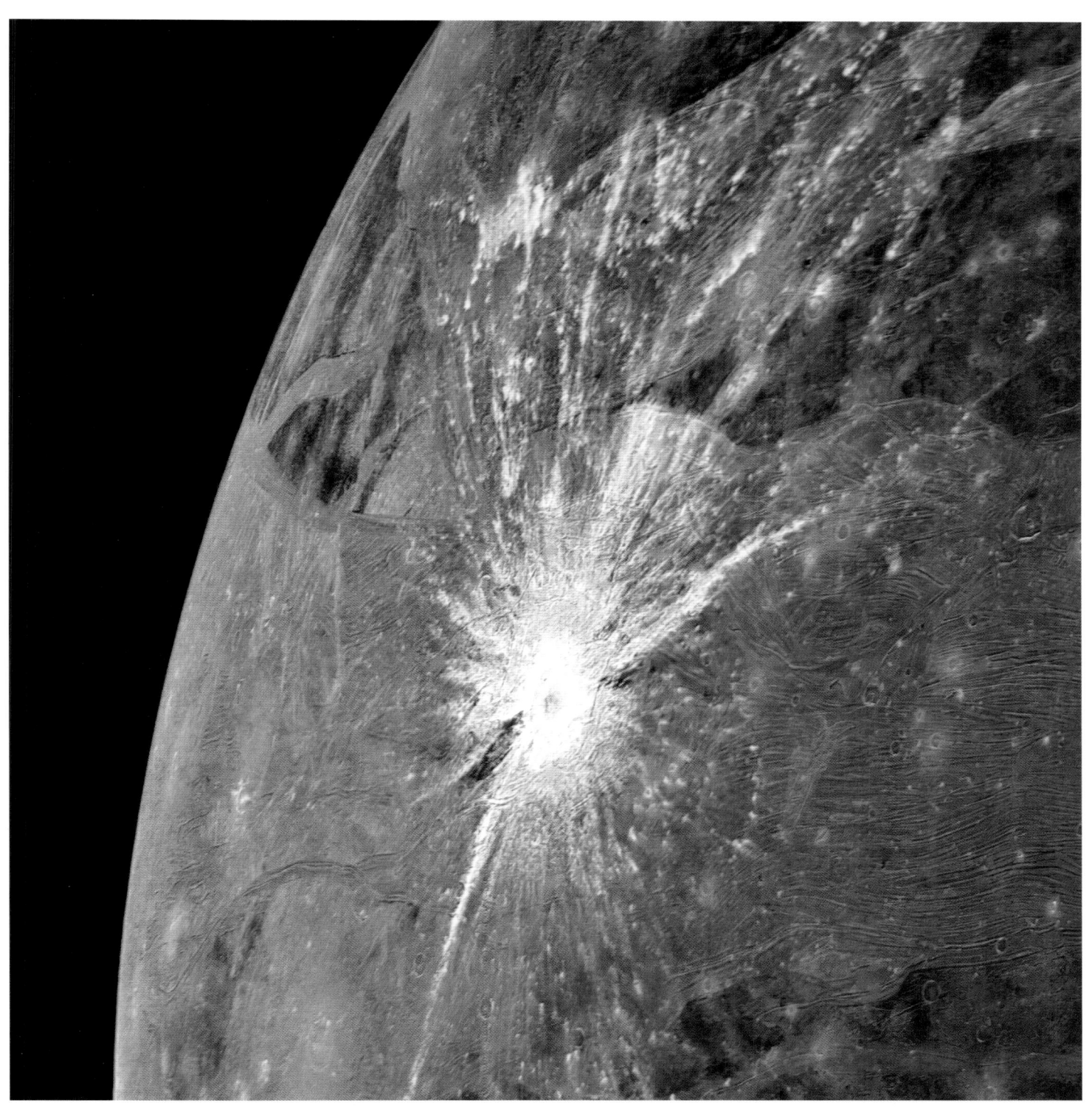

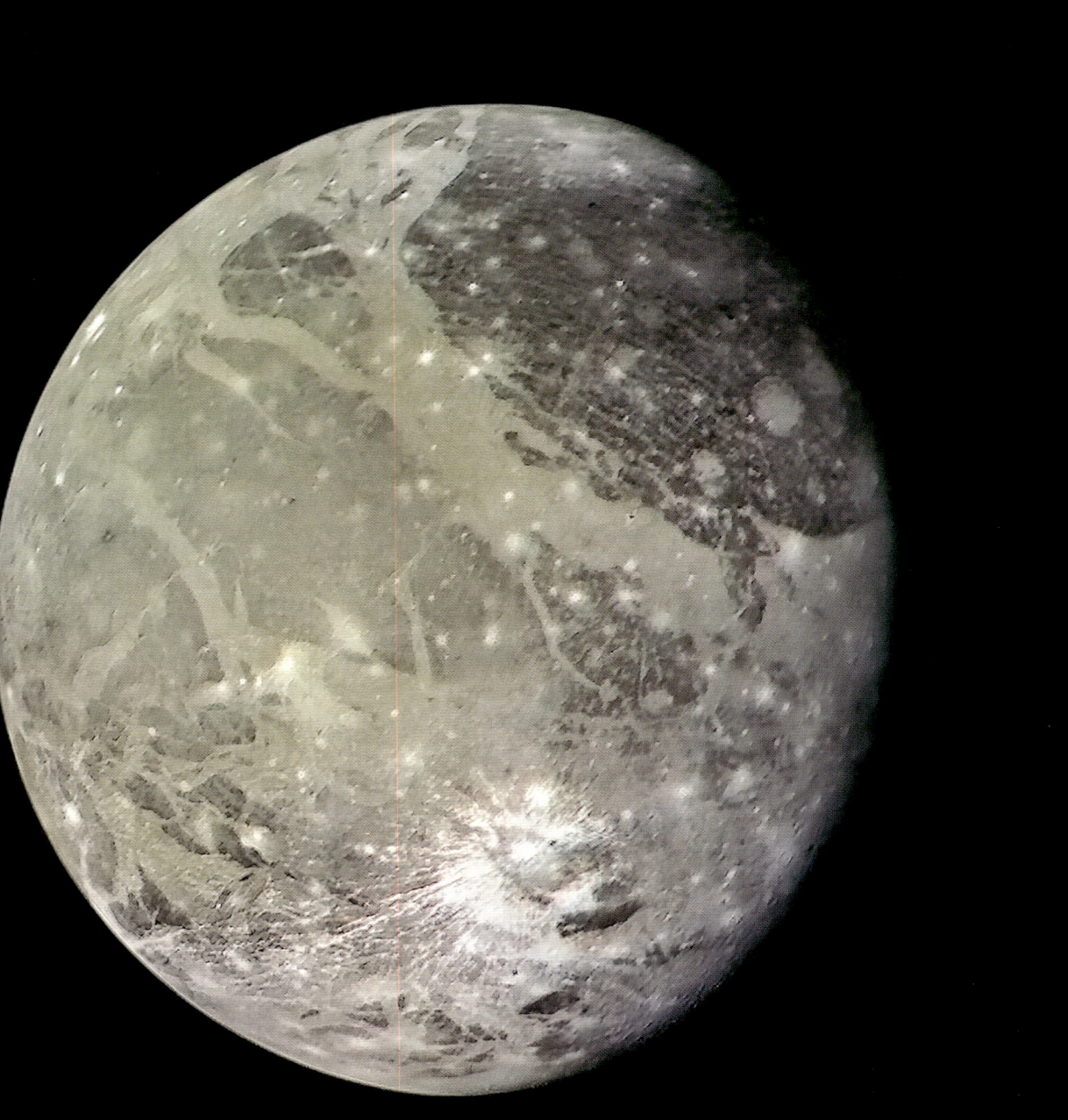

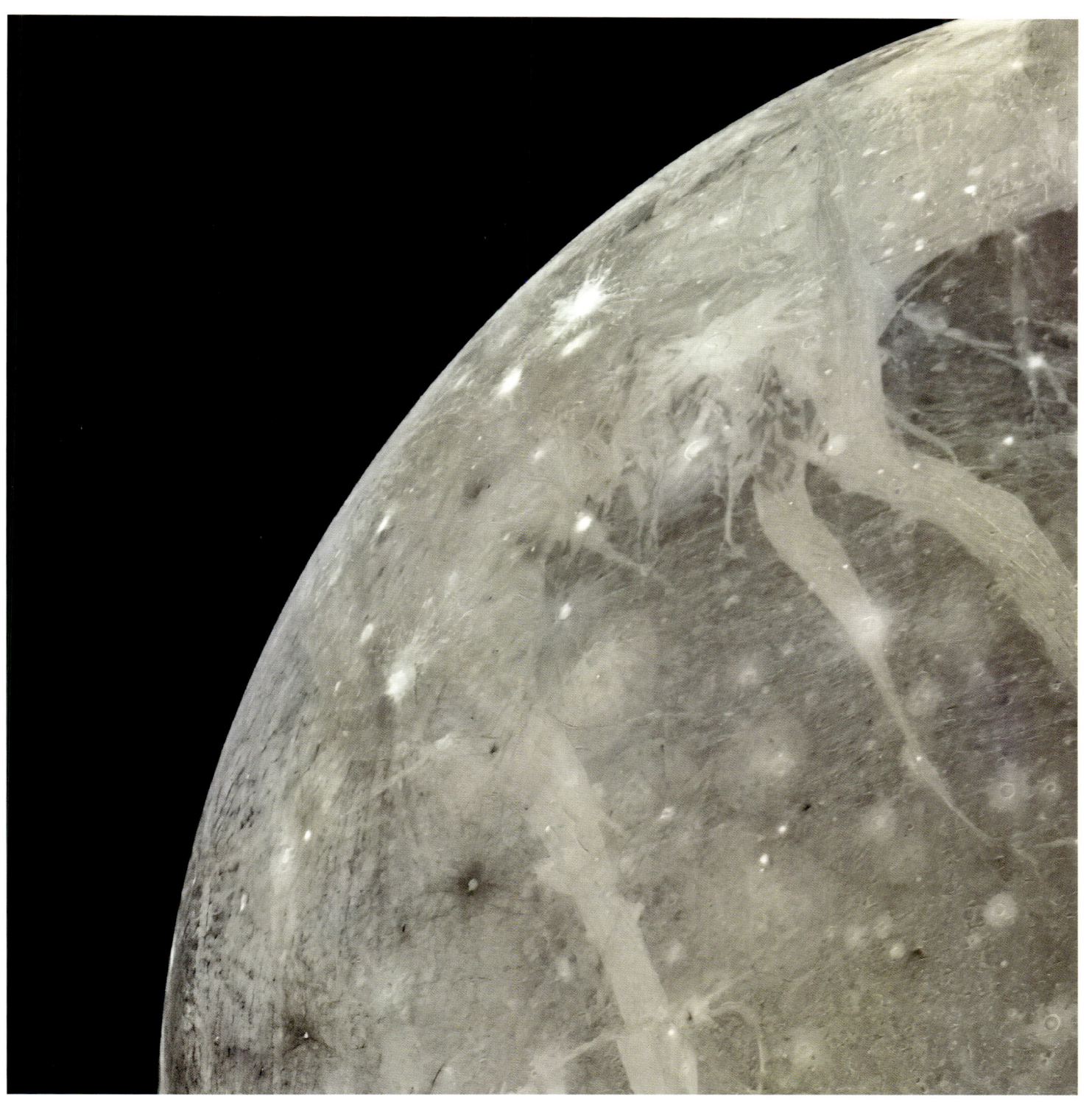

Left: Ganymede, Voyager 2, July 9th, 1979, 00:03:58.

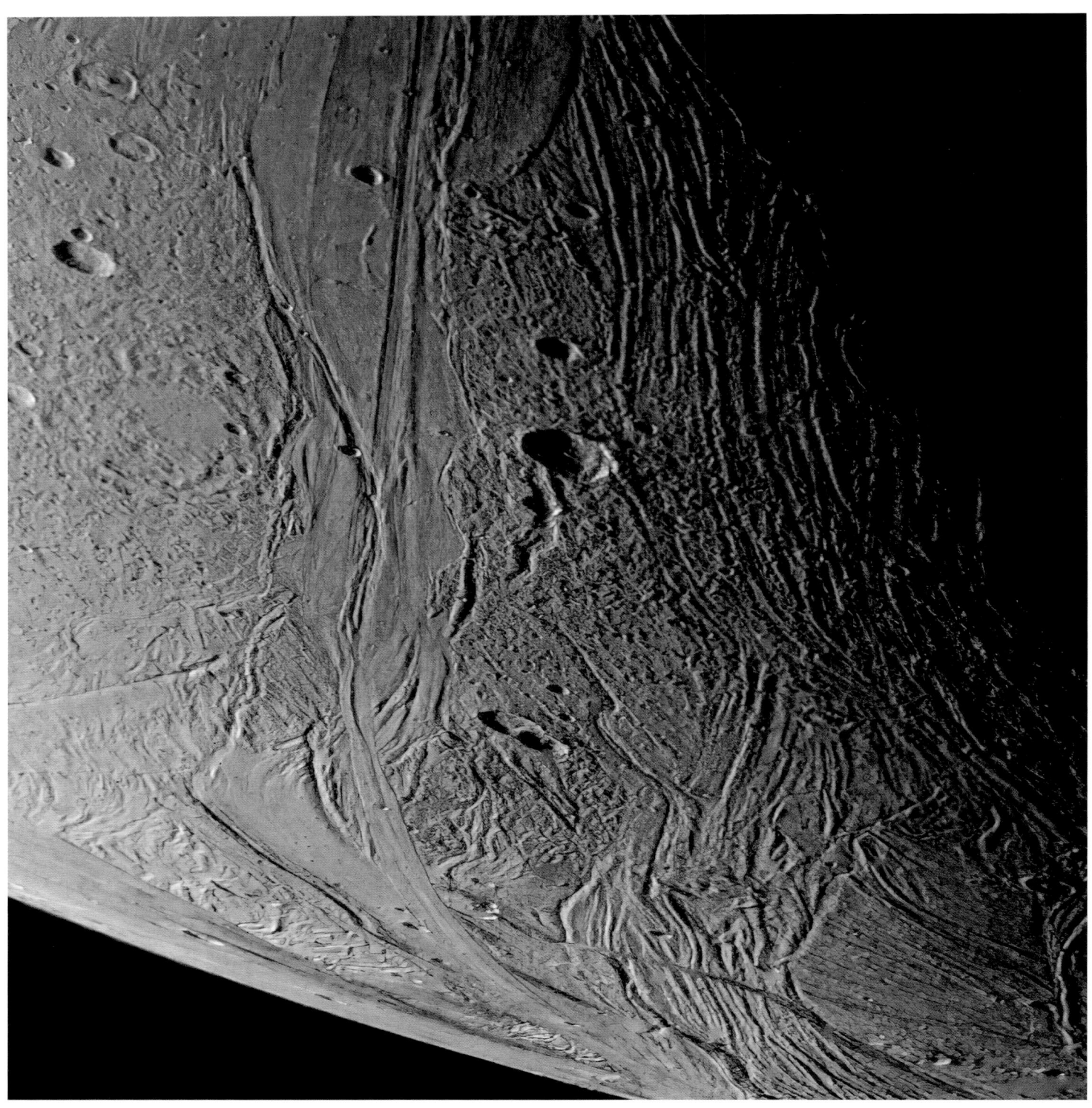

Right: another impact basin, this one is called Gilgamesh.

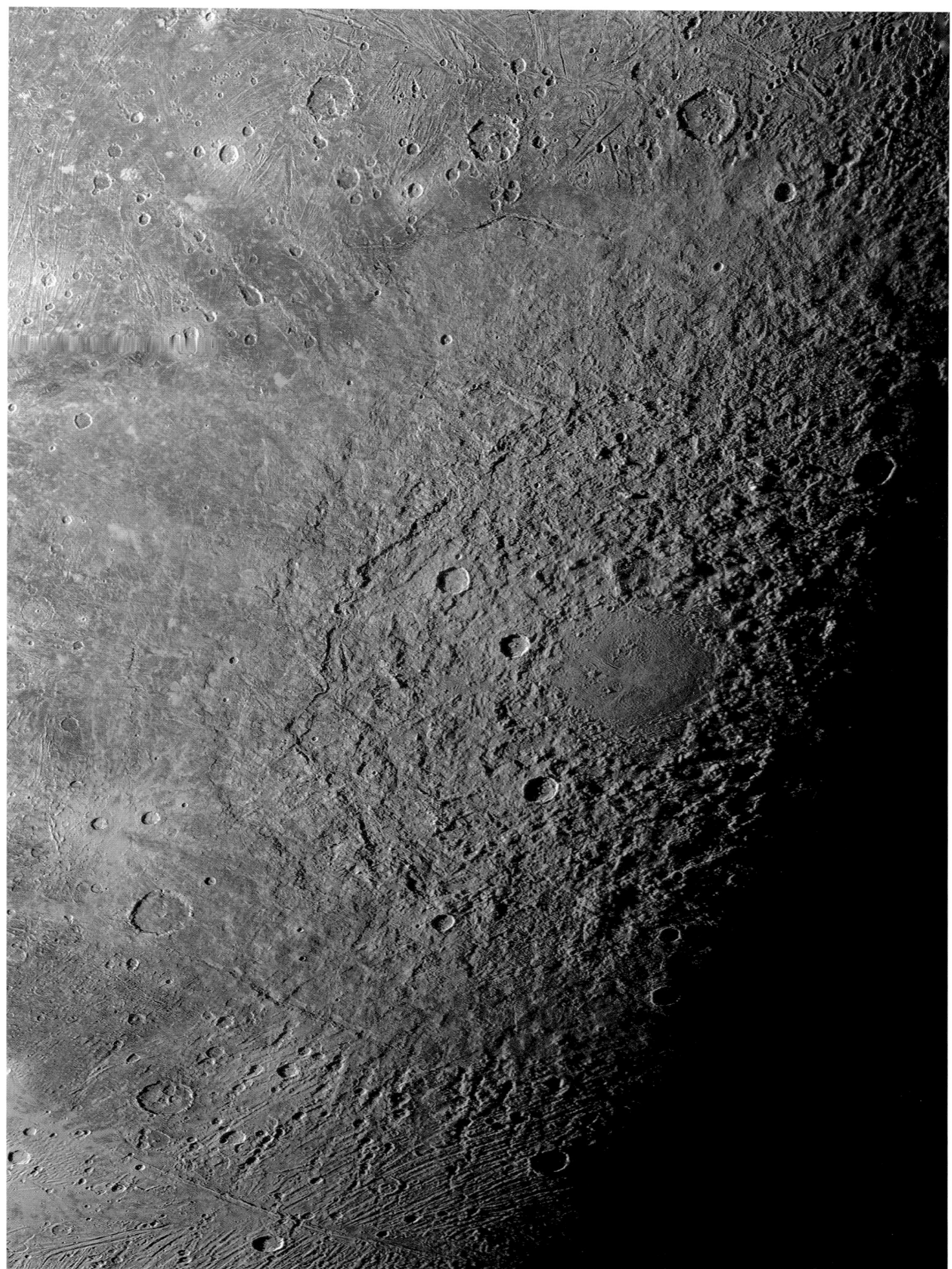

CALLISTO

Callisto is the second largest and the outermost of the four Galilean moons. It takes roughly seven Earth days for Callisto to complete one orbit around Jupiter. Voyager 1 passed by Callisto on March 5th and Voyager 2 on July 7th, 1979. Both managed to return interesting and informative images and data. The pictures showed a gray, desolate, and ancient landscape. Heavily scarred from impacts from the billions of years of pummeling from asteroids and comets striking its surface. NASA researchers calculated that the surface of Callisto is the most heavily cratered object in the Solar System. The crater evidence being a key indicator to how ancient and unchanged the surface is. Callisto is around 4.5 billion years old, roughly the same age as Jupiter itself.

One remarkable image transmitted back from Voyager 1, on March 6th, showing a seriously big multi-ringed impact crater in the northern part of the moon. The central bright region of the Valhalla crater is about 373 miles (600 kilometers) across, with concentric rings rippling outward to a diameter of nearly 2,360 miles (3,800 kilometers). This impact crater discovered by the Voyager 1 craft became known as Valhalla and is considered the largest known multi-ring impact crater in the Solar System.

Callisto is clearly different to its Galilean neighbors. The dead and still appearance of Callisto differs hugely with the hyperactive surface seen on Io. This is believed to be down to the orbital resonance that affects Io, Europa and Ganymede. The gravitational pull between these three moons creates a huge amount of internal heating, leading to much surface change and volcanic action. Callisto is the odd one out, too far out to be affected by the other moon's orbit, making the surface and interior of Callisto static and unaltered. It was thought to be completely inactive until a more recent visit, from the Galileo space probe in 1996. This revealed evidence that Callisto might not be that dead piece of rock and ice, but perhaps may have a thin liquid water layer underneath its thick outer crust.

Right: the circular shape on the limb represents the gigantic impact basin called Valhalla.

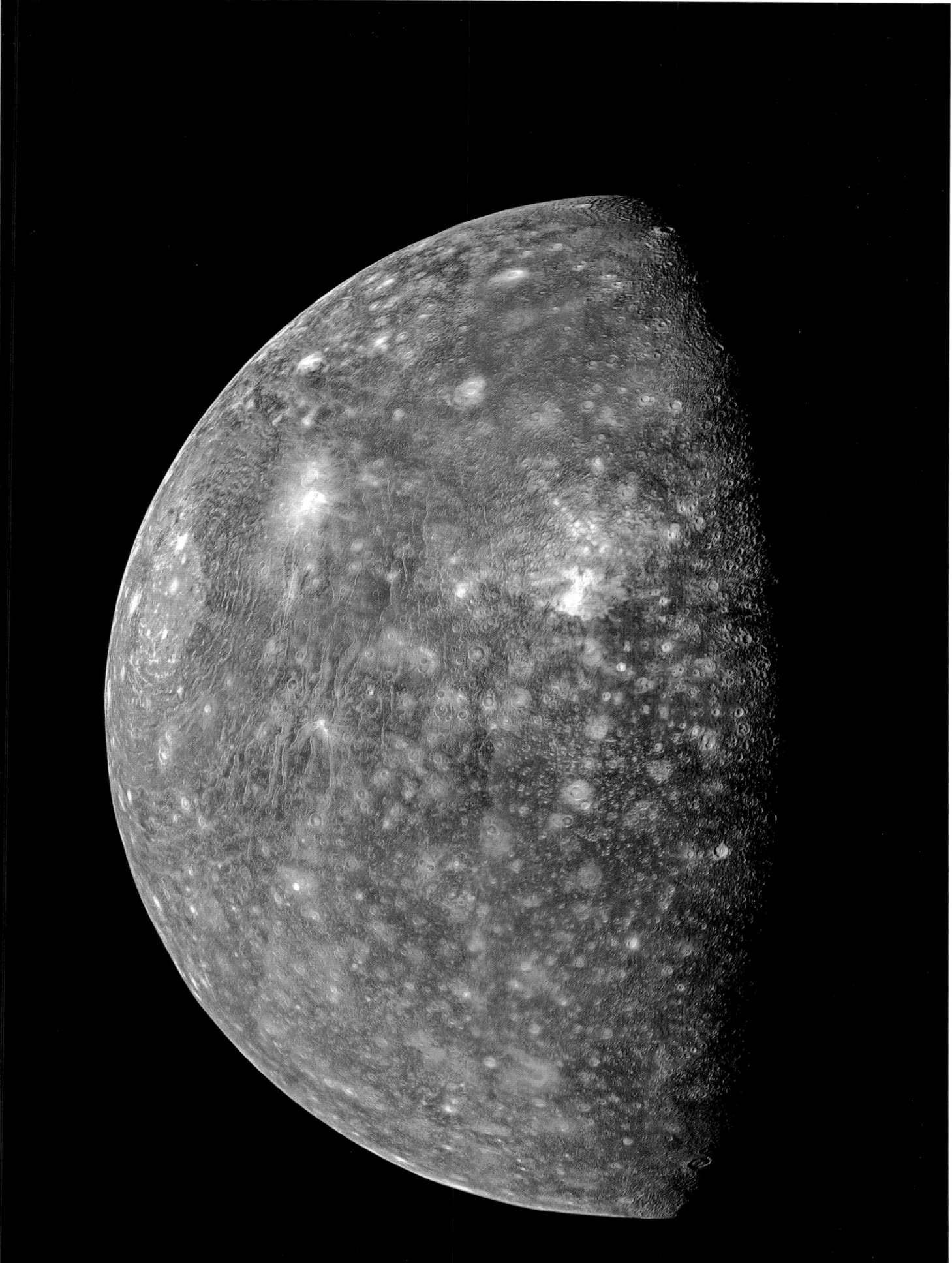

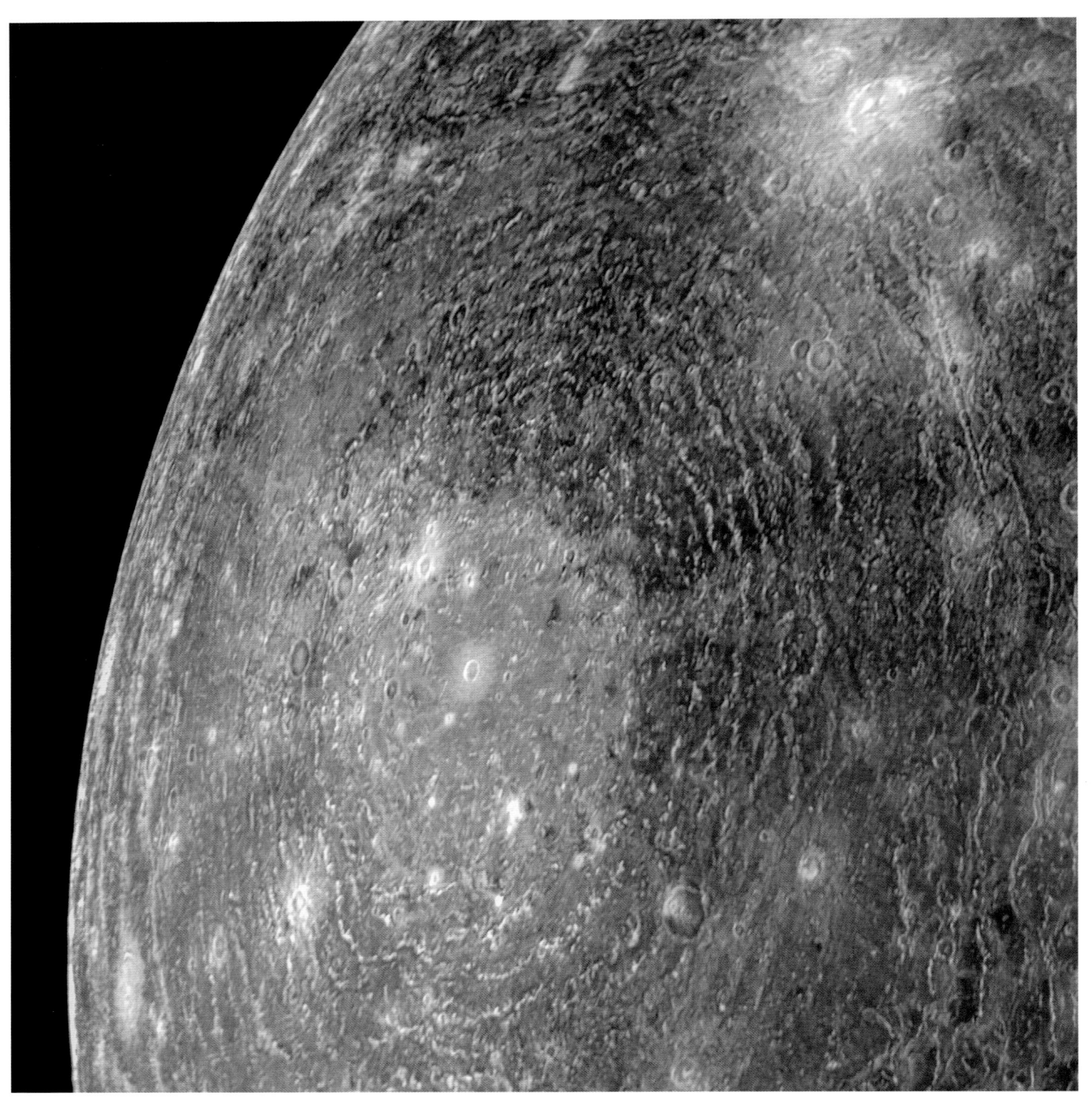

Valhalla up close.

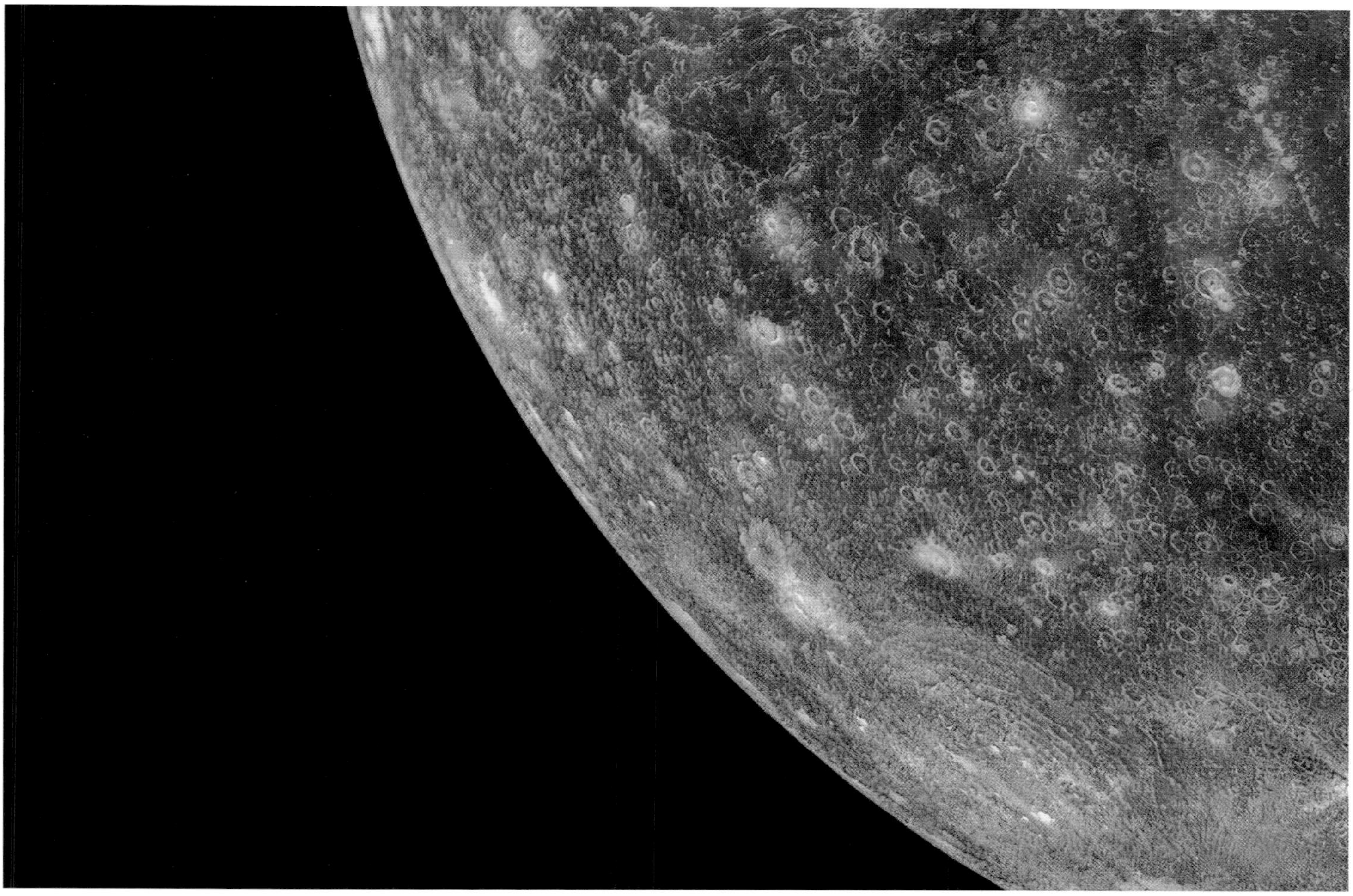

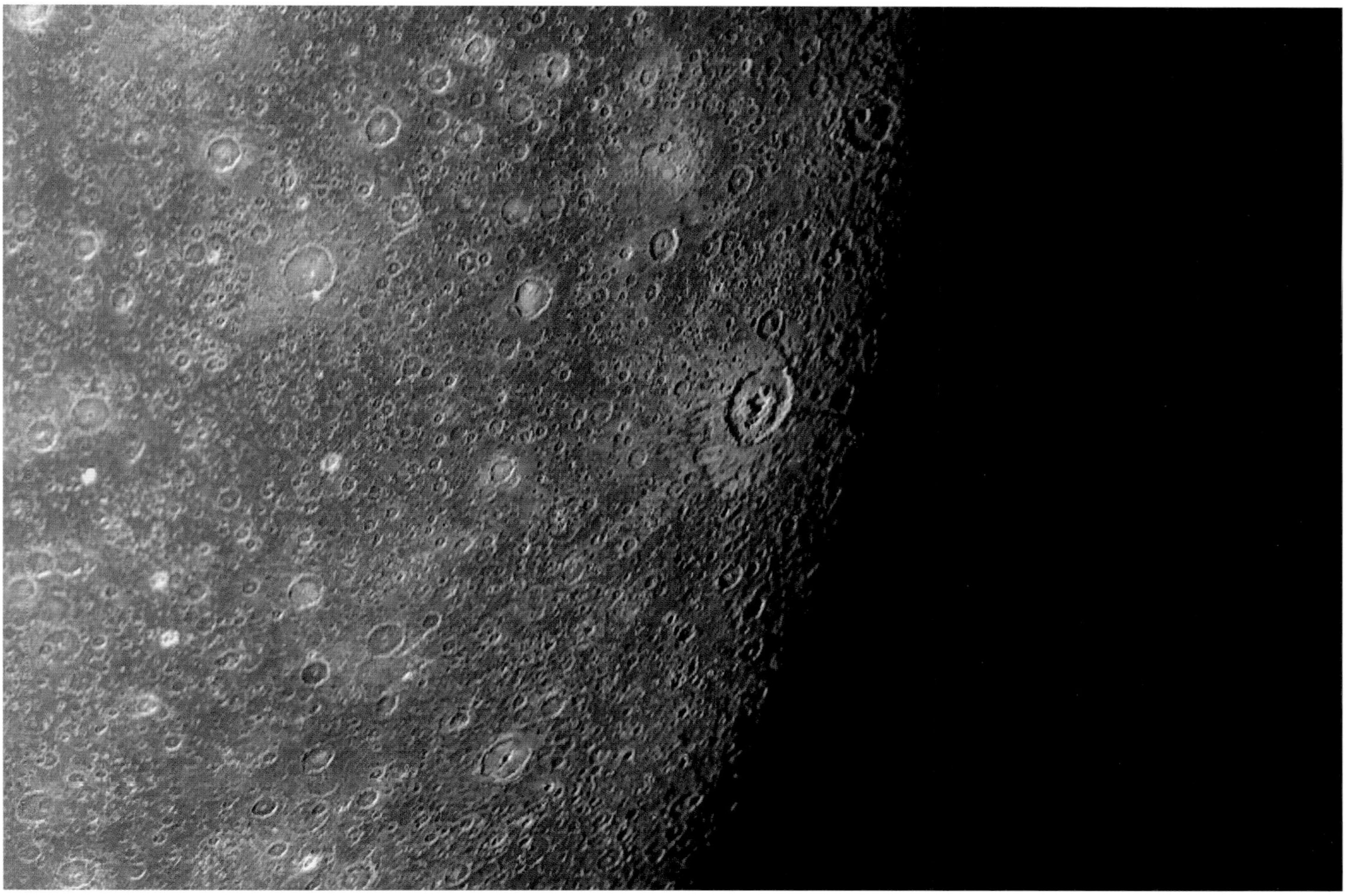

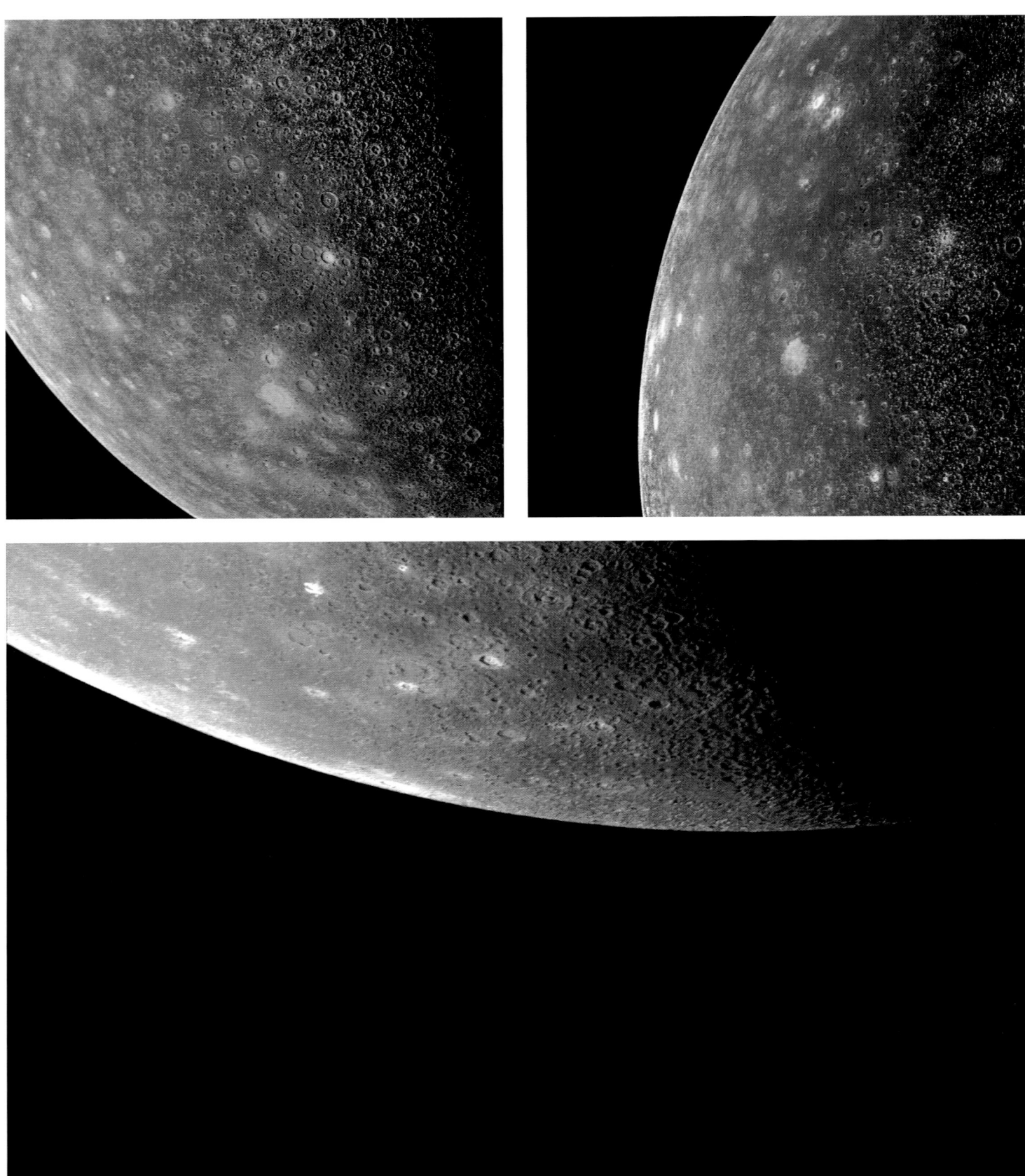

Right: Callisto, Voyager 2, July 8th, 1979, 12:50:22.

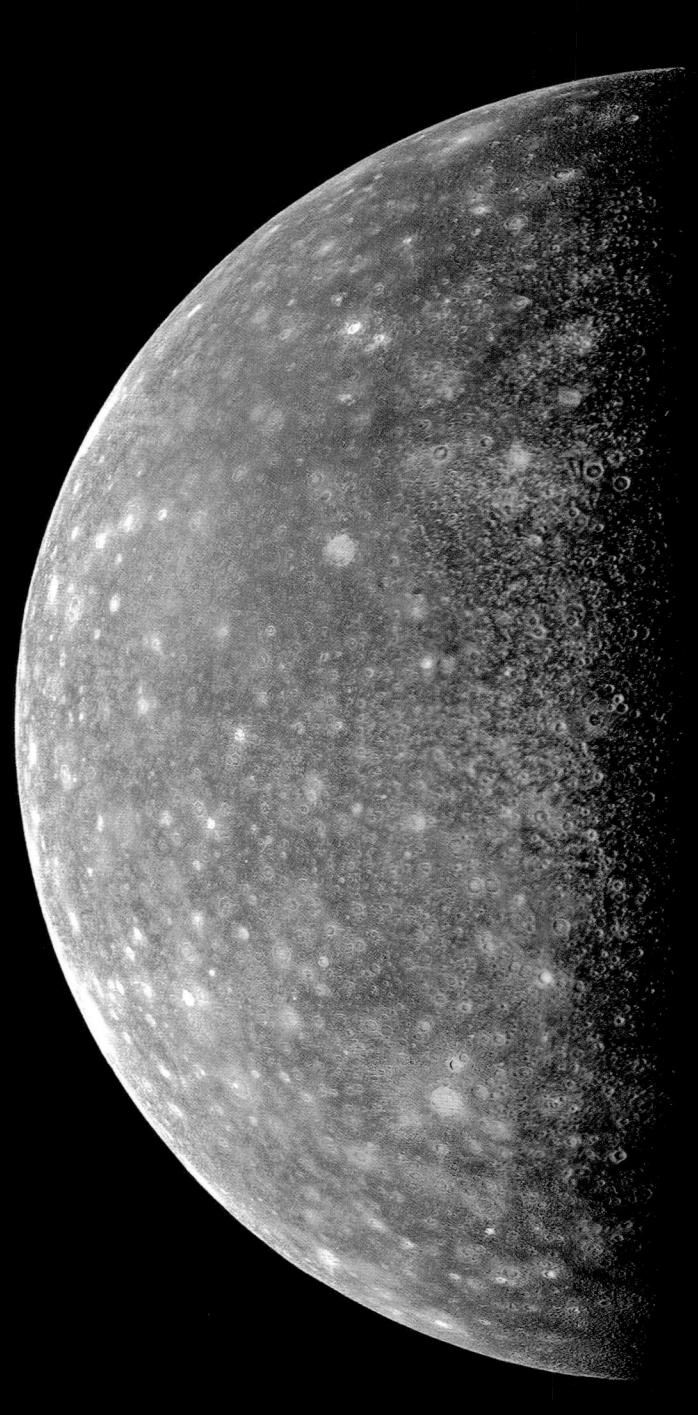

LOOKING BACKWARDS

As Voyager 1 was exiting the Jovian System, the craft turned its camera around to point back sunwards and capture the last "first glimpses" of the planet and its surroundings. This imaging technique uses the shining light of the Sun to help reveal the hidden objects that would be hard to detect if the light was not shining behind these tiny pieces of matter, an imaging trick known as forward scattering.

Although it was on the exit, these images revealed some astounding new findings. Firstly, they showed three new moons orbiting inside the orbit of Io. Adrastea, Metis, and Thebe, all tucked in between the orbits of the previously-known moon Amalthea and Jupiter itself. But the bigger of the new discoveries was that Jupiter—just like Saturn— had its own ring system, although one that is much less profound and visible. These images allowed scientists to determine that the rings are made up from microscopic rock and dust that is being constantly lost out into space, while simultaneously being replaced by new dust particle remnants made from collisions and impacts of comets and asteroids hitting Jupiter's moons. It was shown that Adrastea and Metis were the sources of Jupiter's main ring, and that Amalthea and Thebe contribute to the outer part, known as the Gossamer Rings.

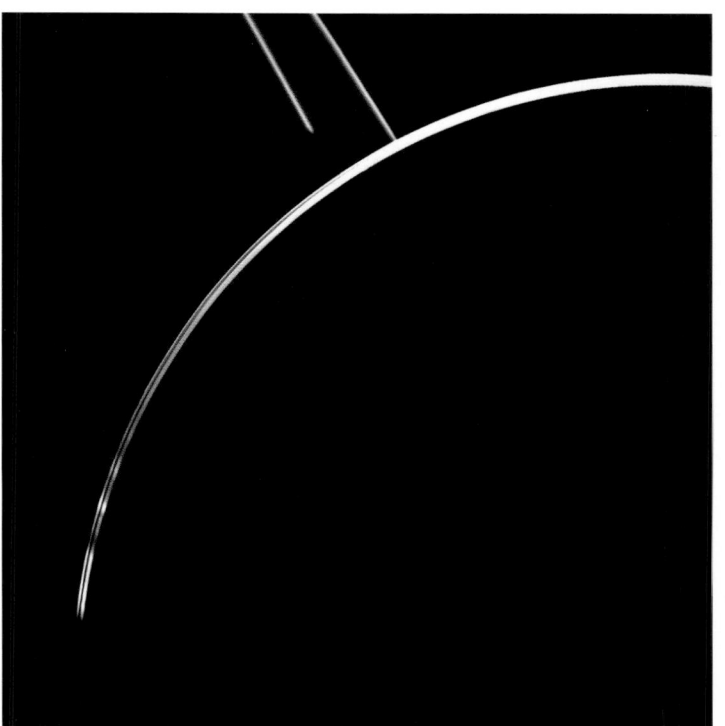

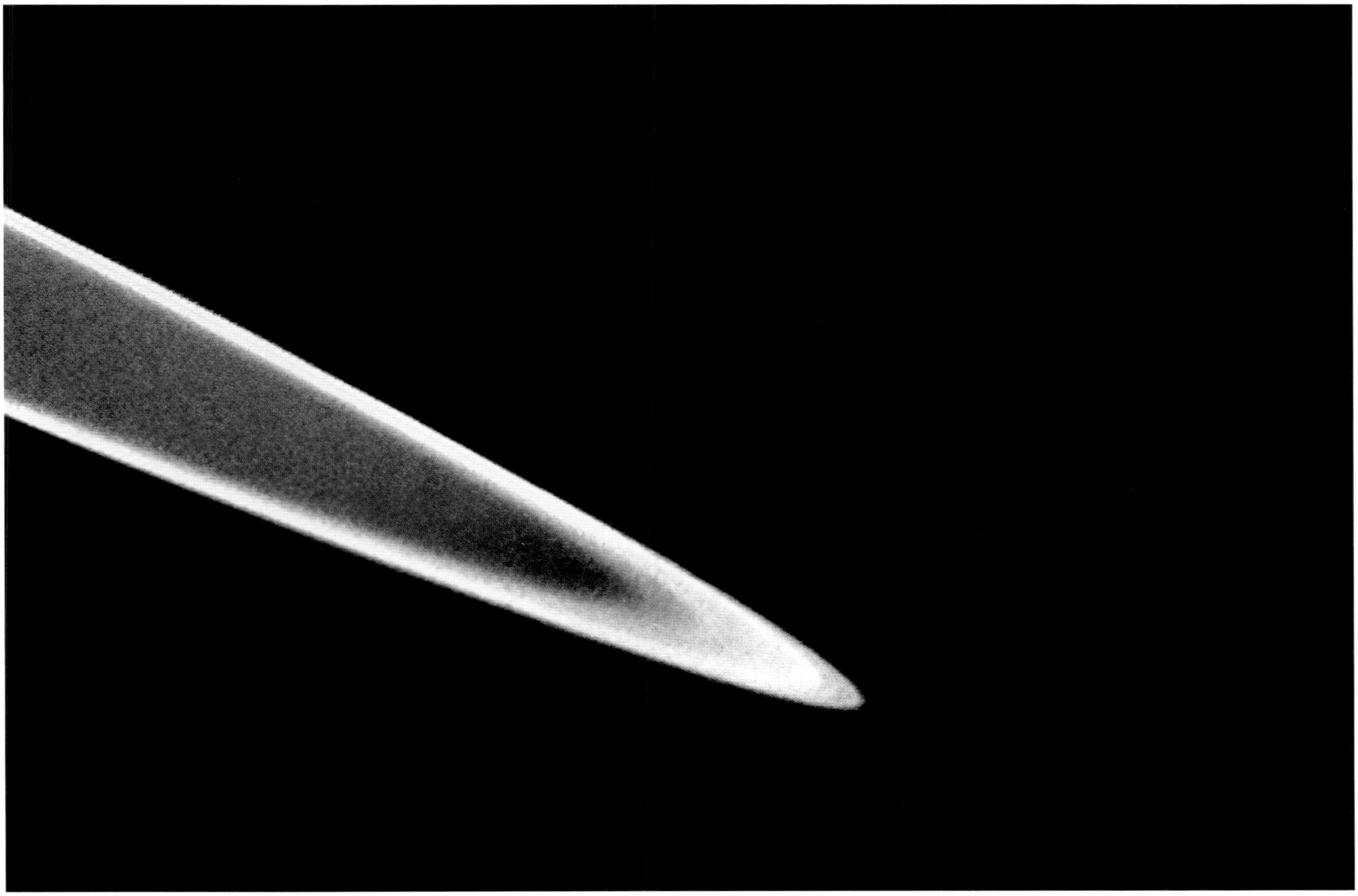

ONWARDS

By the end of August 1979, both Voyager flybys provided a massive leap into humanity's understanding of its biggest neighbor and its equally giant moons. With over 33,000 images returned back to Earth, we were given pictures of some of the most stunning and breathtaking landscapes ever to be captured in the history of photography.

Voyager made remarkable discoveries in its encounter with Jupiter. A faint, barely visible ring system spotted around Jupiter that was completely unknown. A collection of dramatic, changing landscapes, witnessing hyperactive volcanic eruptions created by the Laplace Resonance, a tidal kneading and internal heating of moons, on worlds we originally thought were too cold and distant from the Sun to sustain activity. Besides these discoveries, Voyager also gave us close-up images of the Giant Red Spot, first mentioned to us over 300 years ago, now in stunning detail, showing a storm system bigger than the Earth swirling around counterclockwise in a constant six Earth-day revolution. Jupiter just kept on giving and giving.

The scientific community was very keen to voice their enthusiasm and eagerness to return. With Europa being an extremely high priority to revisit and research further, being one of the lead candidates beyond our world where life may or may have existed in the past.

While the people on Earth entered into the 80s, dancing to disco and playing the Pac-Man video game, our two spacecraft continued humanity's greatest journey in silence. Next stop: Saturn. Voyager 1 started to make its way, speeding up with a boost from Jupiter's gravitational force, giving it a head start in comparison to its twin. Voyager 2's journey would take a longer and slower route, resulting in an estimated time of arrival of 30 months.

Right: Jupiter, Europa, Io and Ganymede all captured in one image. Voyager 1, February 22nd, 1979, 04:37:23.

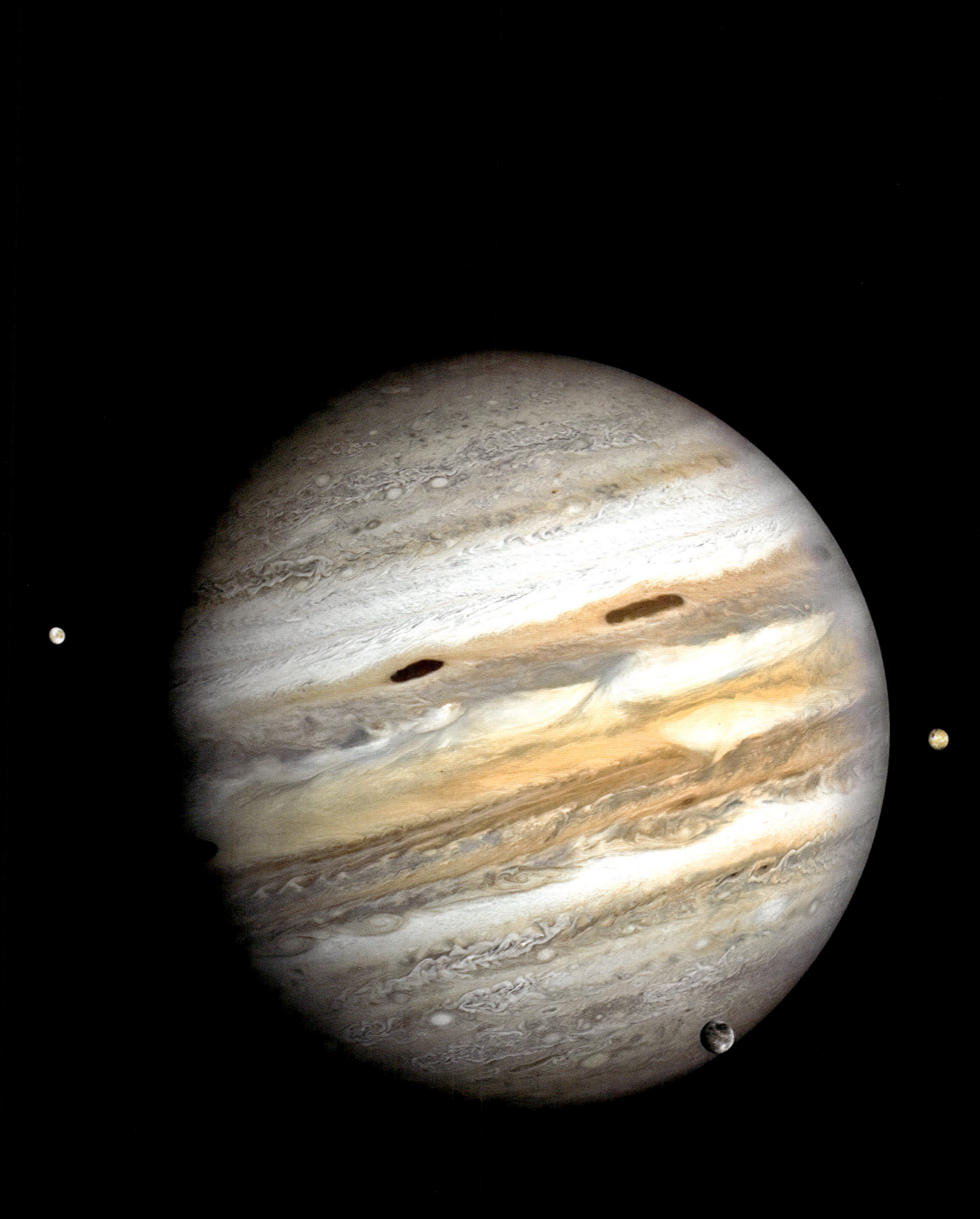

DANCING WITH RINGS

INTRODUCTION

If you ask a child to draw a planet from their imagination, you stand quite a good chance that they will end up drawing a picture that resembles the image of Saturn. A nice round circle, with a simple ring system wrapped around it. That is how powerful and memorable the shape and image of Saturn is to us back on Earth.

There was a great deal of excitement in the fall of 1980, as Voyager 1 was about to send back the most detailed images of possibly the most picturesque and photogenic planet in the Solar System. After the highly successful first flybys at Jupiter, the halls of NASA were filled with anticipation on what kind of photographs and data would be returning to the team.

Saturn is the sixth planet from the Sun and the second largest in our Solar System. It has over 95 times the mass of Earth, with an equatorial radius of 37,449 miles (60,268 km), roughly 9.5 Earths put side by side. Like its neighbor Jupiter, Saturn is also classed as a "Gas Giant." It is a huge sphere of hydrogen-helium gas mix, with a probable rocky core made of an iron-nickel metal and rock, wrapped in an intermediate layer of liquid hydrogen and helium.

Saturn's pale yellow-brown complexion is due to ammonia crystals that are present in its upper atmosphere, creating bands of different colored clouds that wrap around the planet, giving it its varied striped appearance. Wind speeds are exceptionally high, often reaching speeds of 1,100 mph (1,800 km/h). Saturn's non-spherical shape is very noticeable if you look closely at the pictures that show the planet as a whole. The flatter tops at the poles and extruded bulges around the equator are caused by the planet's rotation. The planet's gravity pulls around the most prominent and visually spectacular ring system to be seen in the Solar System, so well defined that it was visible for Christiaan Huygens to clearly observe through his telescope back in 1659.

The Voyager team had been carefully plotting the route that would allow both spacecraft to get a close view of Saturn, and its highly intriguing moon Titan, although they could not push their luck too much. If the probes travelled too close to the surface, it would significantly increase the risk of a potential mission-ending strike from a small stray particle coming from Saturn's ring system.

Right: Saturn & satellites, Voyager 1, August 23rd, 1980, 12:46:56.

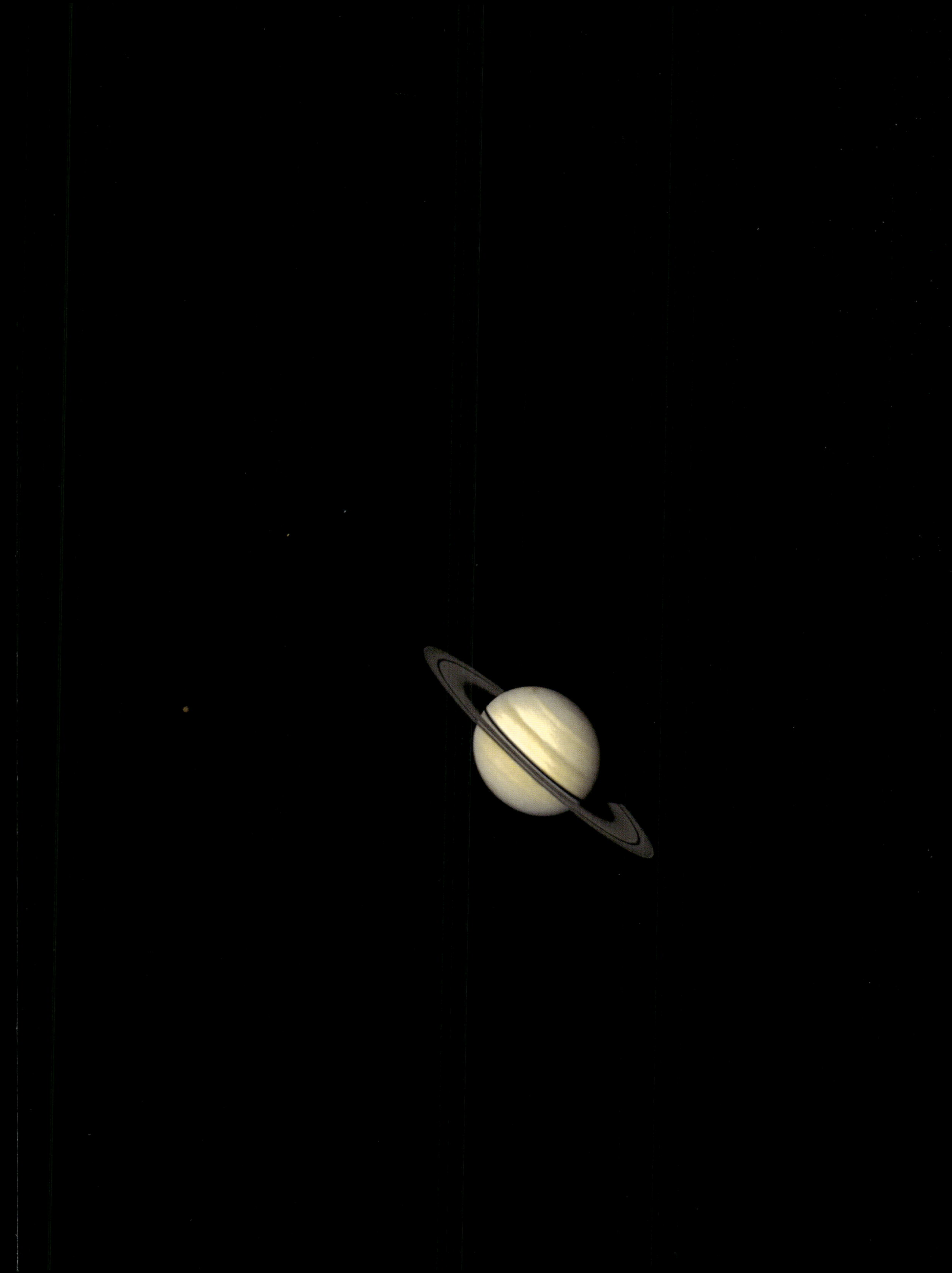

SATURN THROUGH THE AGES

Saturn has been known to humanity for millennia and it followed a similar path of recognition as Jupiter did throughout time. Again, it was the Babylonian astronomers who observed and first recorded Saturn moving across the night sky. It was later that the ancient Greeks referred to the planet as Phainon, a dedication to the Greek god of agriculture, Cronus. Traveling further forward in time, it was finally the Romans who gave Saturn its modern name, Saturnus, after their equivalent God of agriculture.

We started to get a better idea of Saturn's appearance and characteristics at the beginning of the 17th century. Galileo started to hone down his observations of Saturn in 1610. However, the telescope he designed and built could only make out the vague silhouette of Saturn. In his findings, he wrote of strange, non-spherical objects that were pointing out from both sides of the planet. His telescope was not powerful enough to give him a clearer view; he struggled to figure out what gave this strange outline to the planet. Galileo stated that perhaps the shape was created by two moons that sat either side of the planet, or jokingly, suggested that perhaps Saturn had a "pair of ears."

It was not until 45 years later that Dutch astronomer Christiaan Huygens got a better glimpse at Saturn. Publishing his findings that described the ring system for the first time as a "thin, flat disk, nowhere touching, and inclined to the elliptic." Huygens was also to discover Saturn's largest moon, Titan, in 1655. In 1675, Giovanni Domenico Cassini built upon Huygen's discovery, noting that the ring around Saturn was in fact two separate rings. Cassini also discovered four new moons, Iapetus, Rhea, Tethys, and Dione. Another two moons, Mimas and Enceladus, were discovered later by William and Caroline Herschel in 1789.

Even with the ever-increasing advances in telescope technology, revealing more and more about the Saturnian System, there was one big mystery that was troubling astronomers over the centuries. Described as "The Saturn Problem", it was still unclear what this ringed object was that appeared to be floating, at an angle, around the planet. A 27-year old Englishman called James Clerk Maxwell eventually answered "The Saturn Problem" in 1859. He solved the dilemma, not by looking through a telescope, but through the use of existing scientific

theory and the power of deduction to whittle down and eliminate the hypotheses. By working through a number of mathematical formulas, he proved that it would be impossible for the rings to be a continuous ring made from either a solid, liquid, or gas material. This left the hypothesis that the rings are made from many pieces of solid matter as the only possibility.

The last major discovery within the Saturnian System before the age of the space probe exploration was to be found at Titan. In 1944, Gerard Kuiper observed it to have a thick, dense atmosphere. This discovery sparked a huge amount of scientific curiosity, as to whether it could also support life, given that the atmosphere had many similarities to that on Earth. The atmosphere of Titan would become a high priority for the Voyager spacecraft to go and probe in much closer detail.

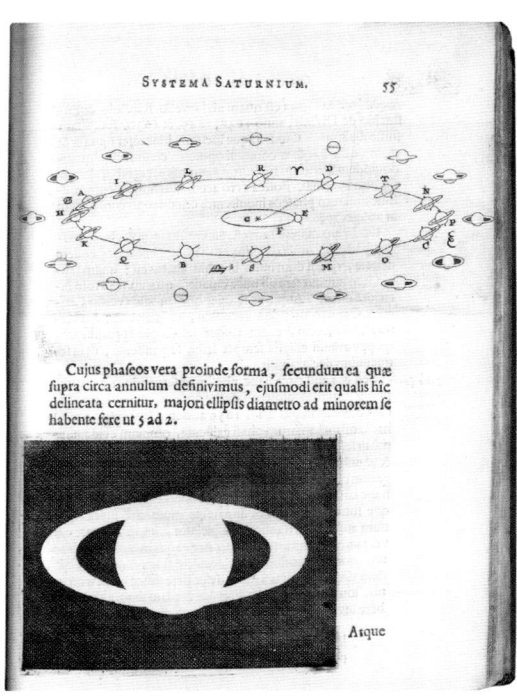

The diagram in Christiaan Huygens 'Systema Saturnium' (1659) that shows how Saturn's appearance changes due to its changing positions of the Earth (E) and Saturn as they orbit around The Sun (G). The bottom diagram is an observation drawing of Saturn and its rings as seen from Earth at their greatest inclination.

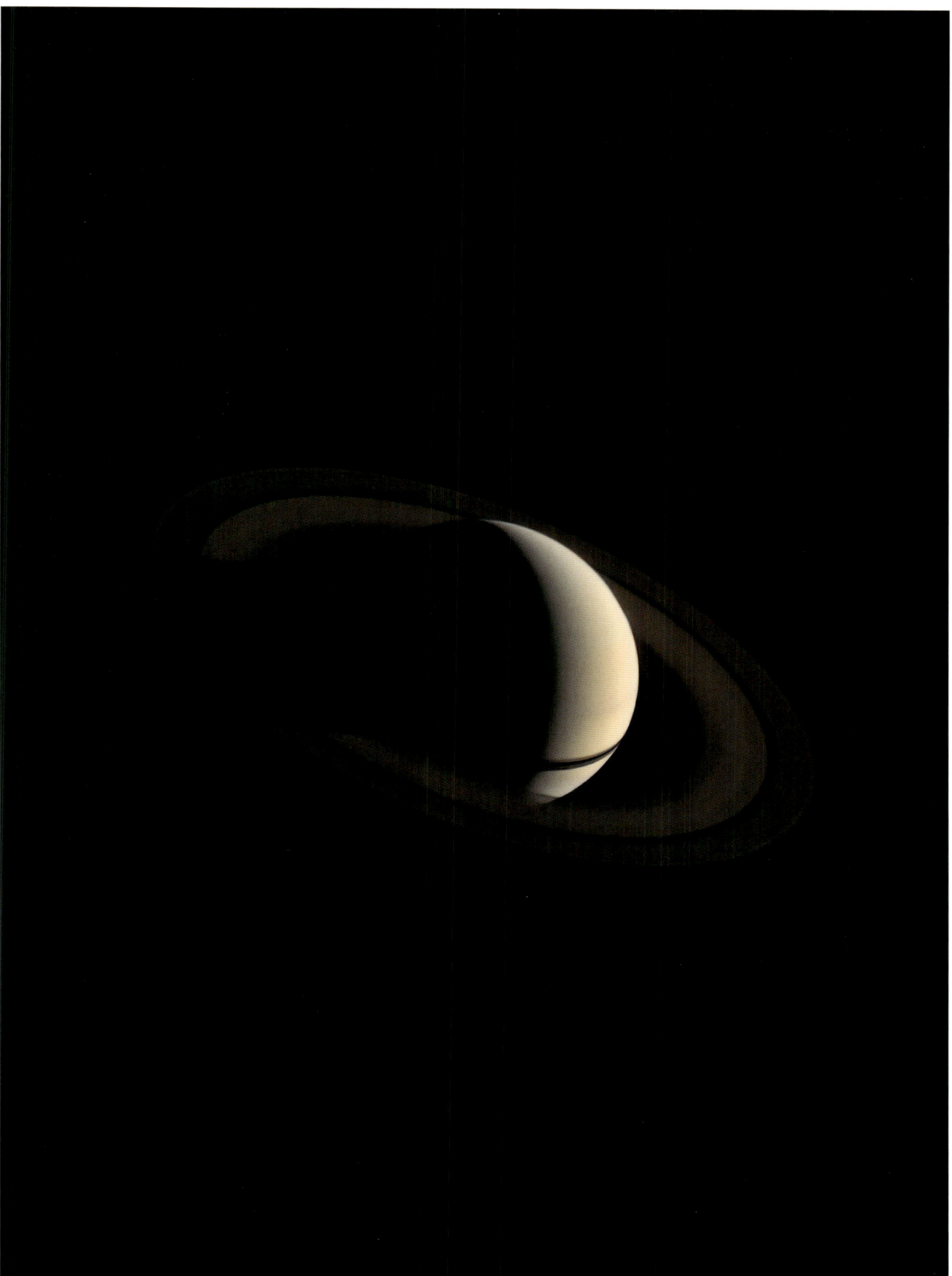

THE APPROACH

The two different routes that each of the Voyagers took when leaving the Jovian System meant that they would arrive at the Saturnian System nine months apart. Voyager 1 arrived in November 1980 and Voyager 2 in August 1981. The reason for the nine-month difference was twofold. Firstly, the team wanted to build in enough time to alter the mission plan for Voyager 2 if the first flyby objectives were not met. Secondly, and most importantly, the slower arrival time to Saturn would suitably align Voyager 2 to head off to Uranus and Neptune. As at this point, the decision on whether to extend the mission beyond Saturn was still not approved.

The intriguing data and images taken of Titan the previous year by Pioneer 11 showed that it was imperative that the Voyager mission got a closer look. The pictures that were sent back by Pioneer showed Titan's atmosphere to be complex and denser than expected. The finding served as a ripple effect of interest to scientists back home, as it raised the question of whether this atmosphere could sustain life. The trajectory was calculated to make sure that a visit to Titan would guarantee quality data, and subsequently putting a close flyby of Saturn and its rings as a secondary goal.

The trajectory of Voyager 1 was designed so that it would get the optimum view of Titan from close-up first, and then travel onwards to Saturn. The Voyager team considered it too risky to pass by Saturn and its ring system first, in case a rogue ring component was to strike the craft.

This meant that Voyager 1 had to make a very specific approach. Starting from a high latitude within the Saturnian System, Voyager 1 pitched downwards to approach Titan and pass by at incredibly close range, around 2,500 miles (4,000 km), in order to get as much quality data as possible. The craft continued to fly downwards, passing behind the moon to gather occultation scan data. After passing by Titan, it continued heading further down towards Saturn's South Pole, avoiding getting too close to its rings. Voyager 1's closest approach to Saturn was on November 12th, passing 40,000 miles (64,400 km) from the top of Saturn's clouds. From this point, the probe would continue downwards, flying under Saturn's South Pole, and then travel upwards on its exit from the pole to capture images of Saturn's rings, Mimas, Enceladus, Dione, and Rhea.

The Voyager 2 trajectory compensated for the lack of close-up images of Saturn's cloud tops and rings. A closer flyby was planned for the second visitation. The probe skimmed the cloud tops from within 26,000 miles (41,000 km) on August 25th, capturing spectacular images of Saturn, as well as detailed observations of its rings and moons.

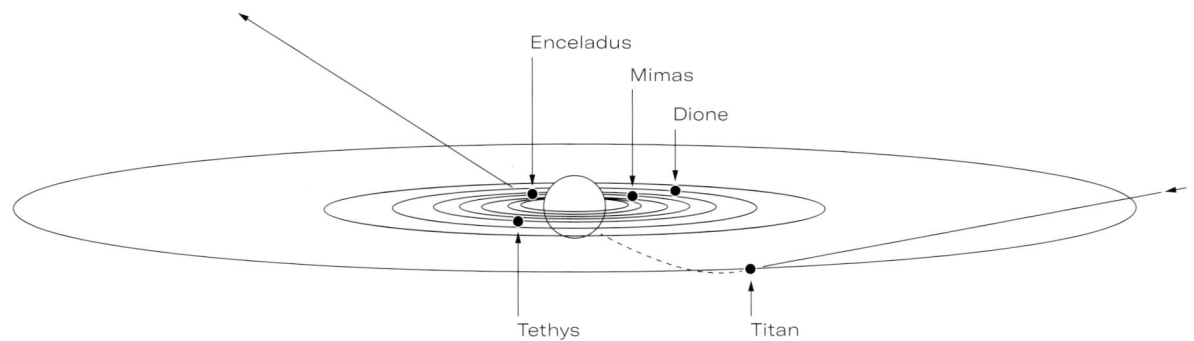

The trajectory approach and exit of Voyager 1 through the Saturnian system.

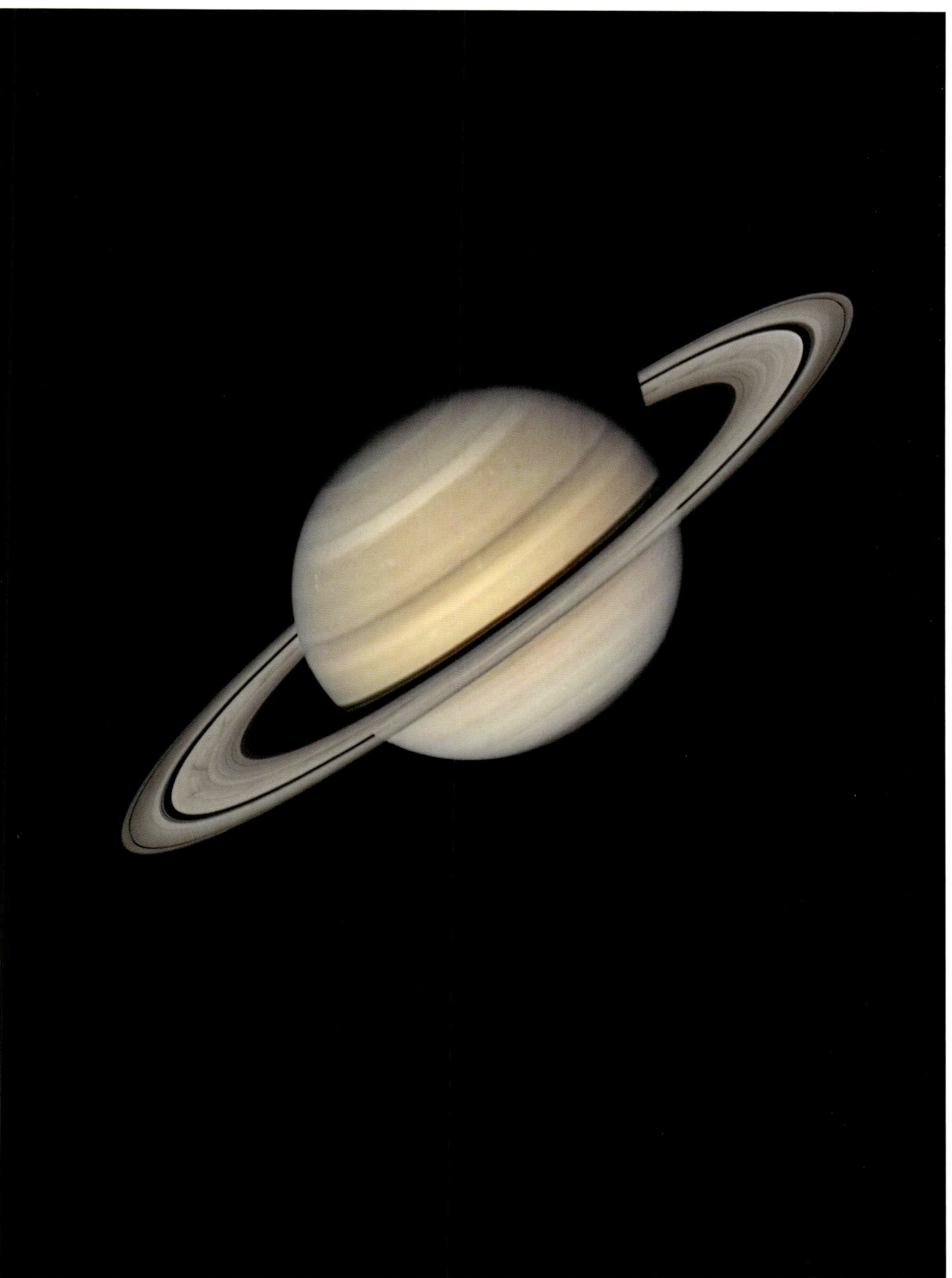

On Titan the molecules that have been raining down like manna from heaven for the last 4 billion years might still be there largely unaltered deep-frozen awaiting the chemists from Earth.

Carl Sagan

TITAN

Titan is Saturn's largest moon and the second largest in the Solar System after the Galilean moon Ganymede. Discovered in 1655 by the Dutch astronomer Christiaan Huygens, and later named Titan by John Herschel, the son of the famous astronomer, William Herschel. The name comes from the mythological Titan, the collective name for the brothers and sisters of Cronus, who the Ancient Greeks named the planet after. It is the only moon in the Solar System that has a substantial, dense atmosphere and is the only known object besides Earth that shows evidence of stable surface liquid.

Earth-based studies and the later observations from Pioneer 11 showed that Titan's atmosphere was similar in composition to that of prehistoric Earth's, before living cells added oxygen to our atmosphere. The low-resolution images taken by Pioneer 11 in 1979 revealed a thick, hazy-orange atmosphere. This covers the moon entirely, hiding from view what kind of surface lay below the clouds. After these first images from Pioneer 11 were received, it made scientists even more eager to see what was below.

Titan is around 50% bigger than our Moon, and larger than Mercury, spanning 3,200 miles (5,150 km) in diameter. The moon is primarily composed of rock and ice, with liquid hydrocarbon lakes found at both the moon's polar regions. The dense atmosphere is at least 370 miles high (600 km), making it considerably higher than Earth's atmosphere. Because of the atmospheric height, Titan was incorrectly considered to be the largest moon in the Solar System for many centuries. It was not until the flyby of Voyager 1 that we discovered the moon was actually smaller than Ganymede. The atmosphere is predominantly made of nitrogen, with areas of hydrocarbon gases such as methane and ethane. Voyager 1 also recorded extremely cold temperatures of around -280 °F (-173 °C) at the surface, just a little warmer than the boiling point of liquid nitrogen.

On November 12th, 1980, Voyager 1 arrived at Titan getting to about 2,500 miles (4,000 km) from the surface. Excitement and anticipation from the Voyager team were high as the Voyager spacecraft ran through its pre-planned flyby well, achieving pretty much all the data and images goals that were set beforehand. The first expected return of data and high-quality close-up images were received back on Earth with a tiny delay. As the images started to download, the excitement among the Voyager team did not last long. The brilliant images of Titan showed a round, featureless sphere, so completely covered in cloud that no landscape or surface feature could be seen.

The return of such plain images was a massive anti-climax for the scientists and astronomers back on Earth, especially given the huge amount of effort, focus, and financial resources spent on prioritizing a close flyby of Titan over other observations of Saturn that could have been made. Unfortunately, both spacecraft were not equipped with the technology that could penetrate through the dense atmosphere and reveal what was hidden below.

Despite the plainness of images, the flyby of Titan was still seen as a great success when it came to acquiring knowledge of the moon. The data provided the scientist valuable insights into the temperature and pressure of its cloud-tops, a more precise knowledge of the composition of the atmosphere and a better understanding of its true size.

The team would have to wait a couple more decades to finally see some (incredibly beautiful) images of Titan's surface. The Cassini probe arrived in 2004, revealing an abundance of liquid lakes on the moon. Some useful data returned from Voyager that indicated that if life exists, or had existed on Titan, these life forms would be truly special, having to depend on chemicals that flow at extremely cold temperatures instead of being water based.

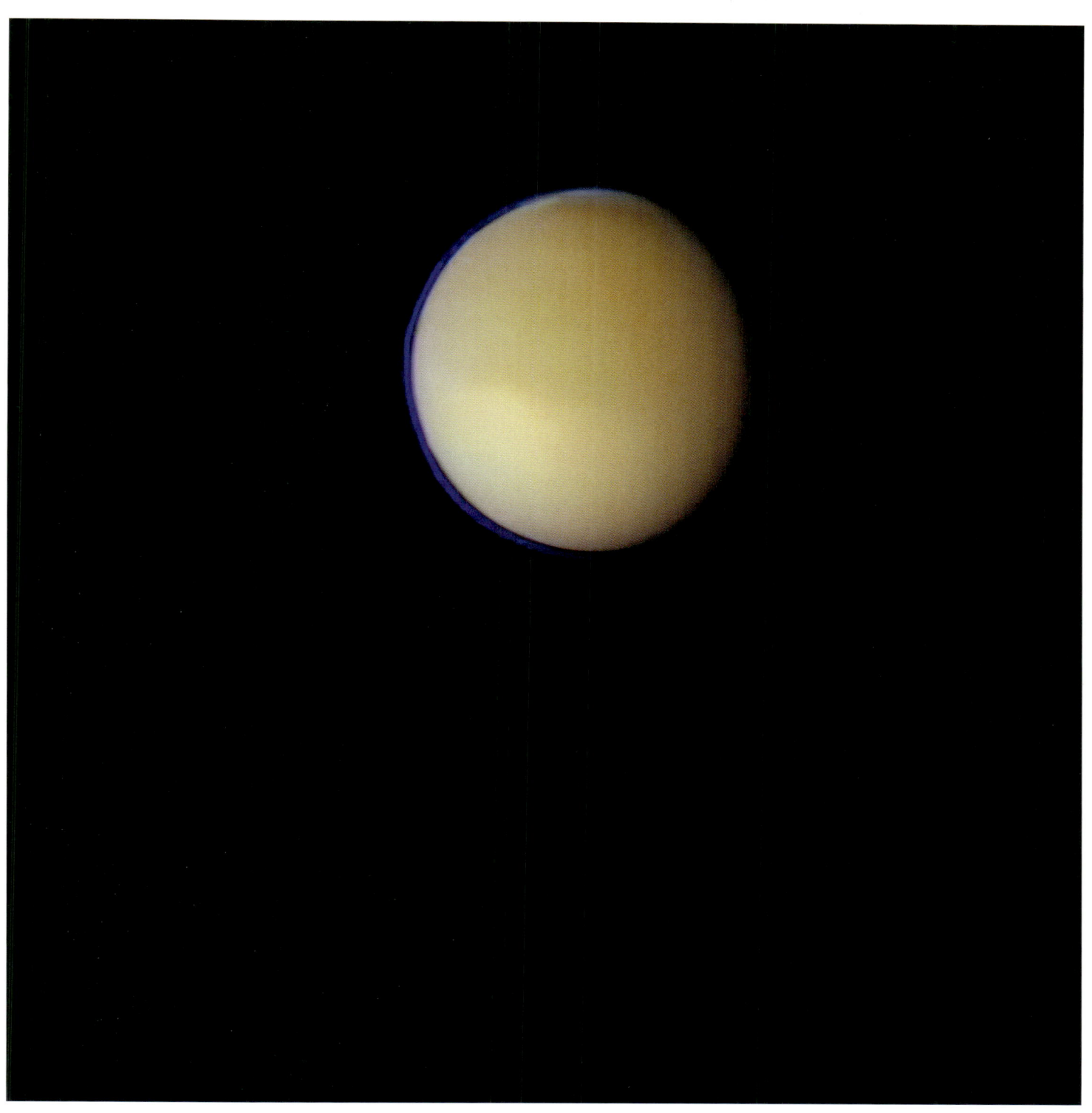

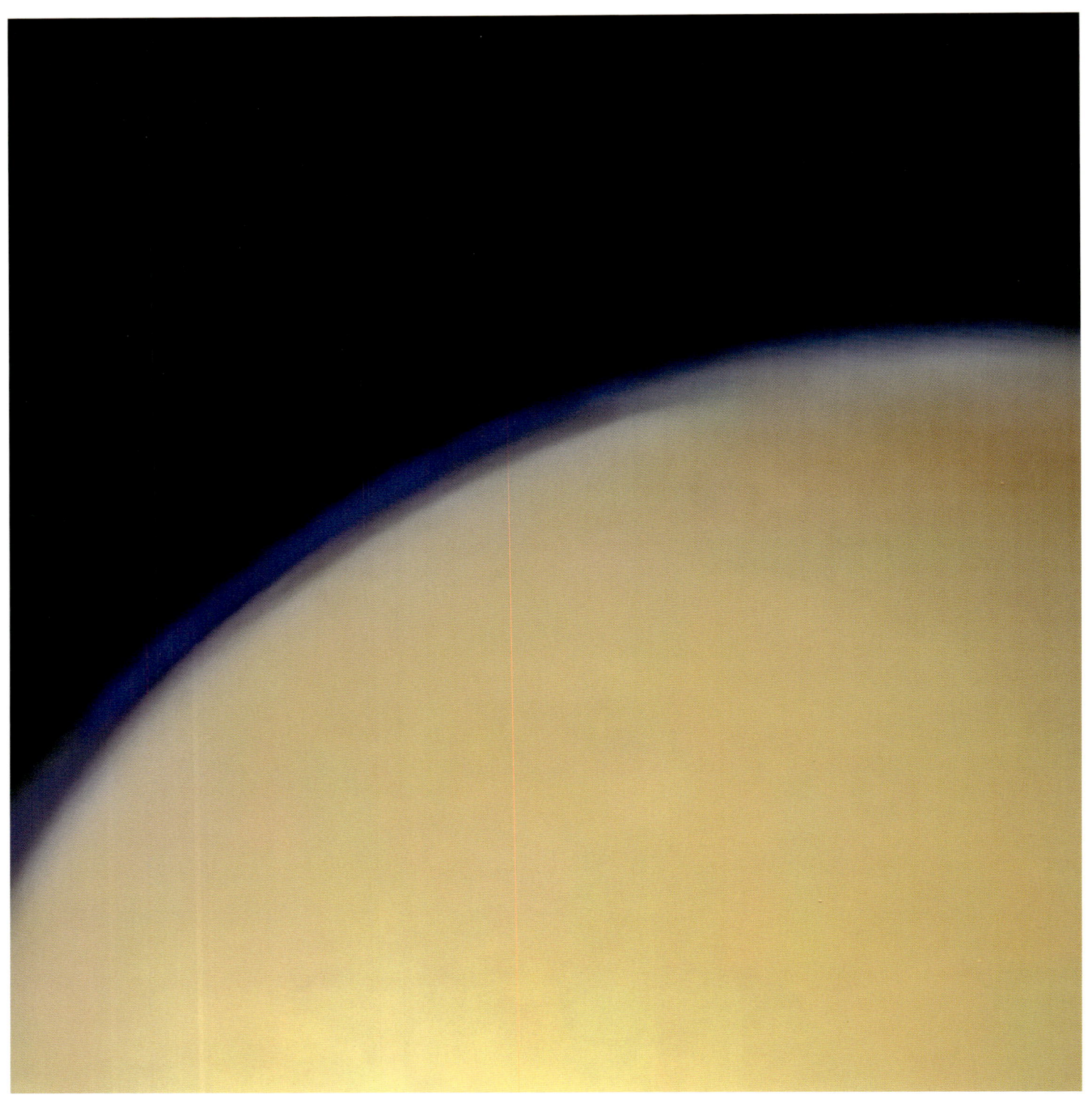

Titan enveloped by a thick layer of haze, which merges
with a darker cloud layer over the north pole.

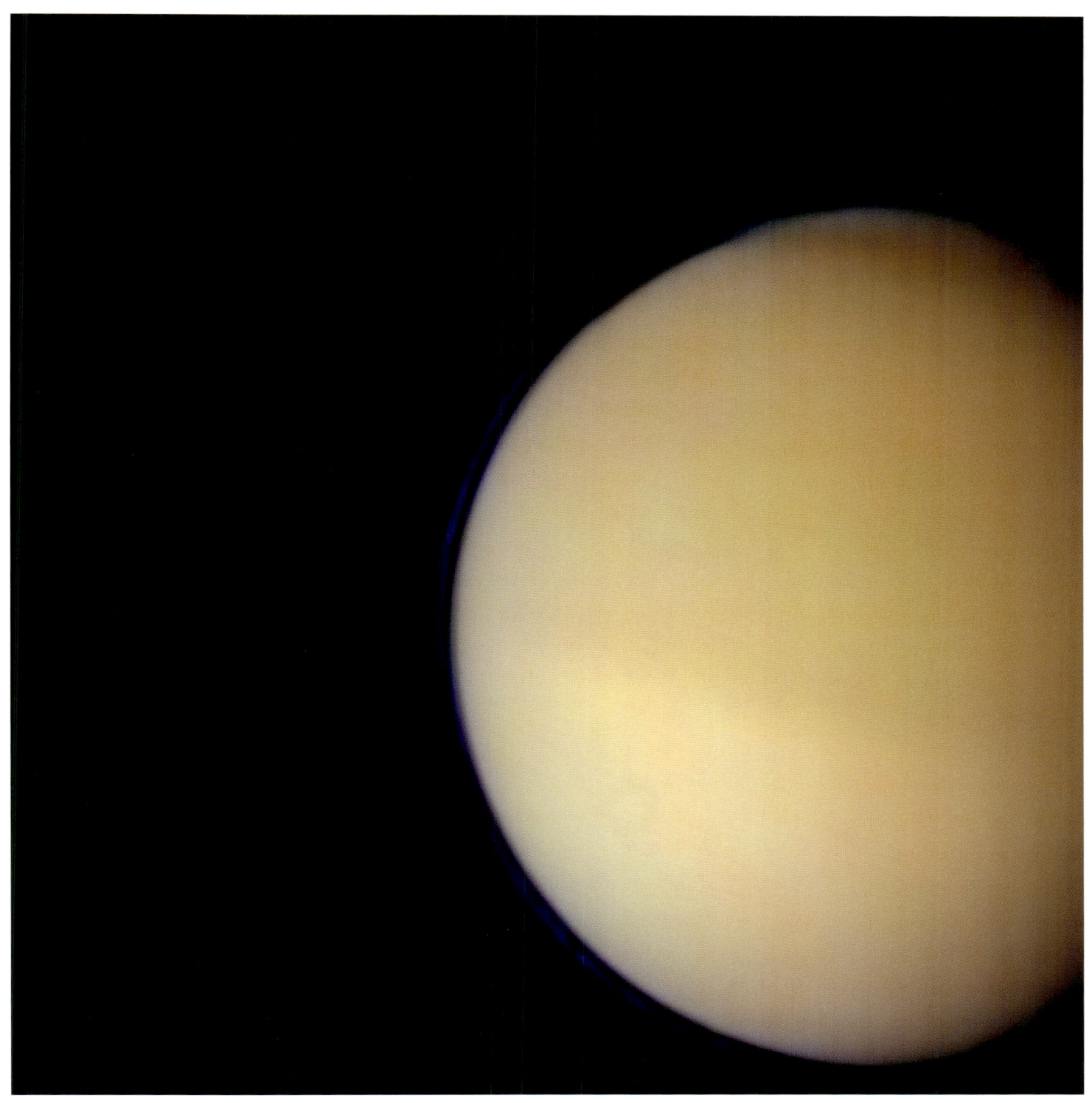

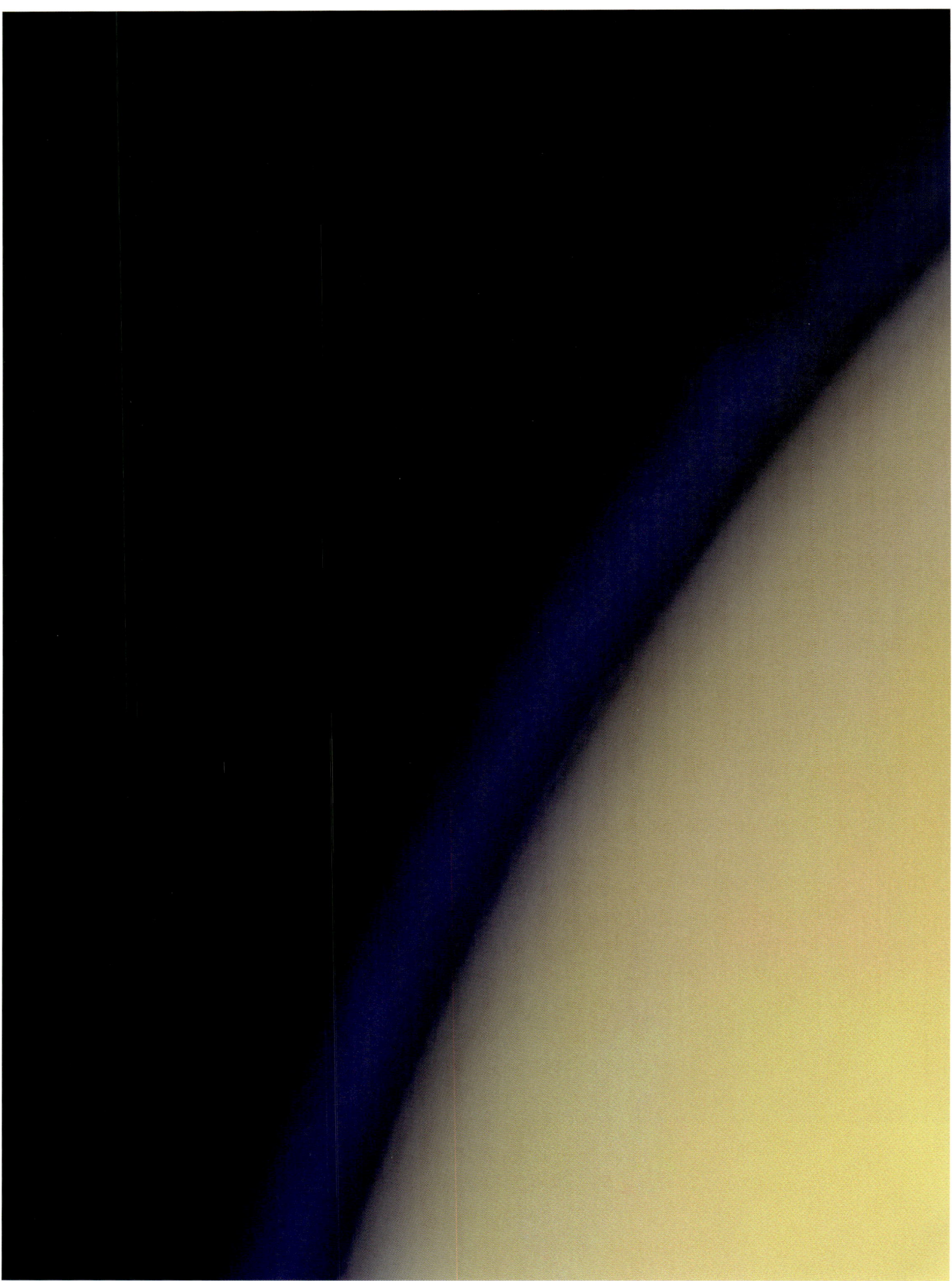

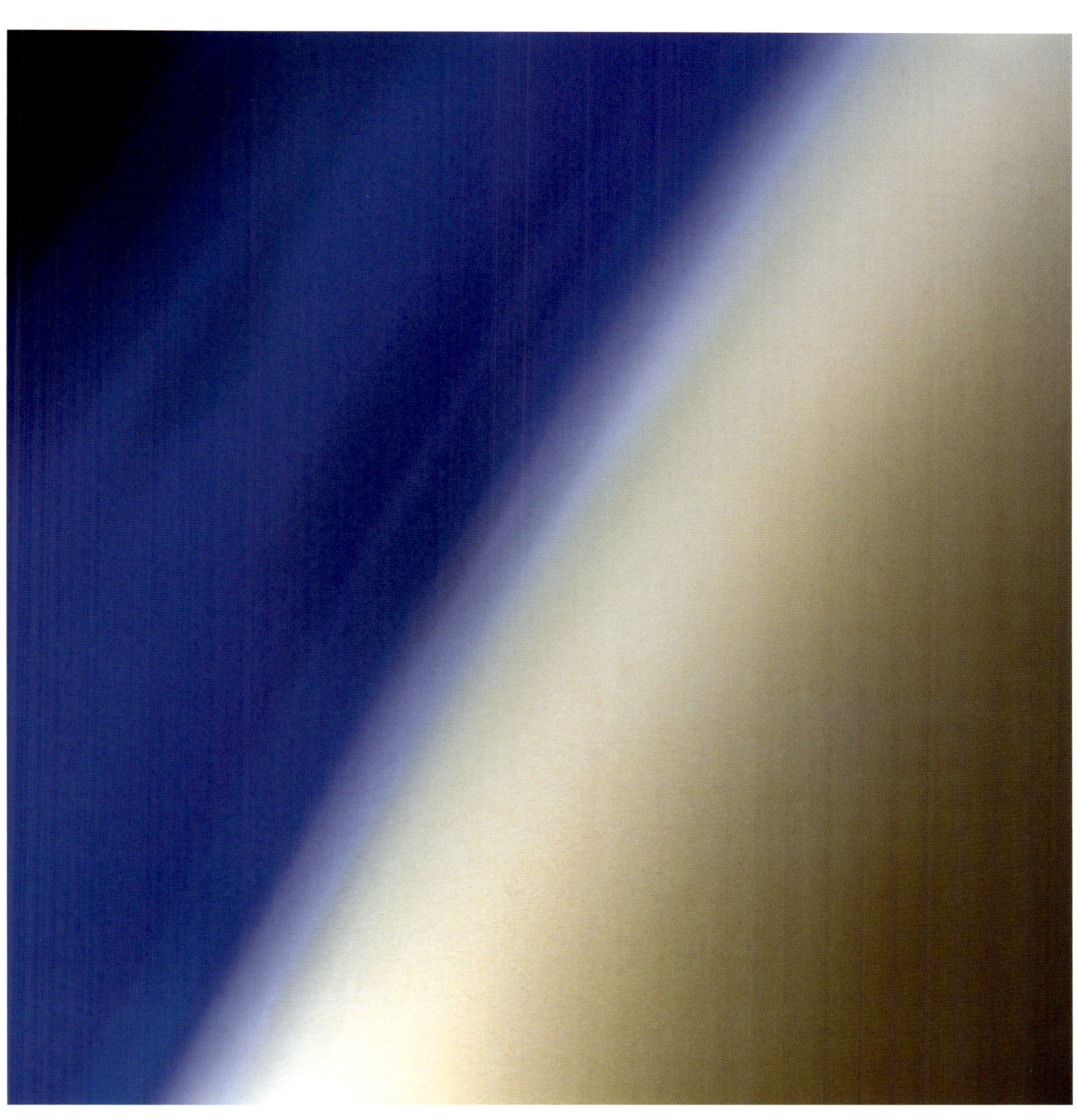

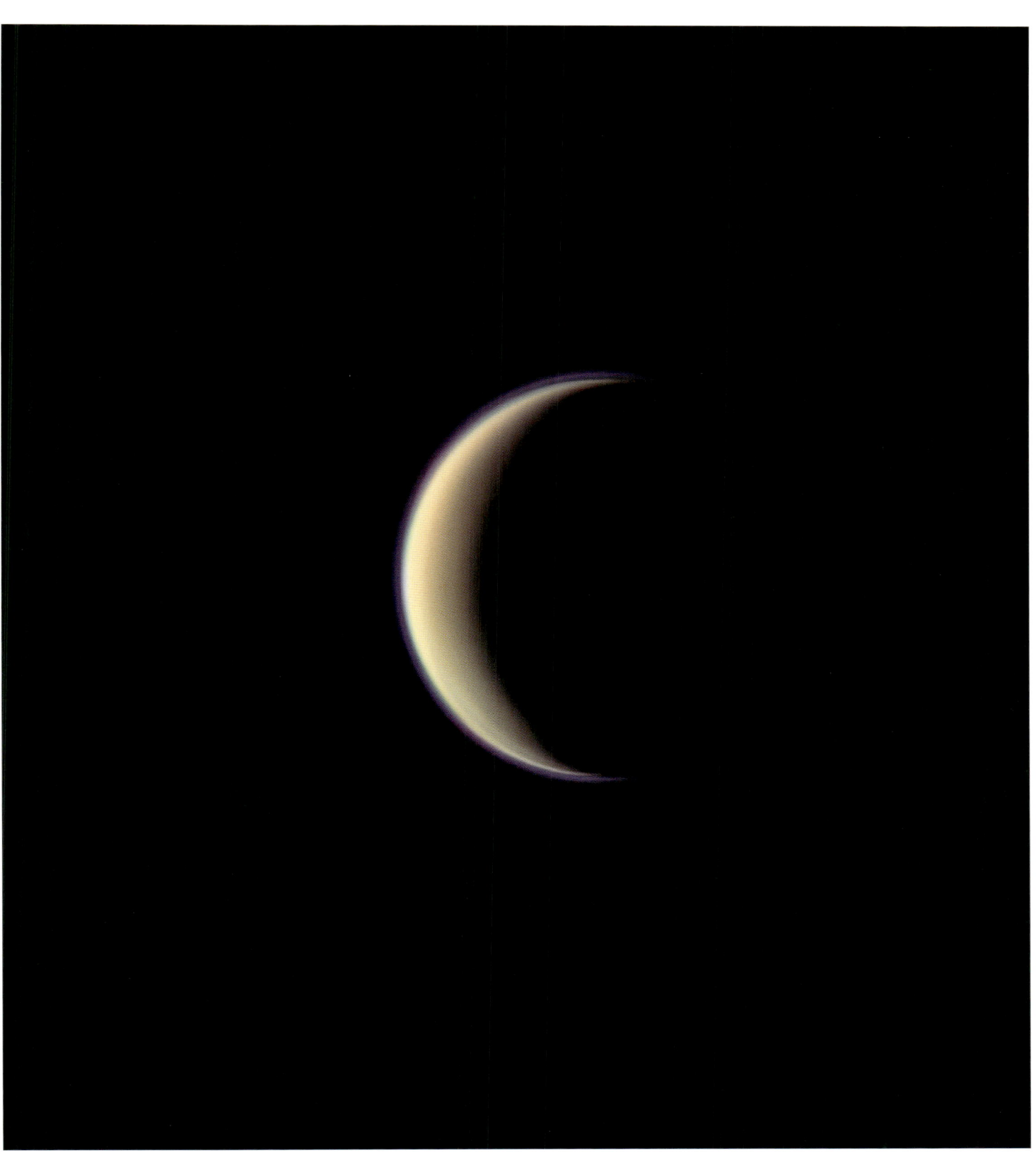

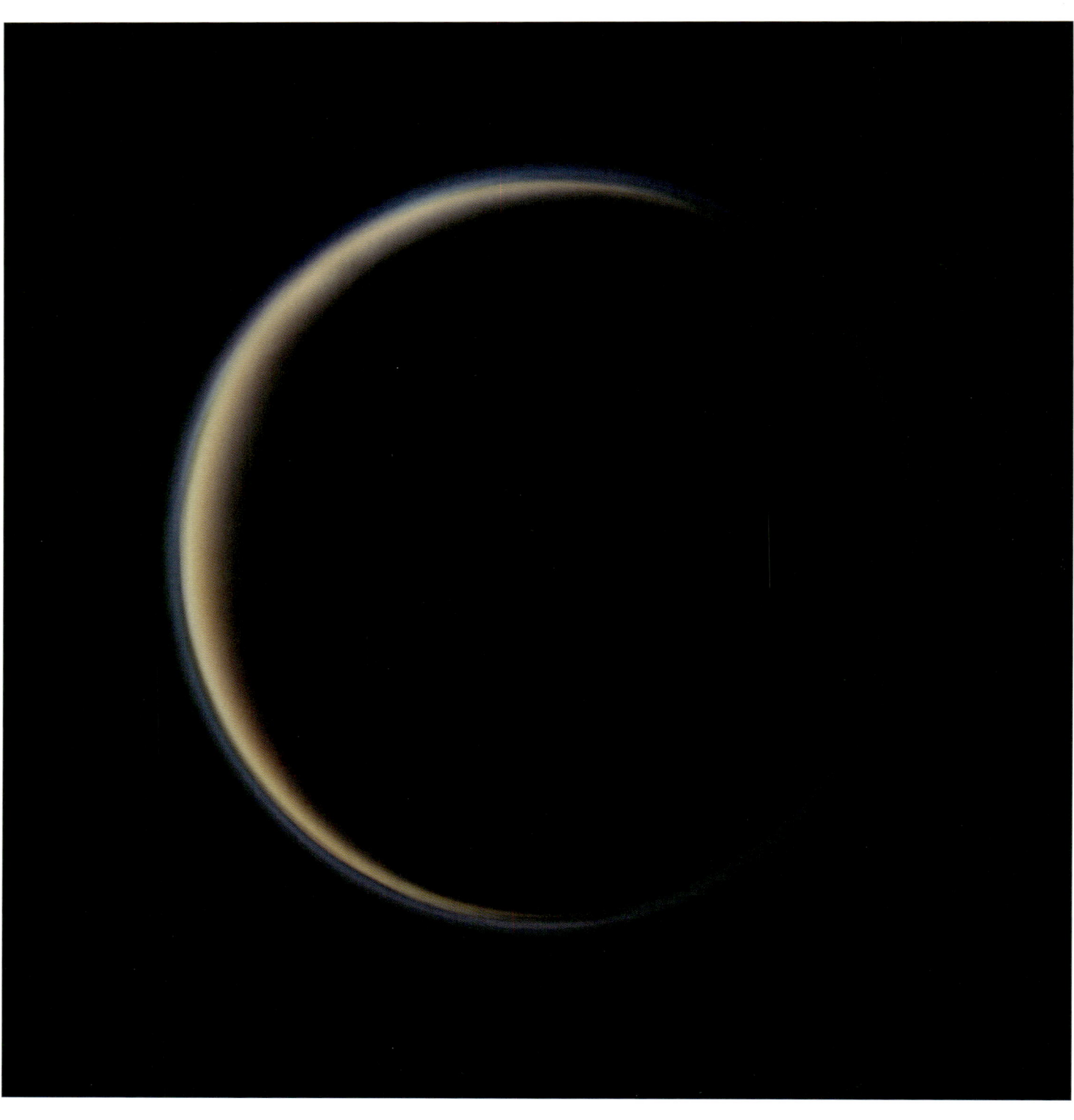

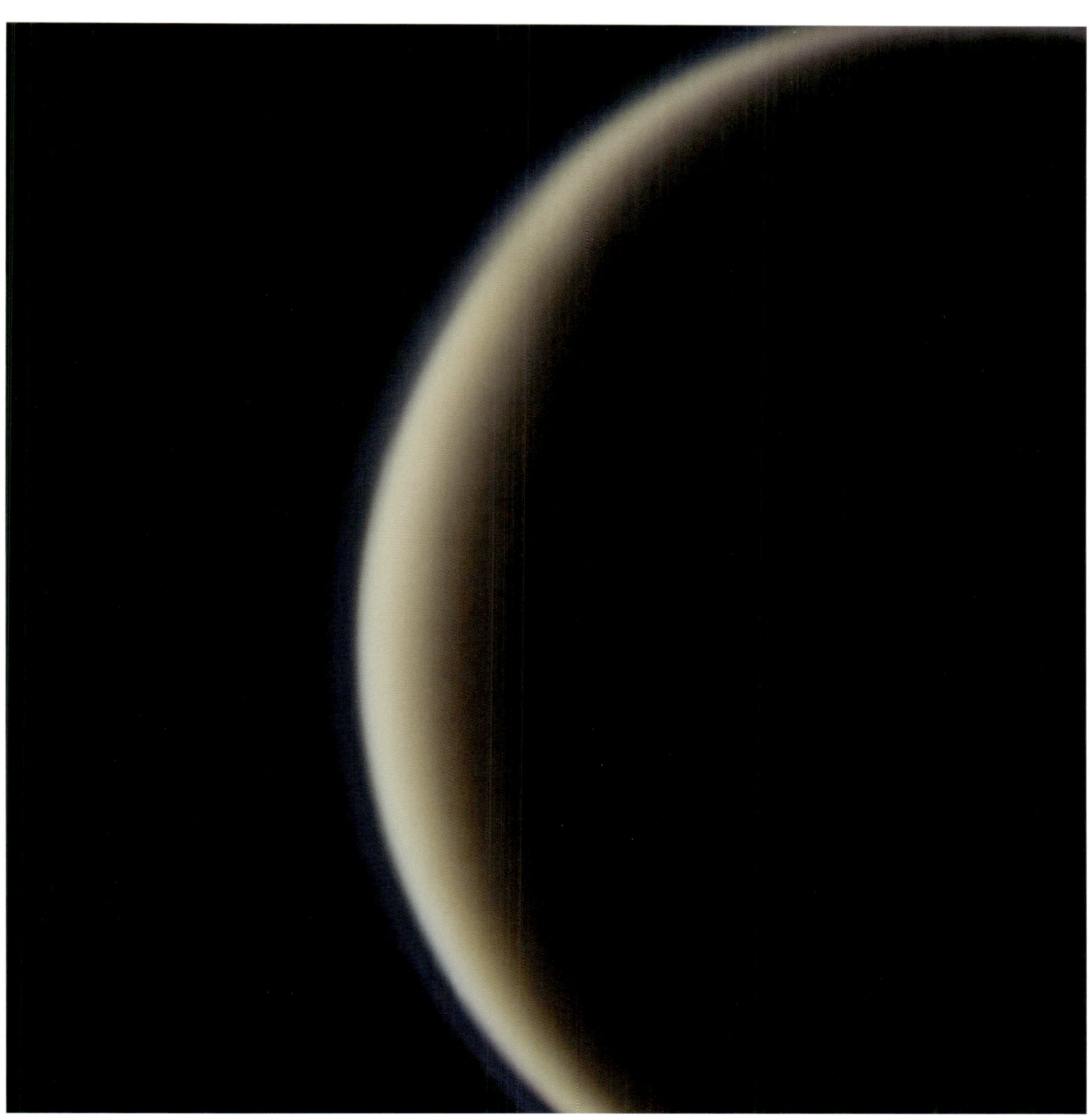

ENCOUNTERS WITH SATURN

Voyager 1 started making its first concentrated observations of Saturn on August 22nd, 1980, while it was still about 67.6 million miles (108.8 million km) away from the planet. Its closest encounter would come 82 days later on November 12th. Voyager 1 recorded that Saturn's day was short by Earth standards, lasting only 10 hours, 39 minutes and 24 seconds.

Voyager 2 started similar distant observations nine months later, and had its closest encounter on August 26th, 1981. Voyager 2's closer approach to Saturn showed images of a similar atmosphere to what is seen on its neighbor and fellow gas giant Jupiter. The photographs showed clear alternating light and dark belts, and circling storm regions. Auroras were seen at both poles and a unique red oval cloud feature surrounded by large, bluish clouds. These auroras were spotted around 55° south of the equator.

The gases inside the red oval give it its distinct look as they absorb more blue and violet light than the neighboring gases. This cloud feature did not change appearance during the months that Voyager 2 was observing the planet.

Right: Saturn & satellites, Voyager 1, October 28th, 1980, 22:31:45.

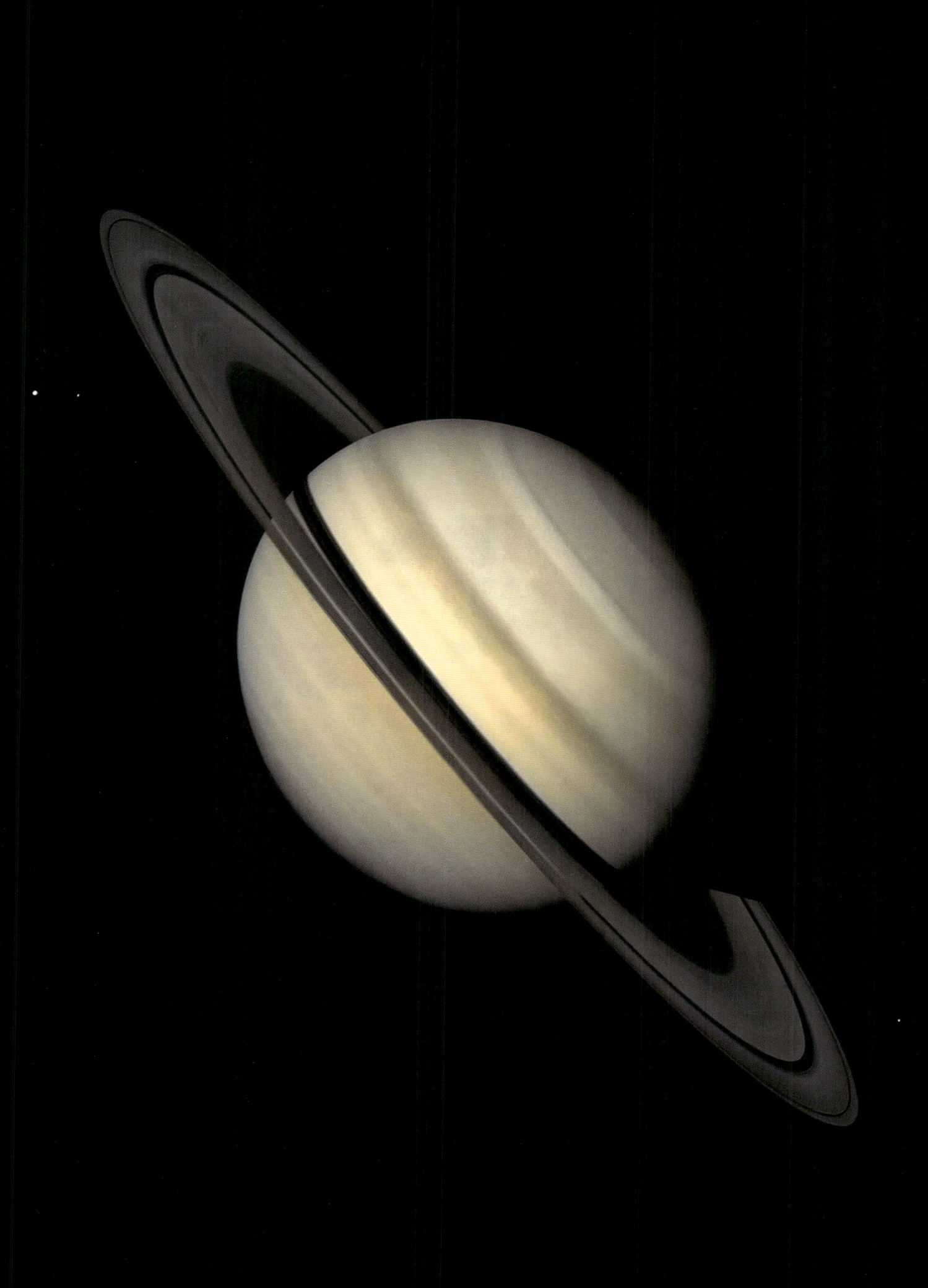

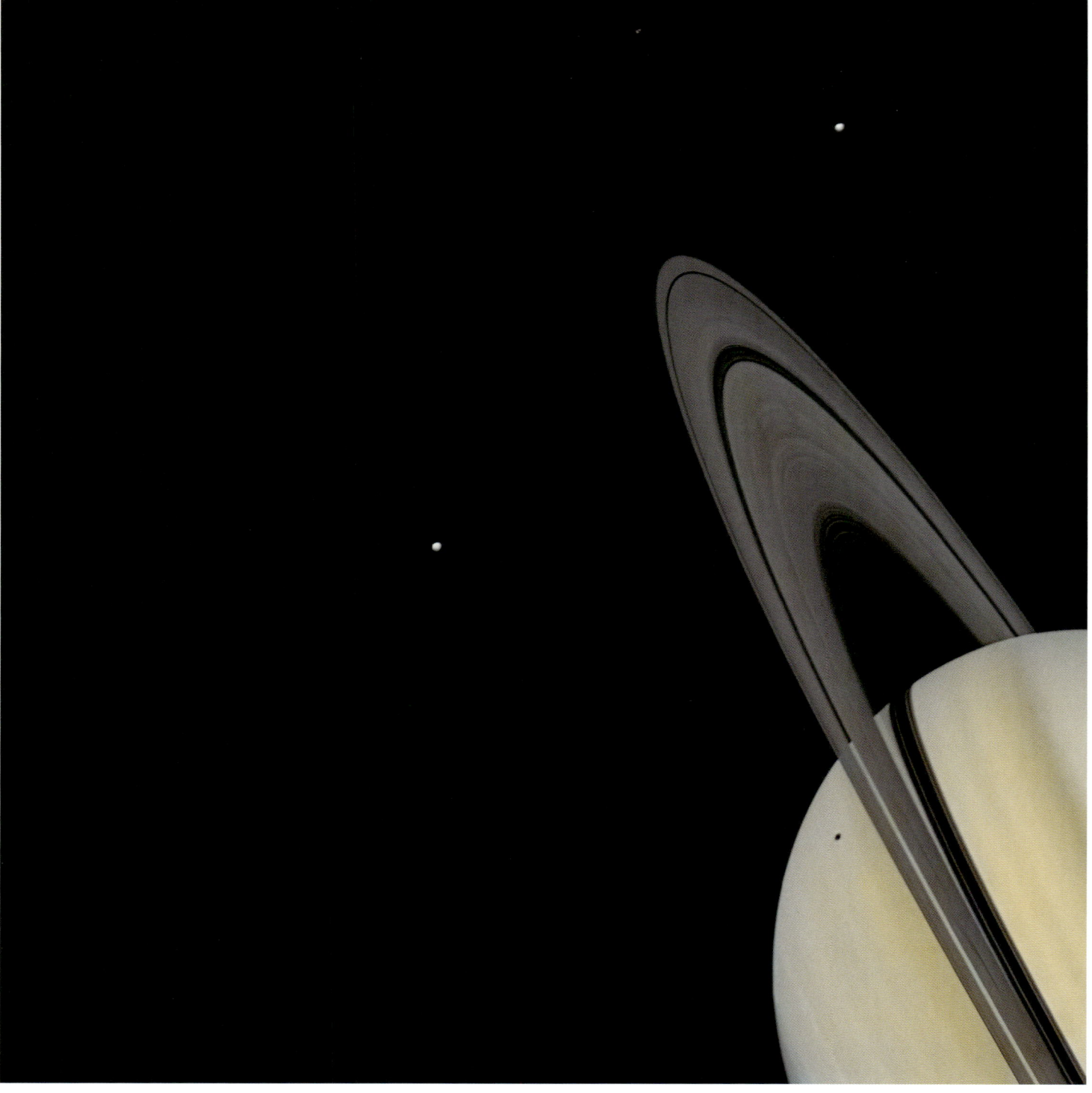

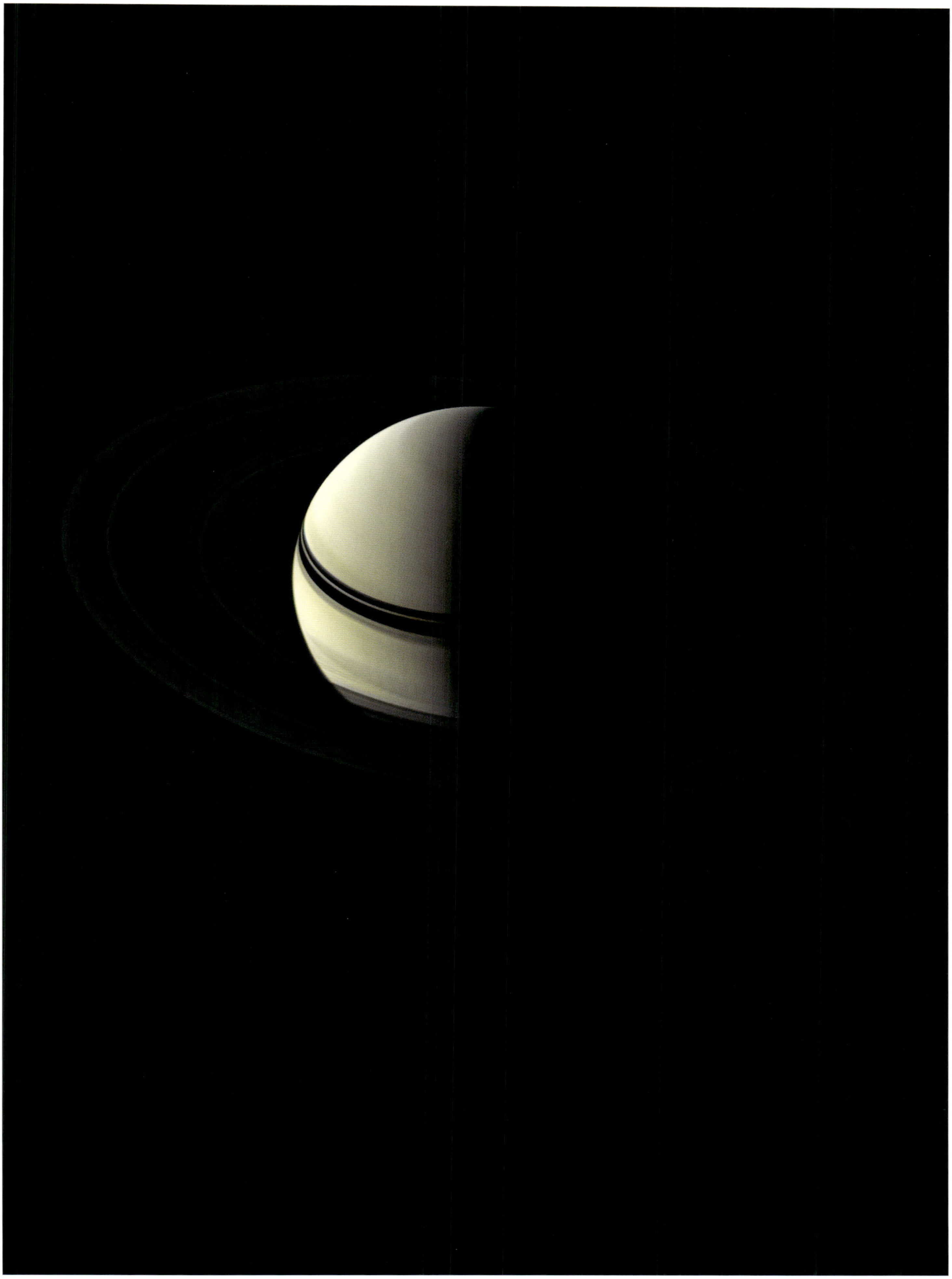

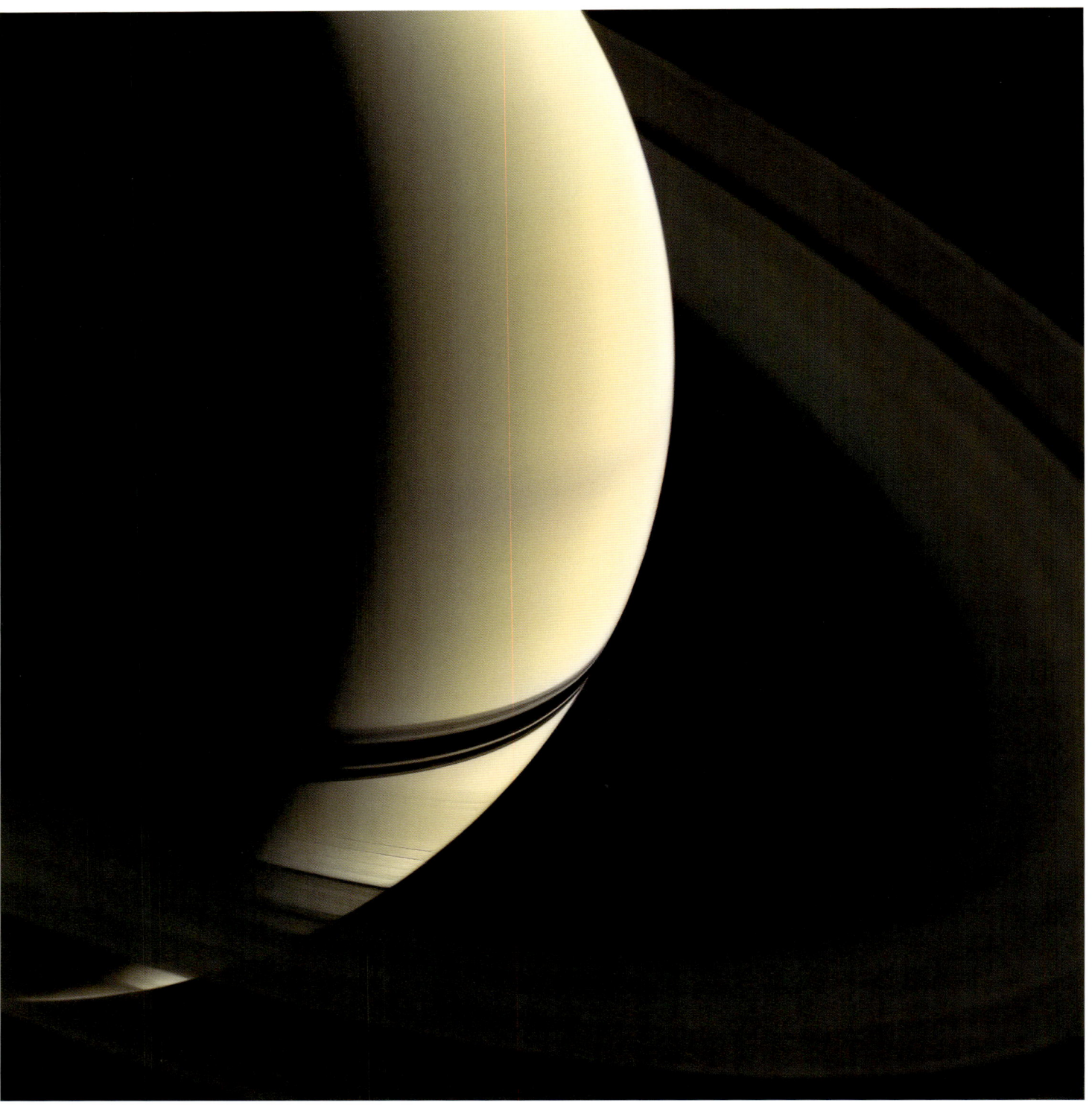

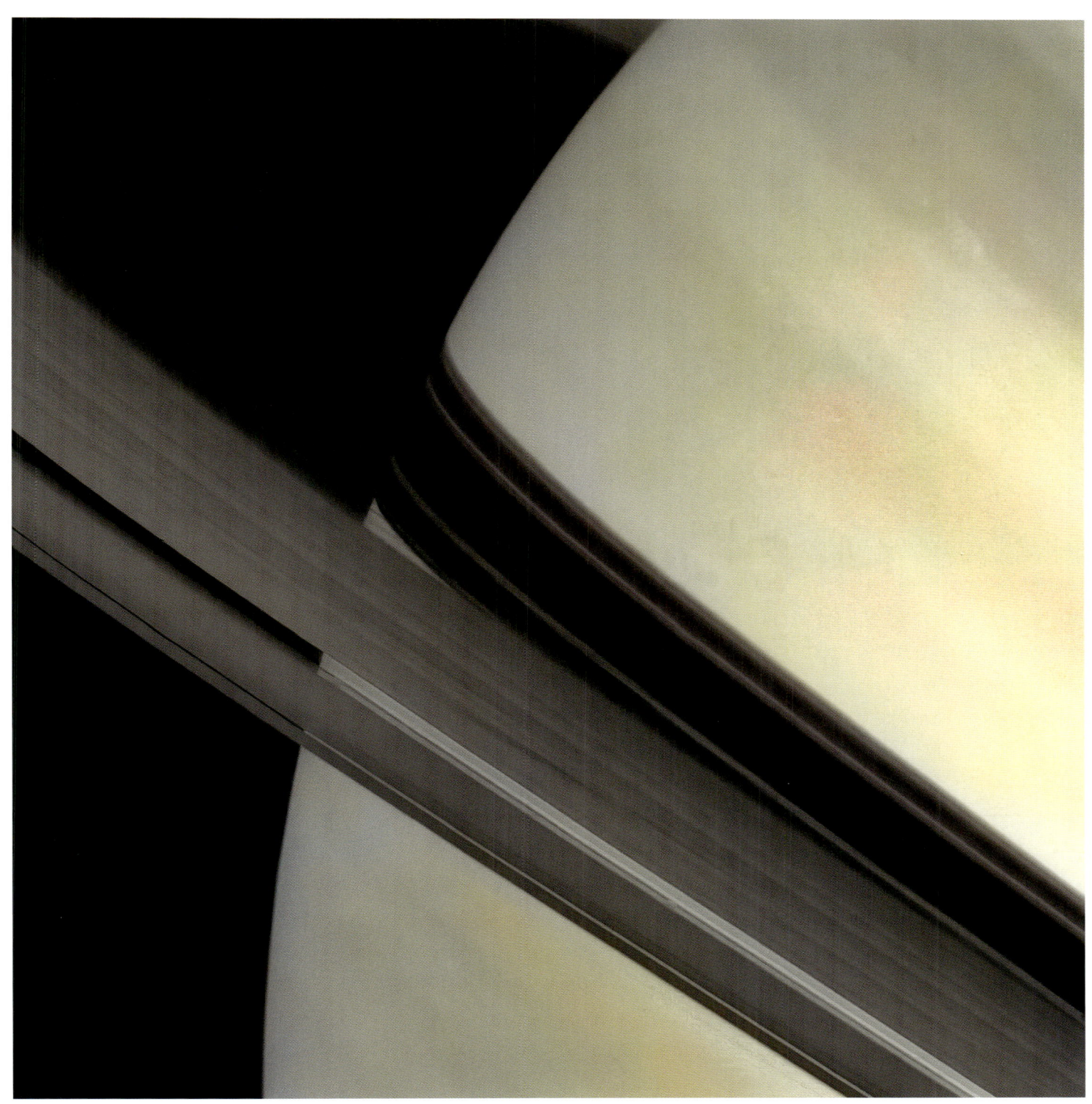

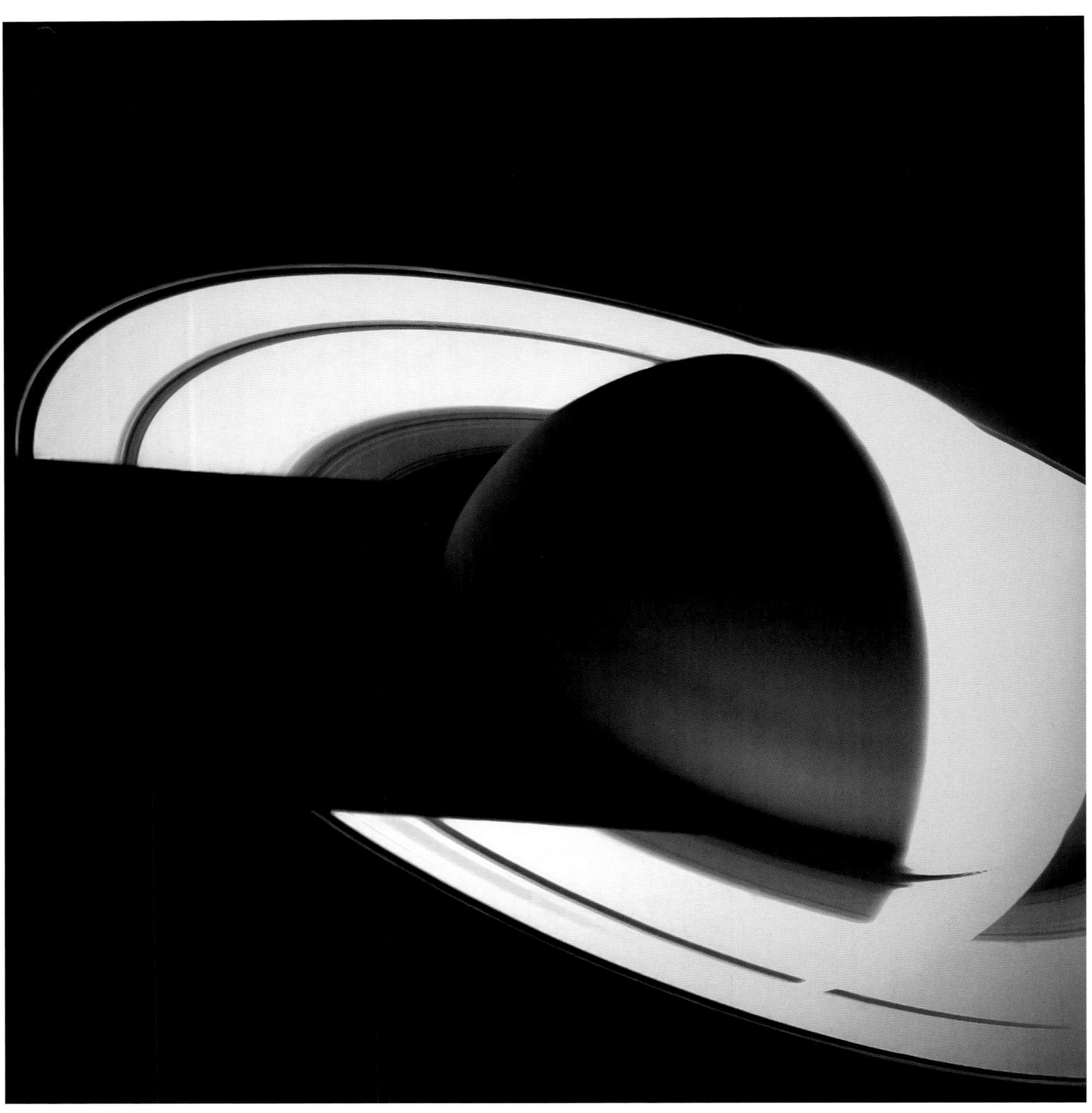

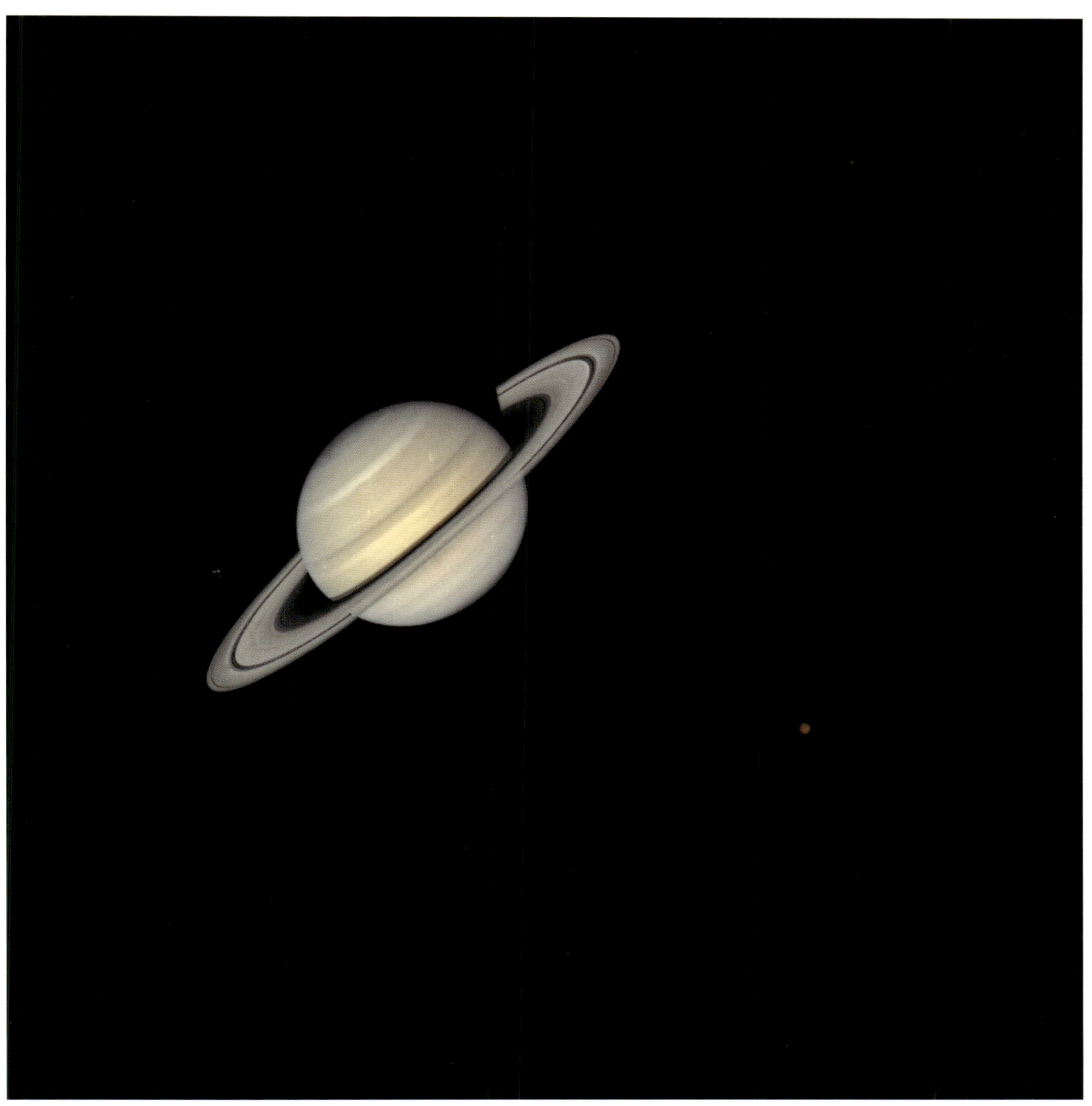

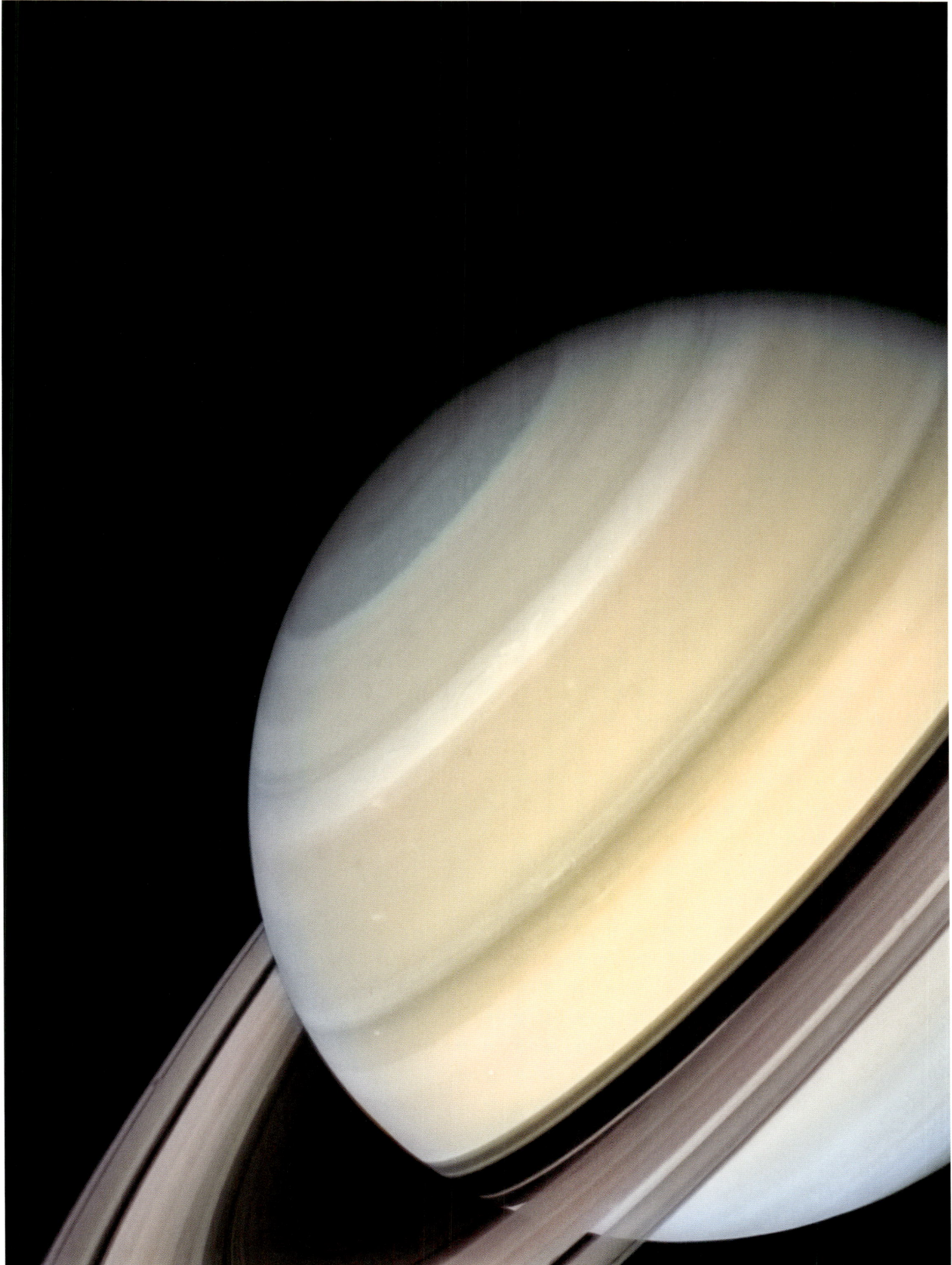

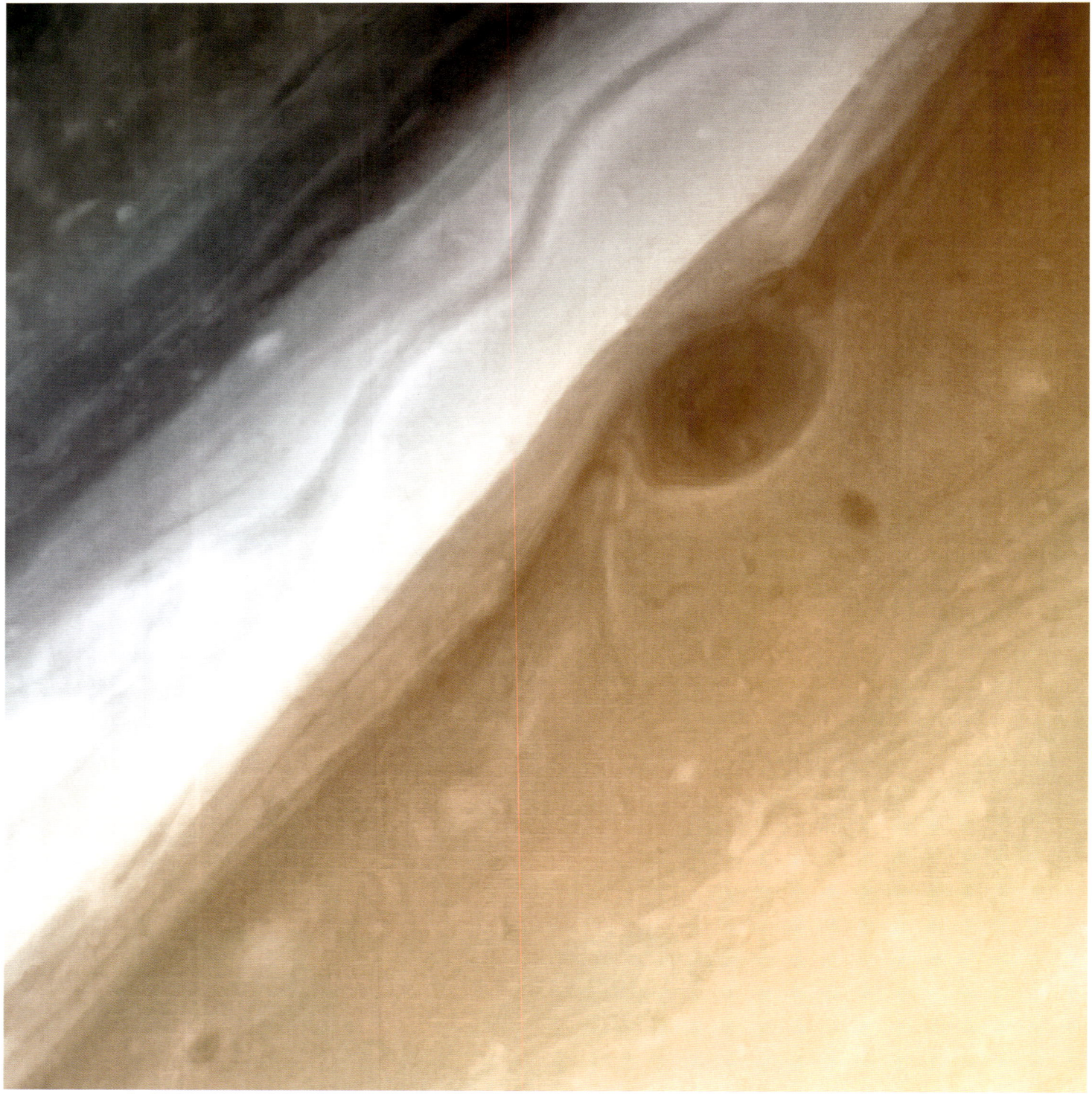

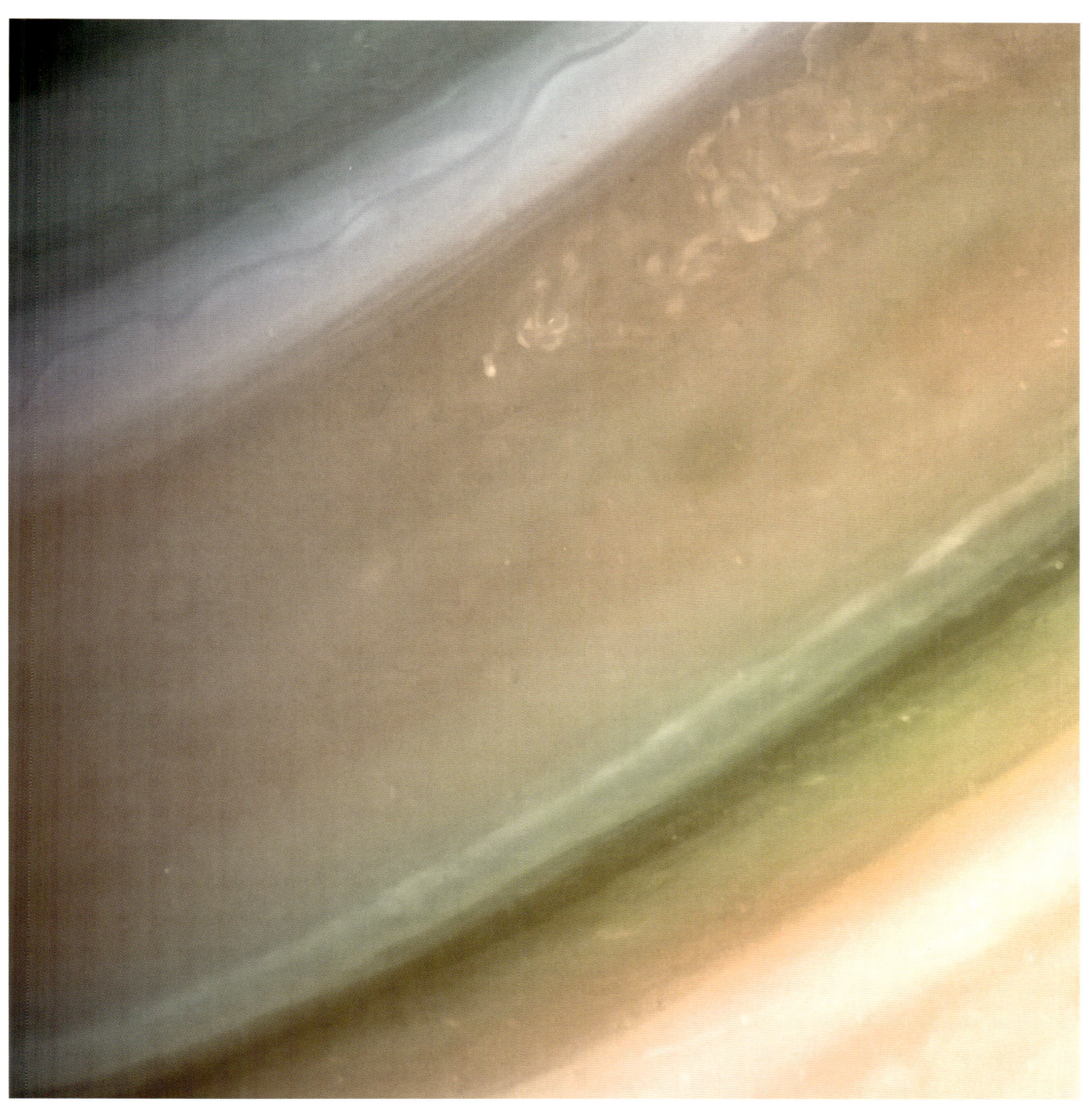

At the mid-latitudes of Saturn one can spot a peculiar curled cloud,
which is seemingly attached by a thin ribbon to the white cloud region
to the north. After monitoring it for seven rotations around the planet,
it seems like the cloud is forming a closed loop. To the east other discrete
clouds are showing.

THE RINGS

Another major objective set for the Voyager Mission was to get a close look and better understanding of the most extensive ring system in the Solar System. First seen in 1610 by Galileo, the rings were later described as such by Huygens in 1655. It was the Italian astronomer Cassini that later discovered that the ring was actually composed of multiple smaller rings.

In 1859, before it was even possible to see the rings in any detail, James Clerk Maxwell proved that the rings were made up of countless "unconnected solid particles." The last major discoveries from the rings before Voyager's visit was from Pioneer 11's approach in 1979. The Pioneer probe discovered a new narrow ring outside the A-ring, which was subsequently named F-ring. The lower image quality on Pioneer still left many questions that the Voyager team was keen to answer.

The sheer size of Saturn's ring system is astonishing. Stretching over 40,400 miles (65,000 km) wide, its thickness varies between 3.3 feet (1 meter) to a couple of miles (a few kilometers) thick in places. Voyager 2 recorded that the ring particles are made from pristine clear water ice, ranging in size from a grain of sand to a four-bedroom house.

Voyager 1 got to within 40,000 miles (64,200 km) of the rings. Unfortunately, a failure in the photopolarimeter prevented the probe from returning images at the planned resolution. The Voyager team had to wait for Voyager 2, which had a much closer observation at 25,500 miles (42,000 km), nine months later. These images returned to Earth, clearly revealing the complete set of rings in detail and showing several new and unseen mysterious features.

Each of the rings orbiting Saturn was named alphabetically in the order that they were discovered. Cassini named the A and B-ring in 1675. Subsequent discoveries added C, D, E, with the F-ring being added last through its discovery by the Pioneer 11 probe in 1979. Voyager 2 sent back images that showed that the A, B and C-rings were elliptical, rather than circular as was previously thought. Each of these rings turned out to consist of hundreds of ringlets that together form into a thicker ring.

The knowledge concerning gaps in the rings also expanded. Besides the two major ones in the A-ring, namely the Encke and Keeler gap, other gaps were also made clear through the images. Within these gaps, it was revealed that a harmonious interplay was taking place between some of Saturn's moons and the ring system. These small moons range from a few dozen to several hundred miles in diameter. Moons like Prometheus, Pandora, Atlas and Mimas keep the gaps clean and use their gravitational field to sweep up any stray objects moving away from the F ring, pushing the particles gently back into line. This "shepherding" activity was shown by Voyager images to be a common feature of many of Saturn's moons, which in turn led to the group of moons being described as the "shepherd moons."

More unseen features were visible within the photographs captured at the rings. Dark finger-like areas that have an appearance like spokes in a wheel were seen orbiting around. The spokes seen in the B-ring baffled the Voyager team, as the existence and rotation of the spokes around the ring is not possible within our understanding of gravitational orbital mechanics. Today it remains a mystery how these spokes can exist. A modern theory on this matter suggests that the spokes are a seasonal feature, caused by electrical disturbances from lightning bolts striking below in Saturn's atmosphere, or coming from meteor impacts within the ring system themselves.

Another seemingly impossible occurrence was spotted within the F-ring. A series of three narrow rings appeared to be braided and intertwined with each other. Again, like the spokes seen in the B-ring, the braiding of particles also goes against our understanding of orbital mechanics. Modern, more detailed images taken by the Cassini spacecraft showed that these braided areas were actually made from separate objects shaped in a particular way that collectively gave the appearance of braiding.

It is still unclear exactly how the most extensive ring system in the Solar System came into existence. An early theory proposed by French astronomer Édouard Roche suggested that the debris in the rings came from a moon that strayed too close to Saturn, and was pulled apart and broken up into pieces by tidal forces early on in the lifetime of Saturn.

New data received from the Cassini probe suggests a different theory. It showed that the rings were actually very young, between 10 to 100 million years old, which would rule out that Saturn had broken up a moon or comet during the early formation of the Solar System 4.6 billion years ago. Further data showed that the ring system around Saturn might not be a long-lasting feature, as measurements suggested that the rings appeared to be shrinking, with debris falling out and crashing into Saturn's atmosphere. Given the current loss rate, the rings would disappear in around 300 million years' time.

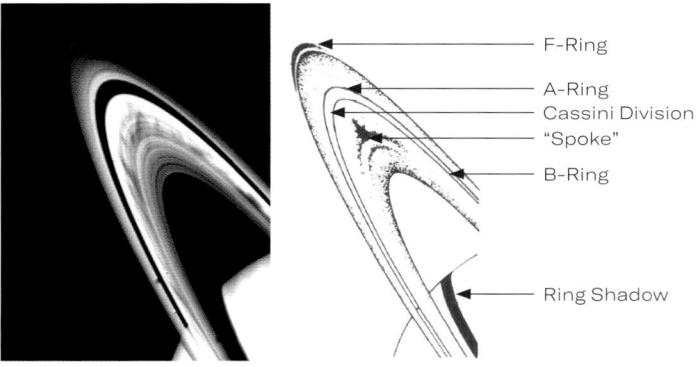

F-Ring
A-Ring
Cassini Division
"Spoke"
B-Ring

Ring Shadow

Diagram showing the different rings, divisions and "Spoke".

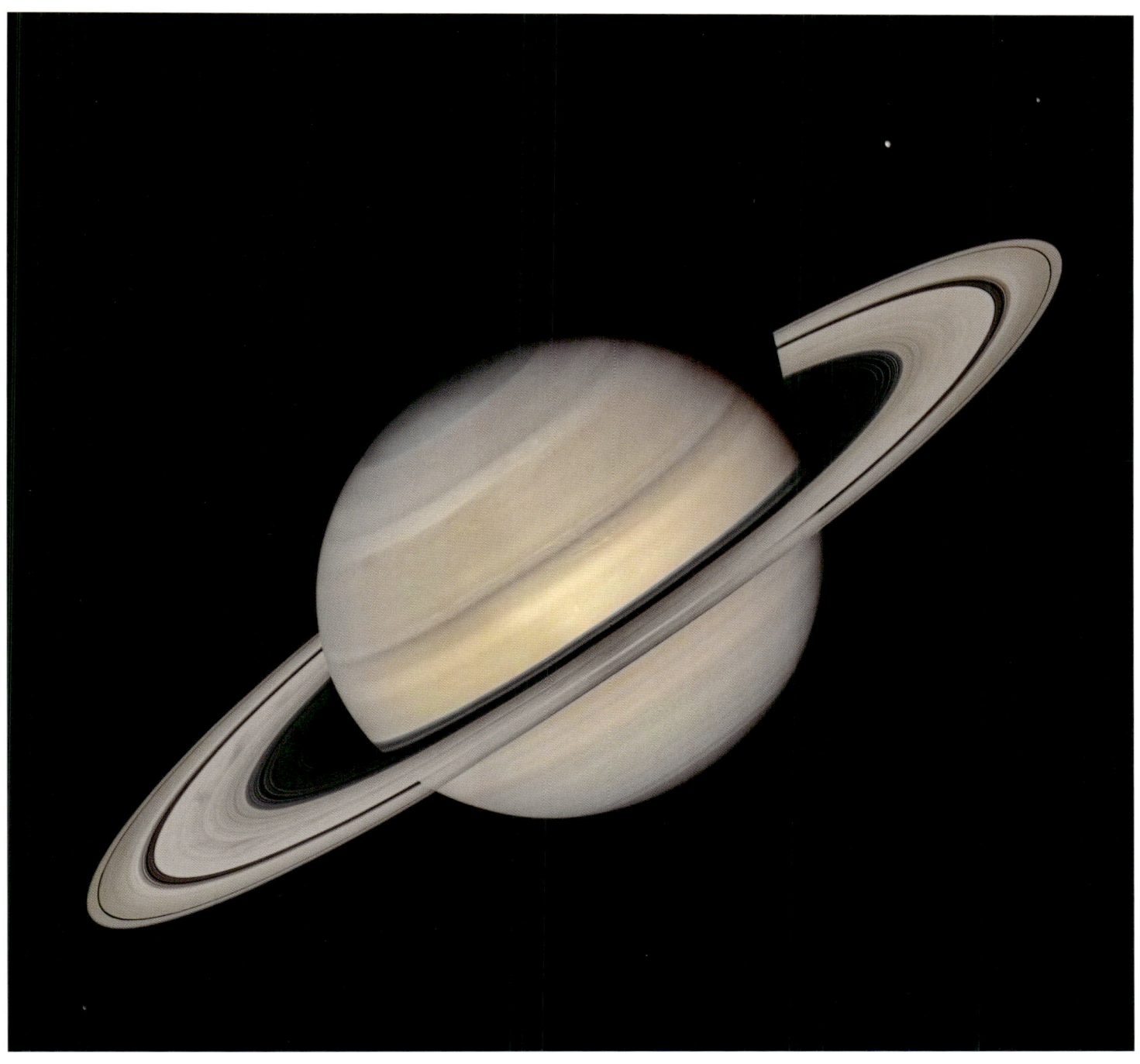

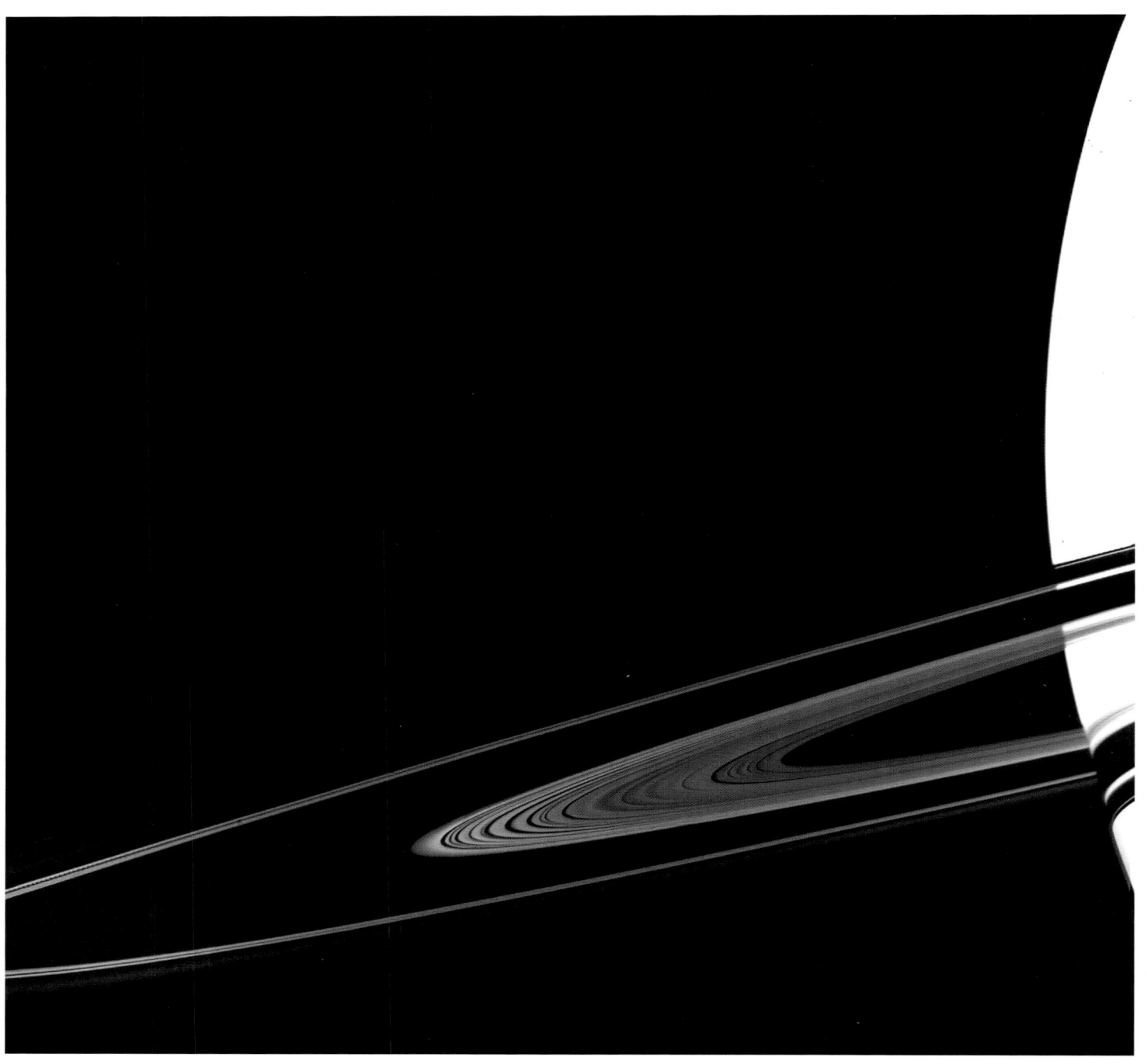

Right: Saturn Rings, Voyager 2, August 15ᵗʰ, 1981, 12:16:47.

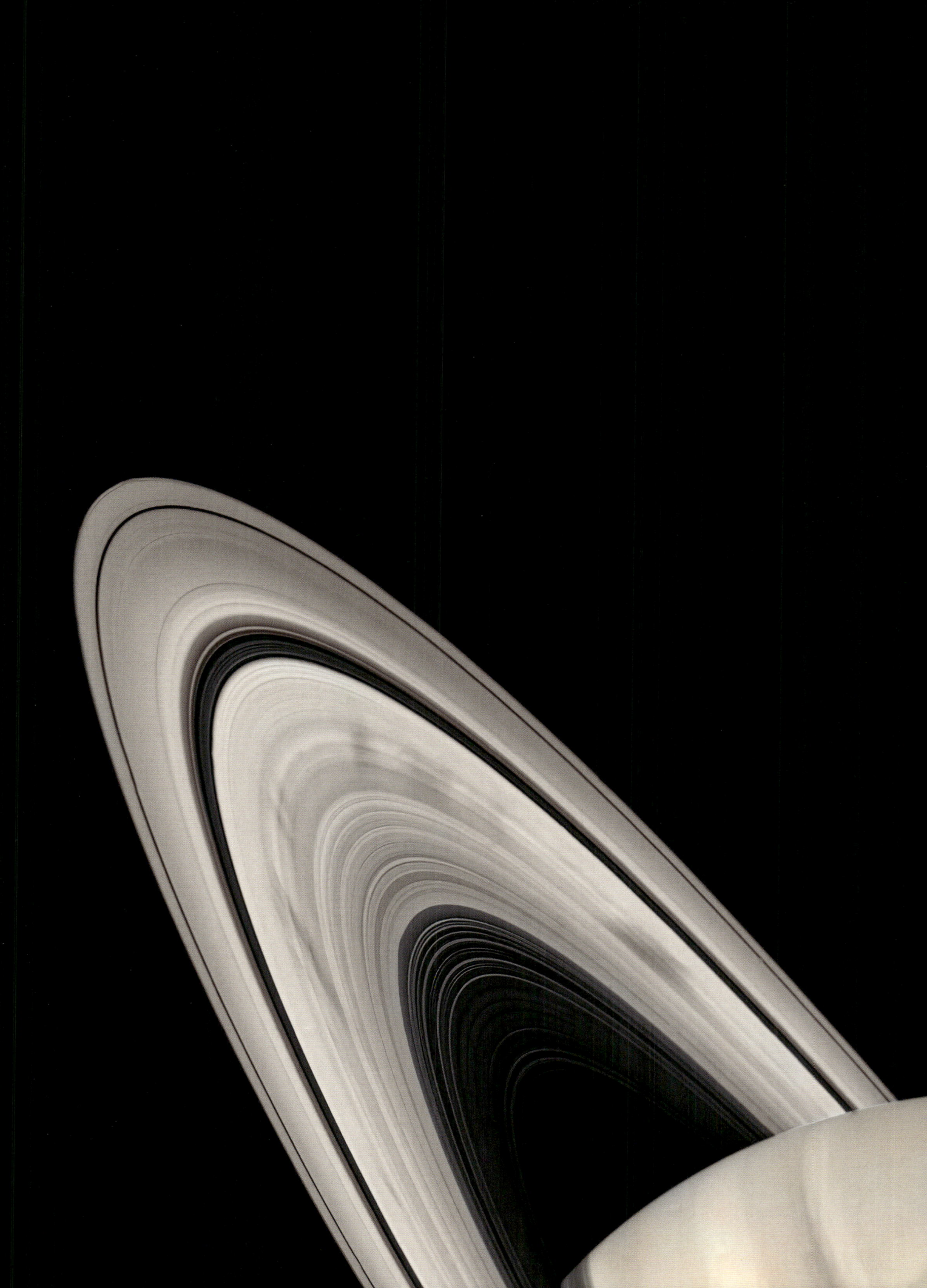

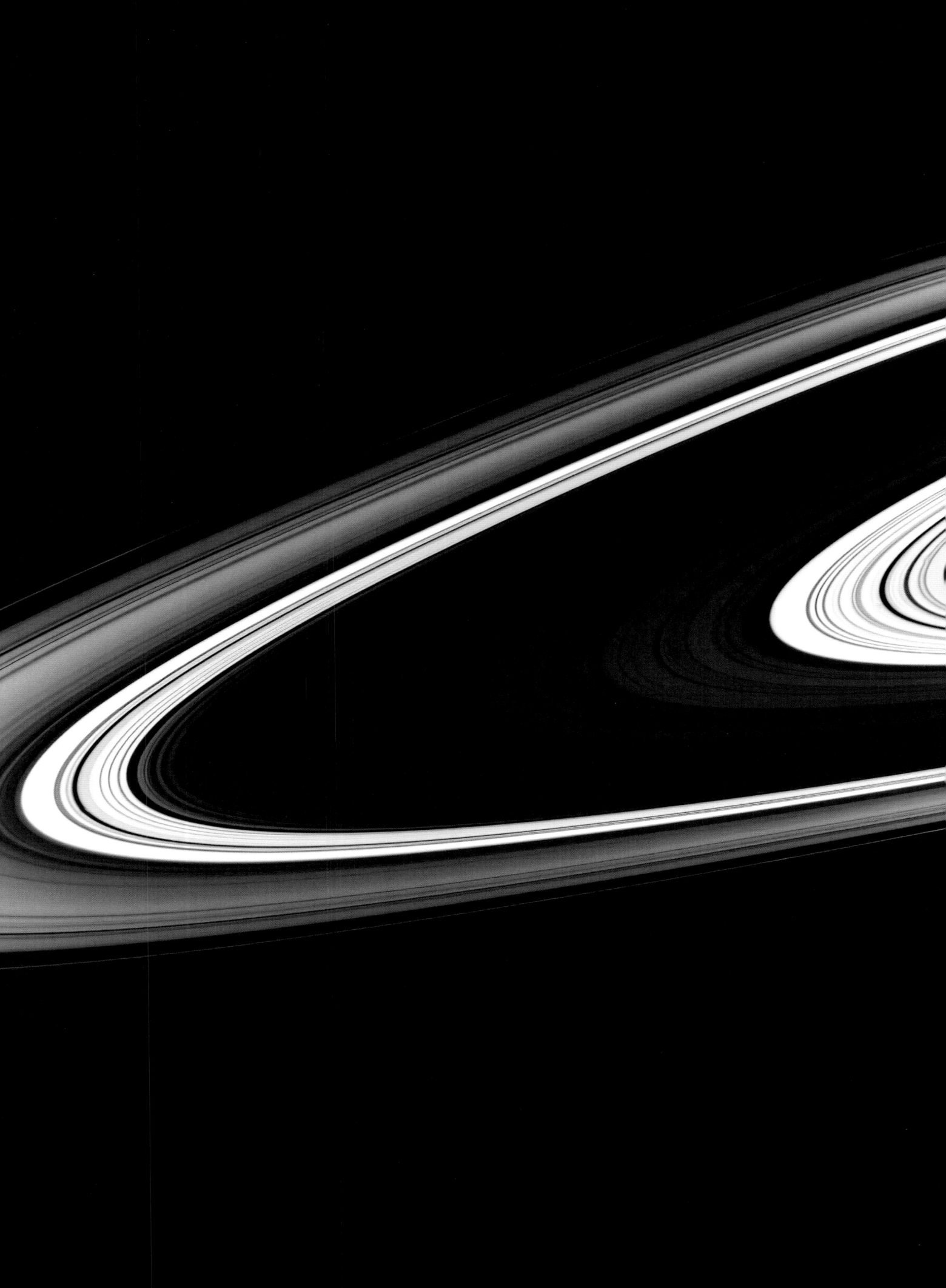

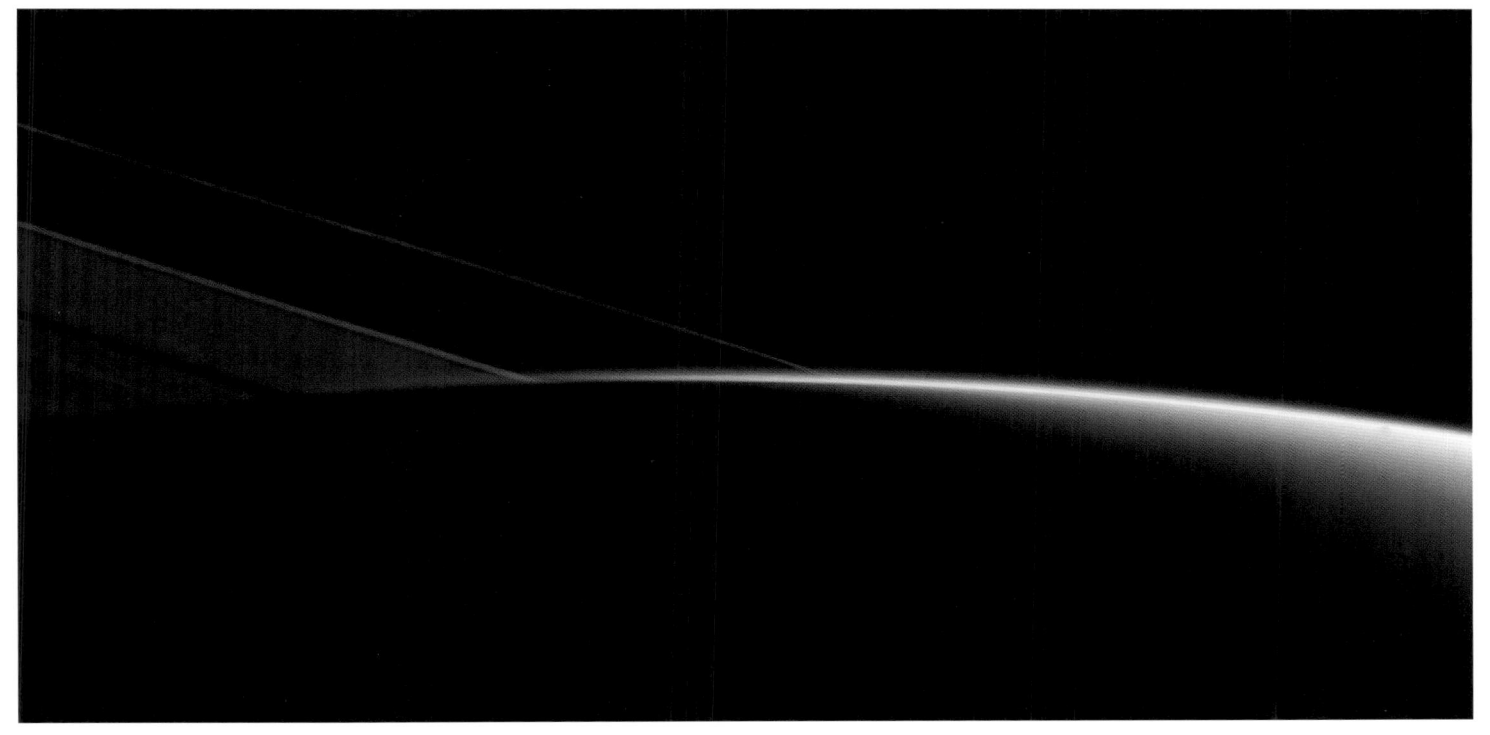

Left: Saturn Rings, Voyager 1, November 12th, 1980, 13:31:42.

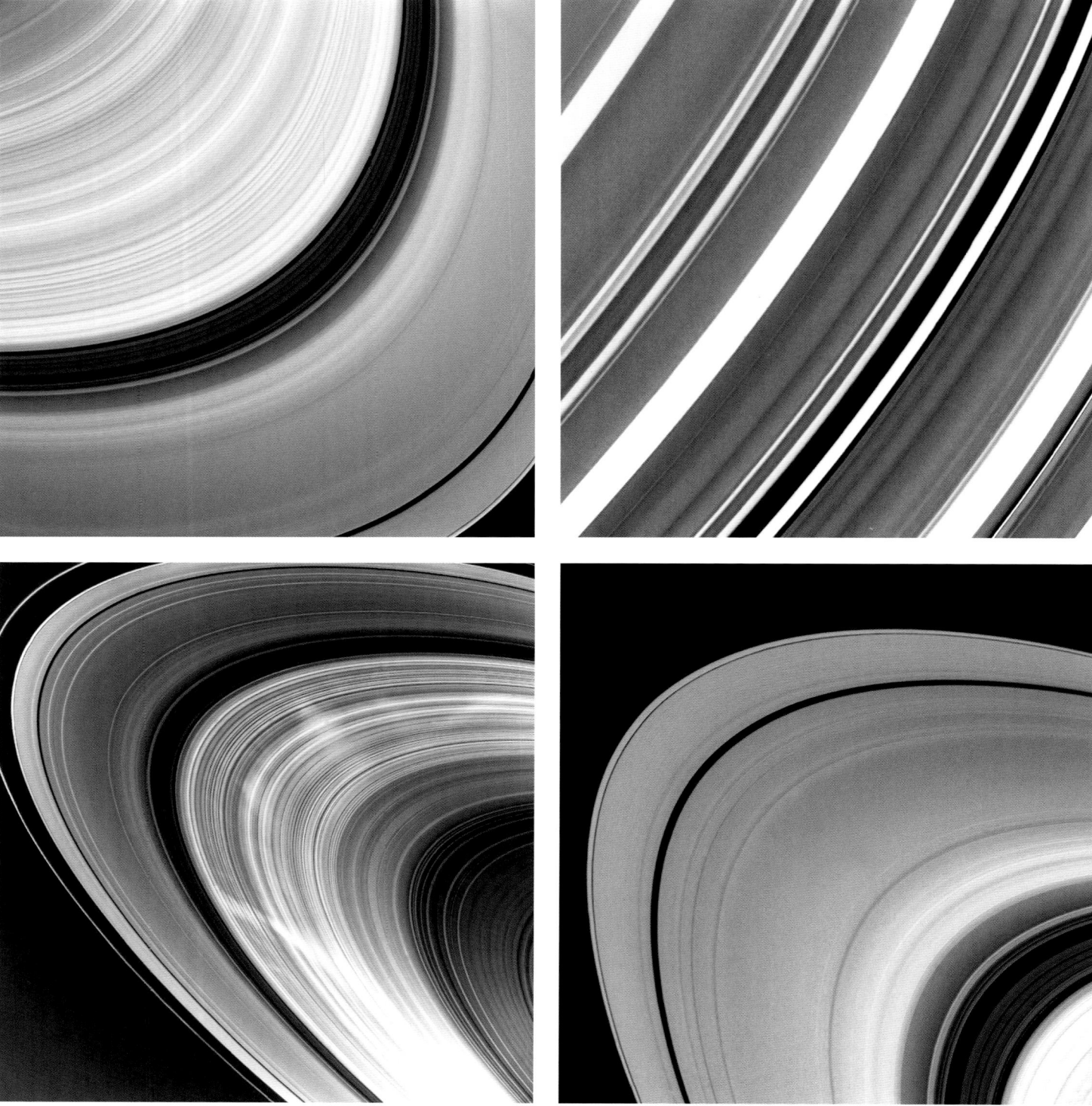

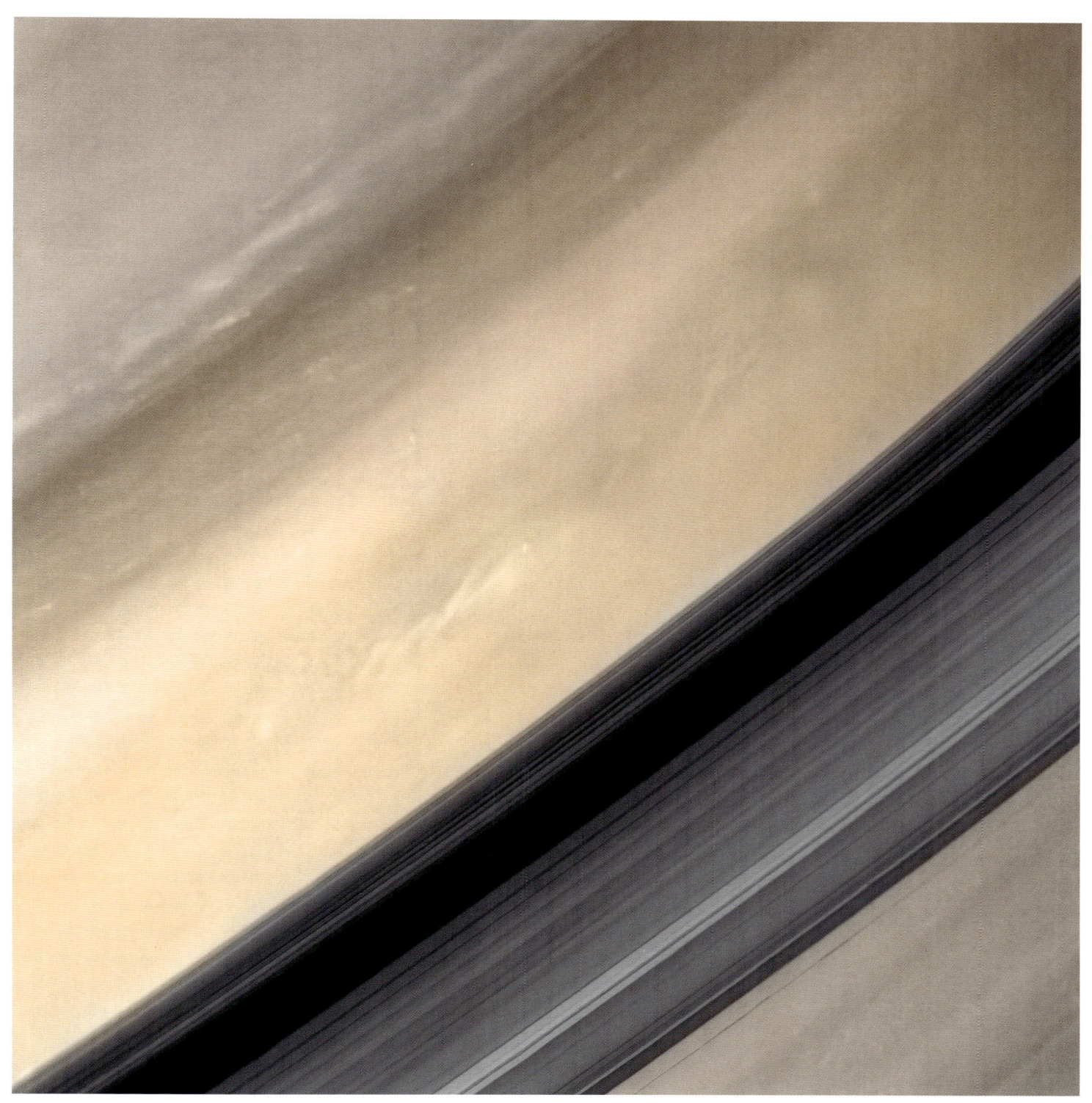

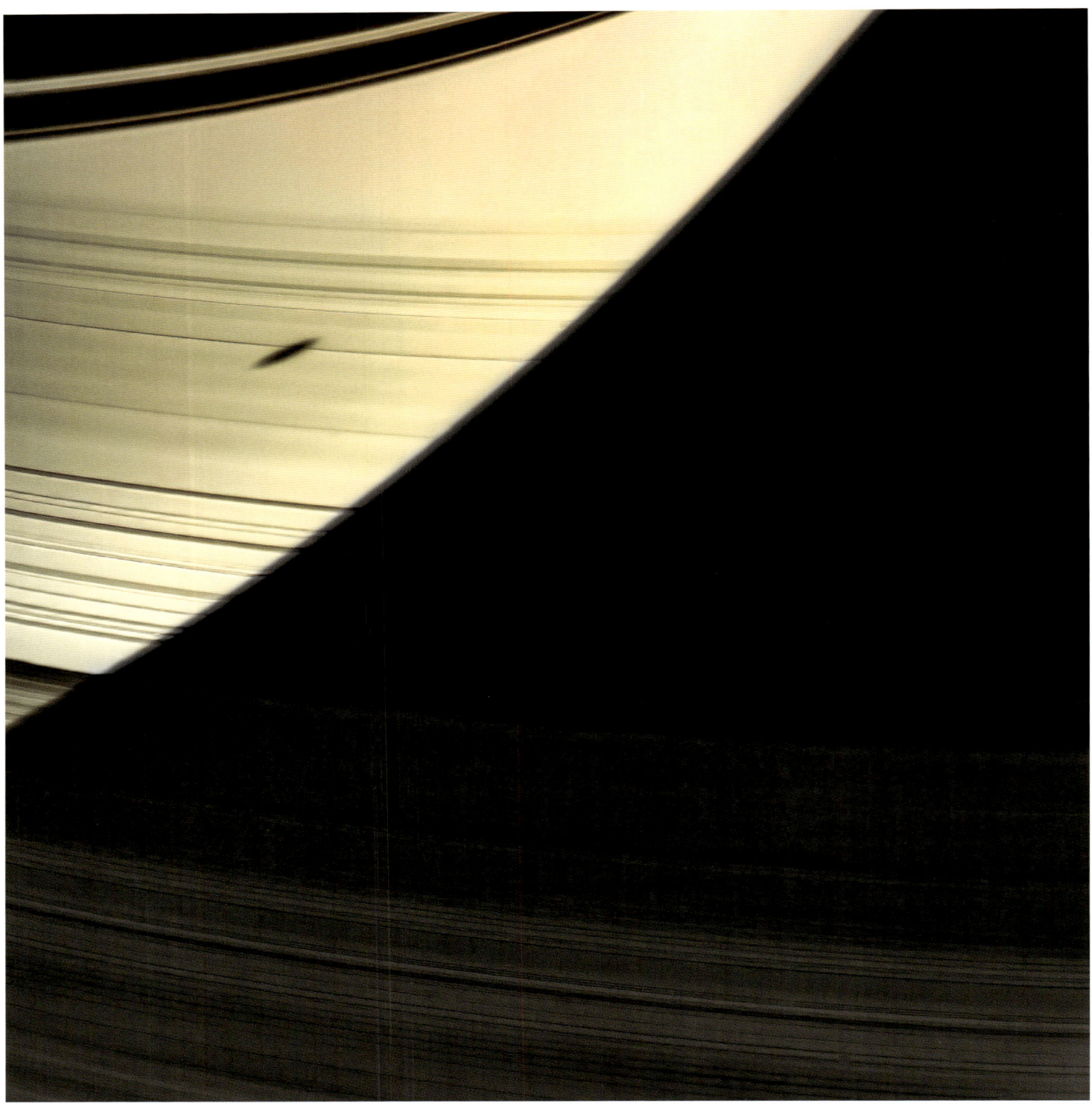

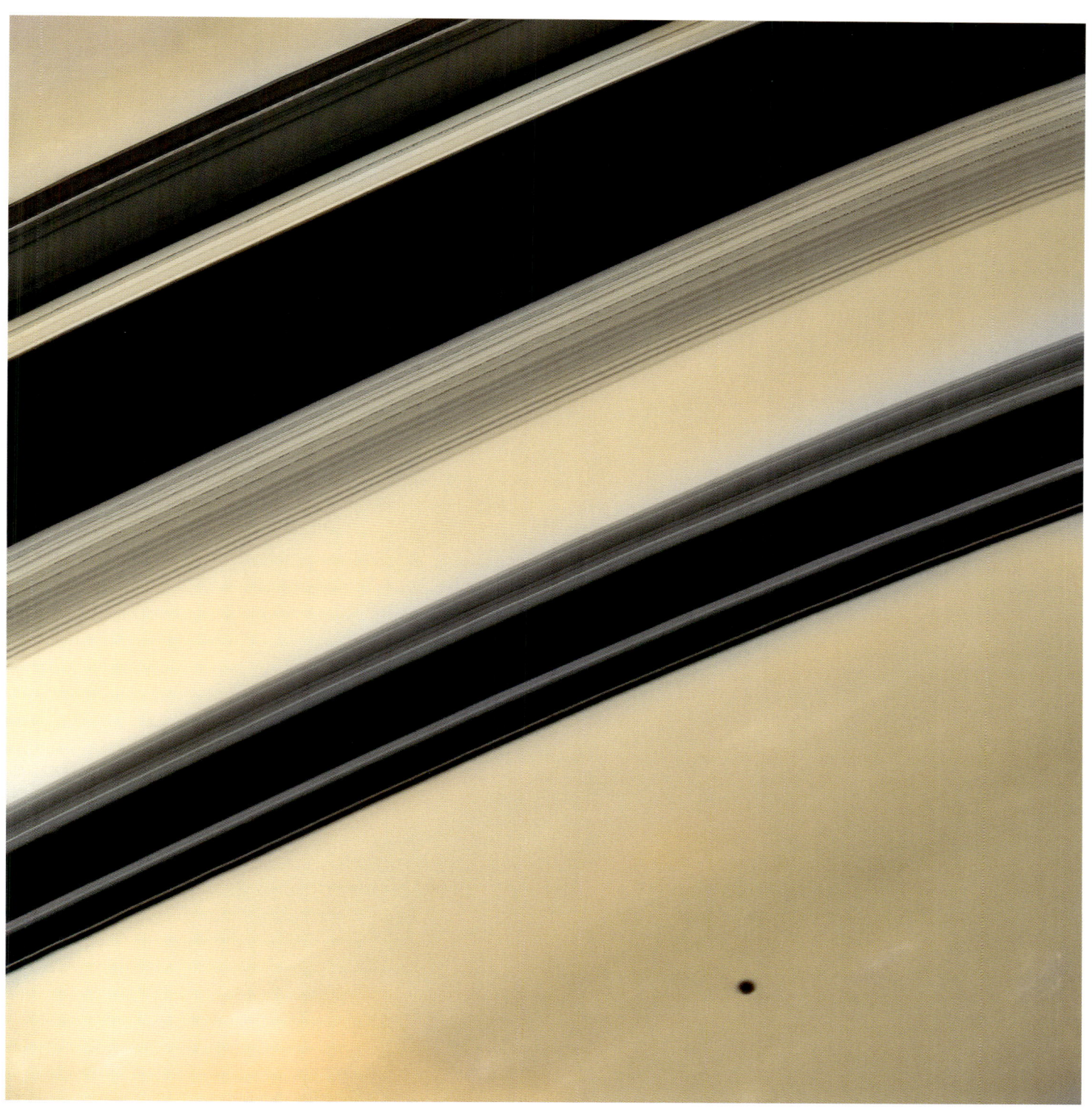

THE COLLECTION OF MOONS

Saturn has a large and diverse collection of moons. To date, 82 moons have been confirmed to orbit around the planet, ranging from tiny moonlets that span 30 odd yards (around 27 m) across, to Titan being bigger than Mercury. However, only 13 out of the 82 moons are of a significant size, having a diameter greater than 30 miles (50 km) or more.

Although the Voyager 1 encounter with Titan was an initial disappointment, the later images and data sent from the Voyager probes of Saturn's other large moons returned interesting data. The probes also discovered three new moons during the flyby: Atlas, Prometheus, and Pandora. Pan was discovered years later when the Voyager data was studied again in 1990.

Mimas, Enceladus, Tethys, and Dione collectively form a group that is categorized as the "inner large moons" that orbit within Saturn's E ring. Mimas is the smallest moon found in the group. It has a strange egg-shaped appearance, which was caused by the effects of Saturn's gravity shaping it in an unusual way. A huge impact crater that wraps one-third around the moon was seen. The Voyager images also showed that there appeared to be no past or present geologic activity internally, as the surface was fully covered in smaller impact craters.

Enceladus is the second smallest moon from the inner moon group. Flying past it in August 1981, Voyager 2 had a fairly close view of this moon. The returning data showed a young surface that had hardly any impact craters. Internal heating was predicted to be present, caused by the gravitational resonance influence of Enceladus's interaction with its neighbor Dione. The highly reflective and pure ice surface makes it one of the brightest known objects in the Solar System.

Tethys has a diameter of 660 miles (1,060 km) across, and was discovered by Cassini in 1684. The moon is predominantly composed of water ice, showing a surface that is old and heavily cratered. Voyager images revealed a huge scar-like valley that is about 60 miles (97 km) wide, and stretches an enormous 1,250 miles (2,000 km) long, stretching nearly three-quarters of the way around the moon.

Images of Dione also showed another old, pockmarked moon. Heavily cratered, it has a large long valley that wraps around the whole side of the moon. Like Enceladus, Dione showed evidence of past tectonic activity with signs of fractures and other features caused by cryovolcanic activity.

Another group of moons orbit beyond the E ring. This group is called the "larger outer moons" and it consists of Rhea, Titan, Hyperion, and Iapetus. Rhea is the second largest of Saturn's moons. It has an icy body that, like many of the other moons, shows an old and heavily cratered surface. The images brought back by the Voyager craft showed a surface that resembles the surface of our own Moon and that of the planet Mercury. Hyperion and Iapetus were only studied by Voyager from a distance. From the images of Hyperion, the team was able to just about make out an irregular shaped moon, and a peculiar, tan-colored icy surface that looks quite a lot like a sponge. Hyperion orbits around Saturn in a highly chaotic rotation, which means that it has no clear equator or a north and south pole. Iapetus is the third largest of Saturn's moons. Voyager images and data revealed that the moon has an odd walnut shape that has a two-tone colored surface. Iapetus orbits 2.2 million miles (3.5 million km) away from Saturn, by far the most distant orbit of any of Saturn's large moons.

The Voyager probes also imaged Epimetheus and Janus. These two moons behave together in a unique way. They essentially share the same orbit around Saturn. The two moons swap orbital positions every four years as the inner orbit that one of the two moons orbits around, travels slightly faster than the moon traveling on the slower, parallel outer orbiting track. The faster moon overtakes the slower one. During this pass, the movement pulls the slower moon into the inner and faster lane. The faster moon, as it overtakes, moves out to the slower outer orbiting track, performing a fascinating orbital dance that happens nowhere else in the Solar System.

Right: Rhea, Voyager 1, November 13th, 1980, 06:59:45.

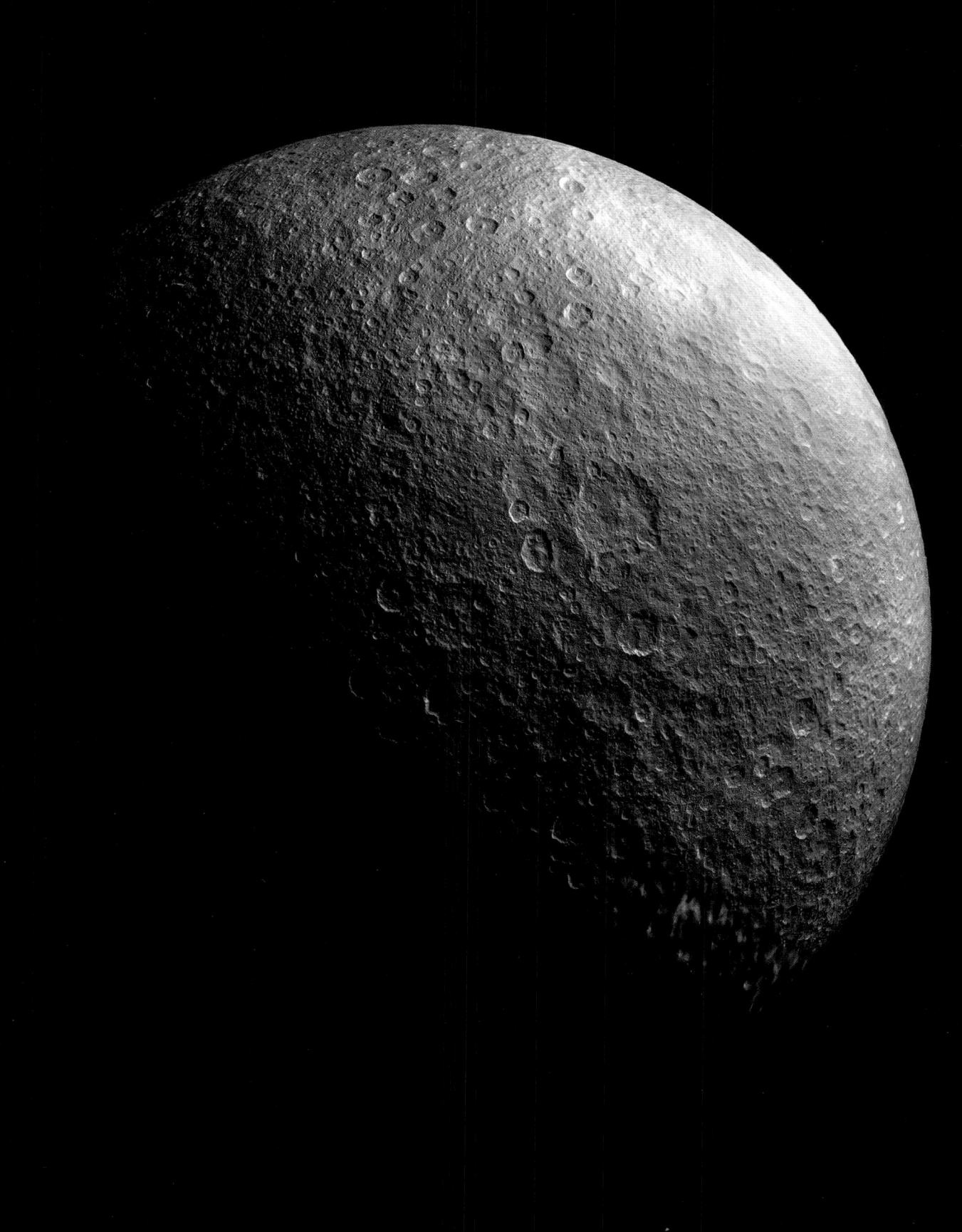

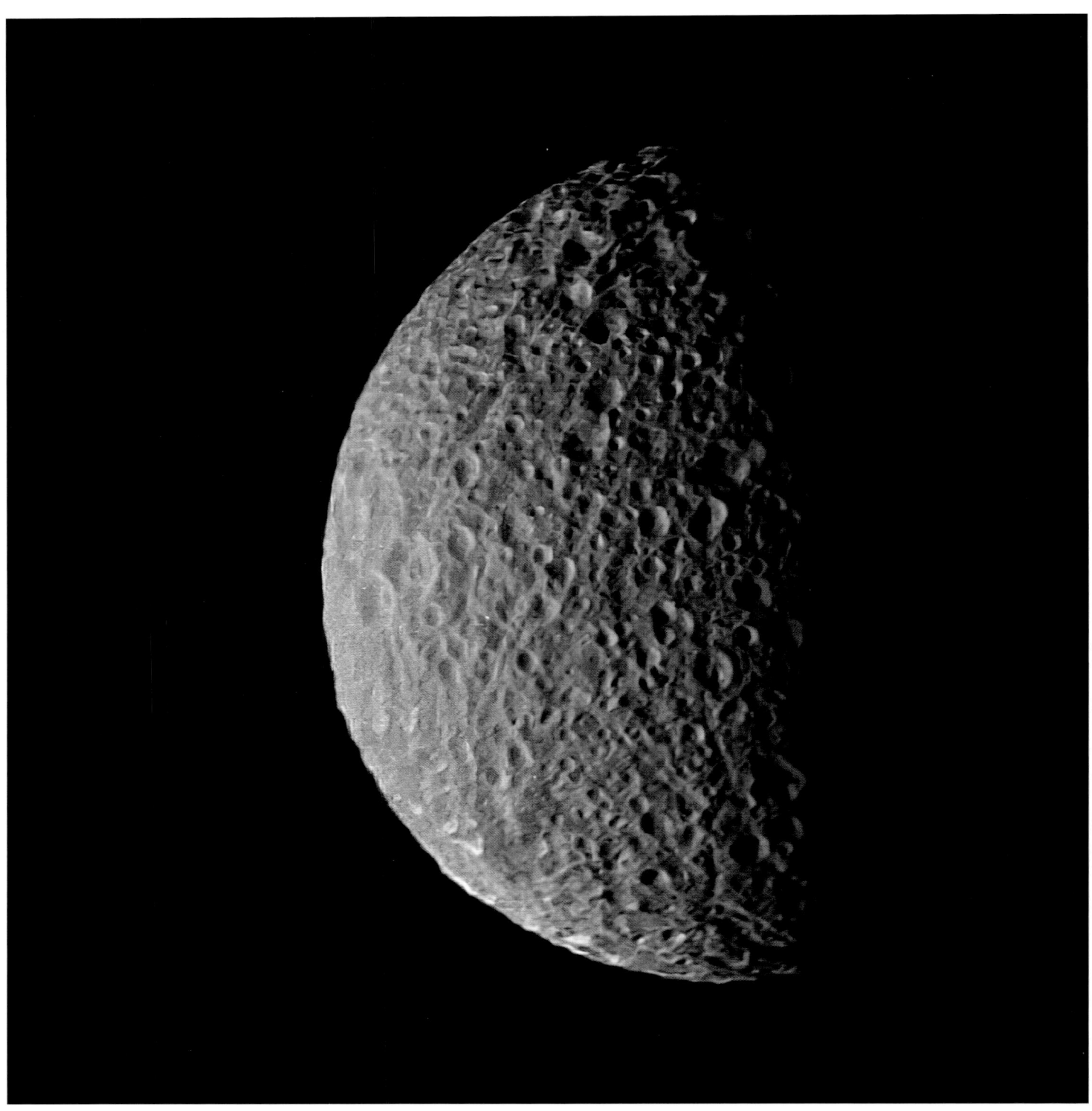

Right: the giant crater Herschel makes Mimas resemble the Death Star.
Mimas, Voyager 1, November 12th, 1980, 15:37:21.

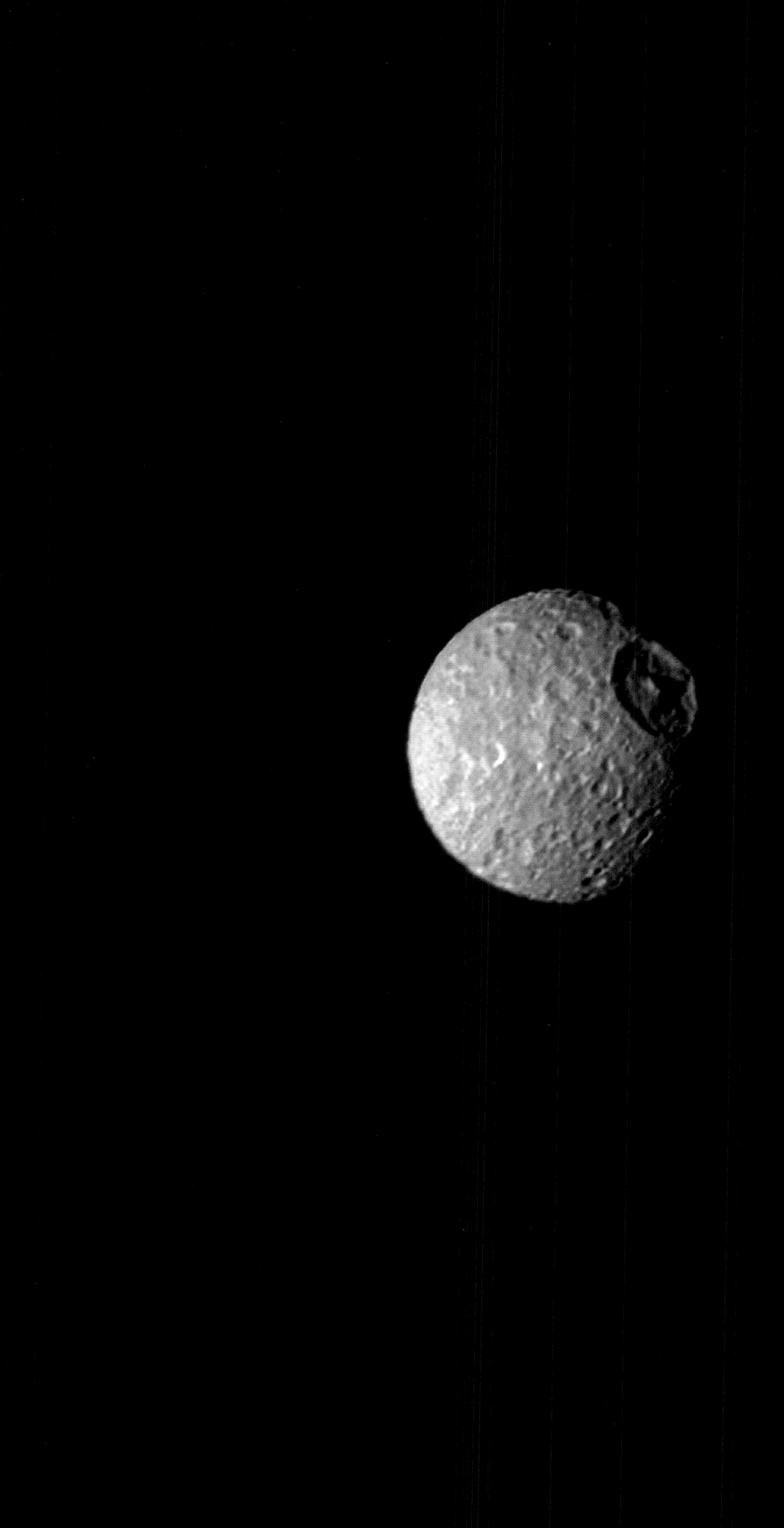

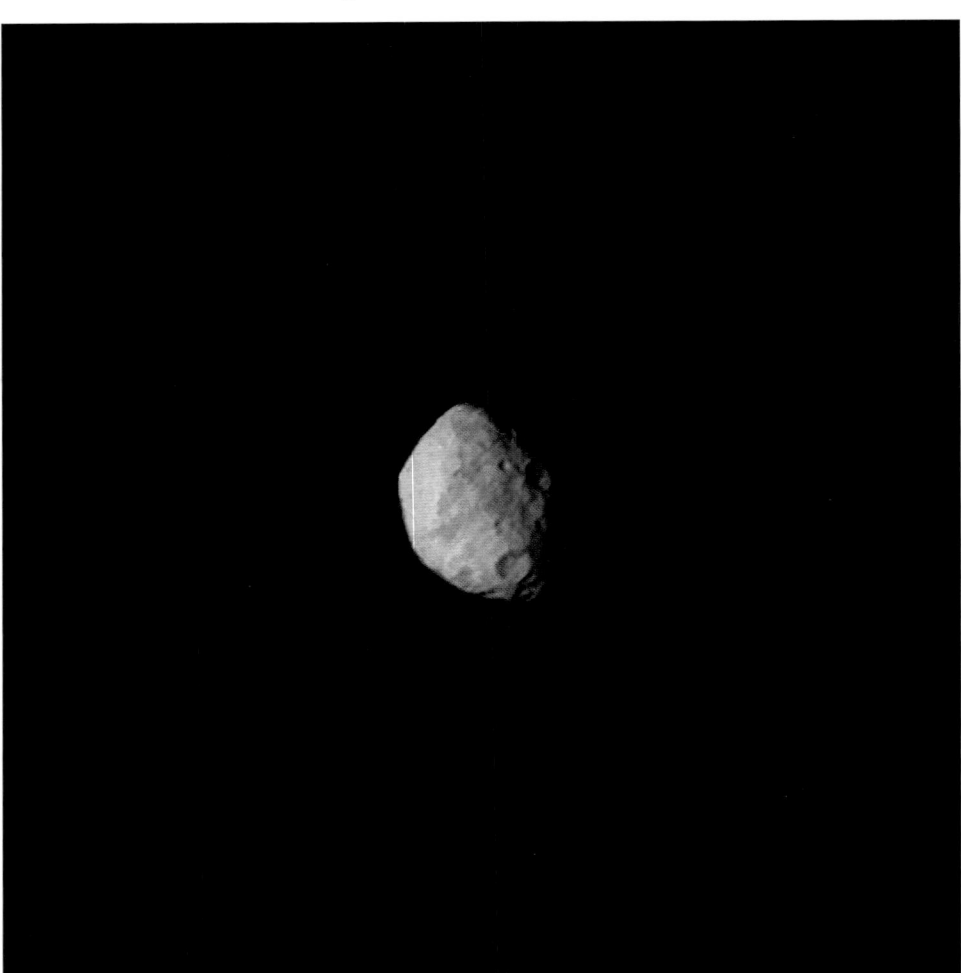

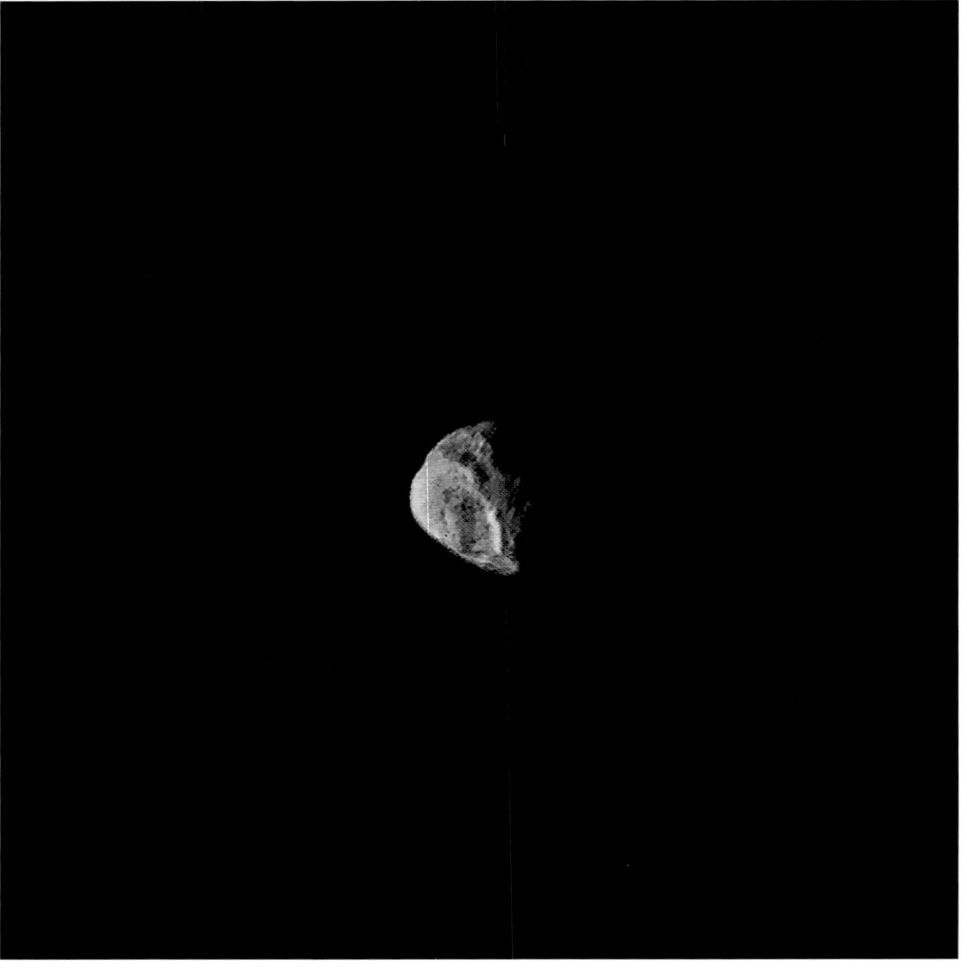

The moon Hyperion is a strange breed. Its spin axis has a chaotic orientation in time, which makes it impossible to predict how the moon will be spinning in the future. The moon seems to tumble as it orbits Saturn. Scientists so far have only found a few bodies with chaotic spins as such.

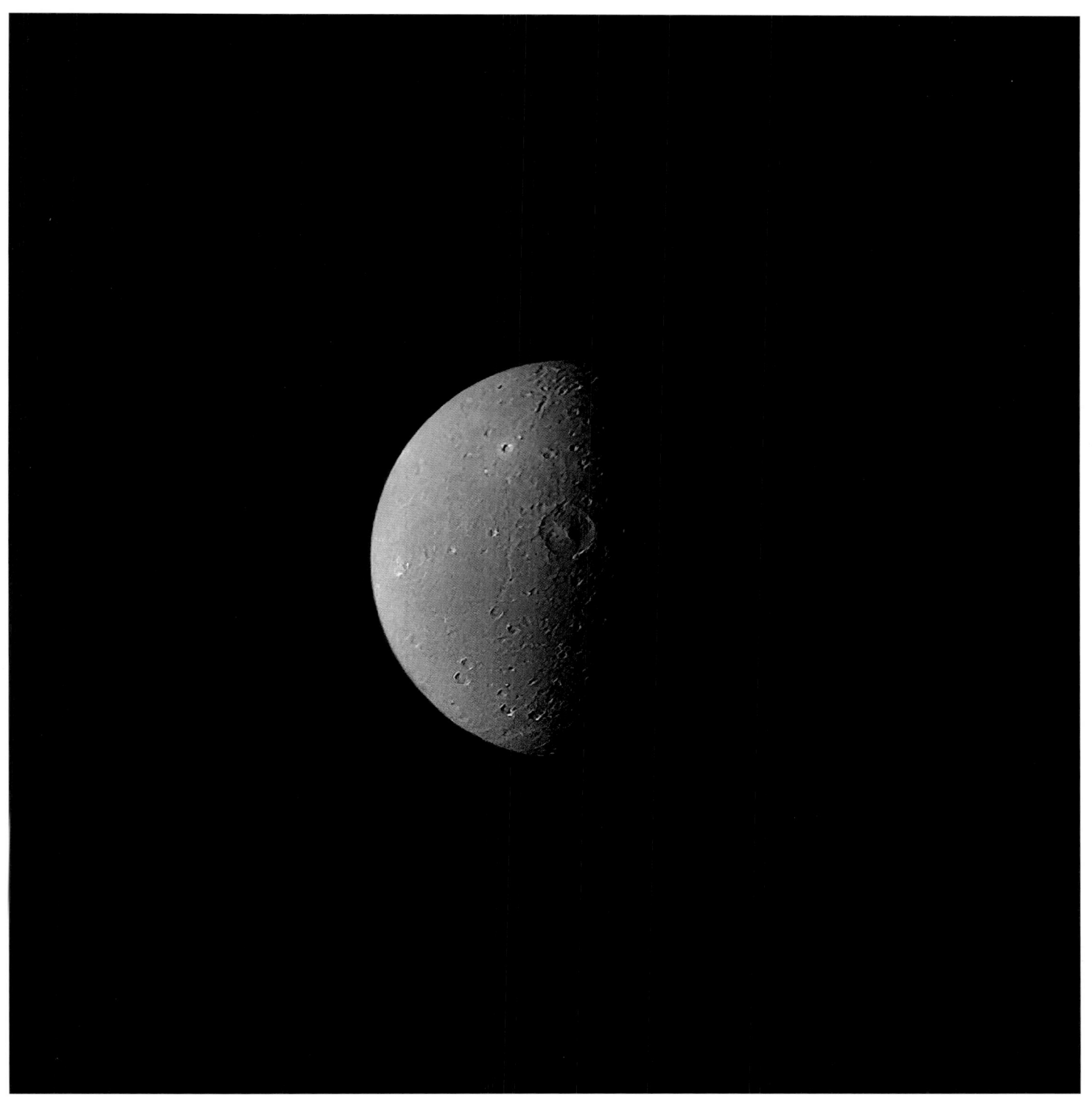

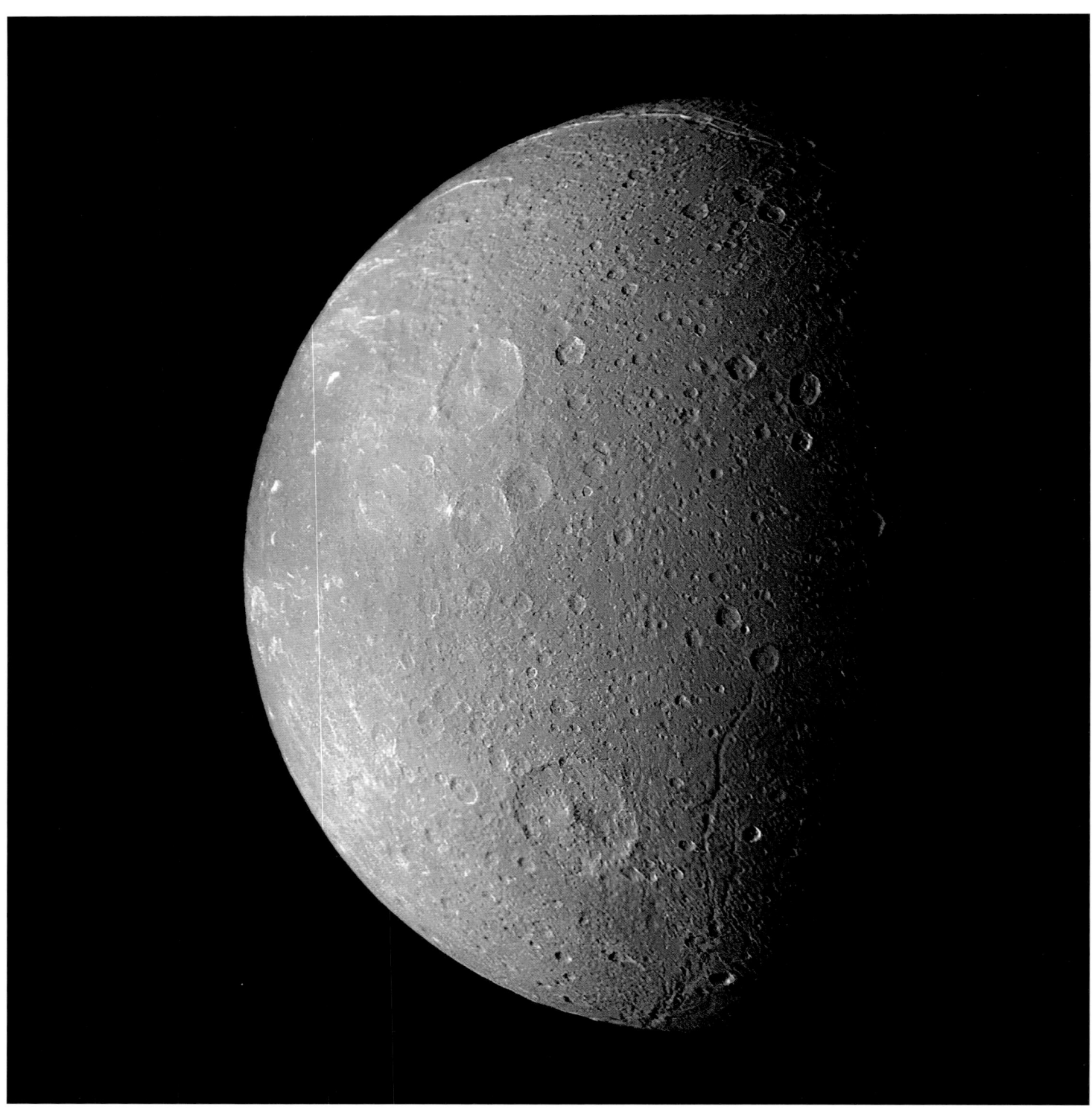

This is Saturn's moon Dione. Its many impact craters show the record of collision of cosmic debris. With a diameter of roughly 62 miles (100 km) the largest crater shows a well-developed central peak. The bright rays that are seen represent material ejected from neighbouring impact craters.

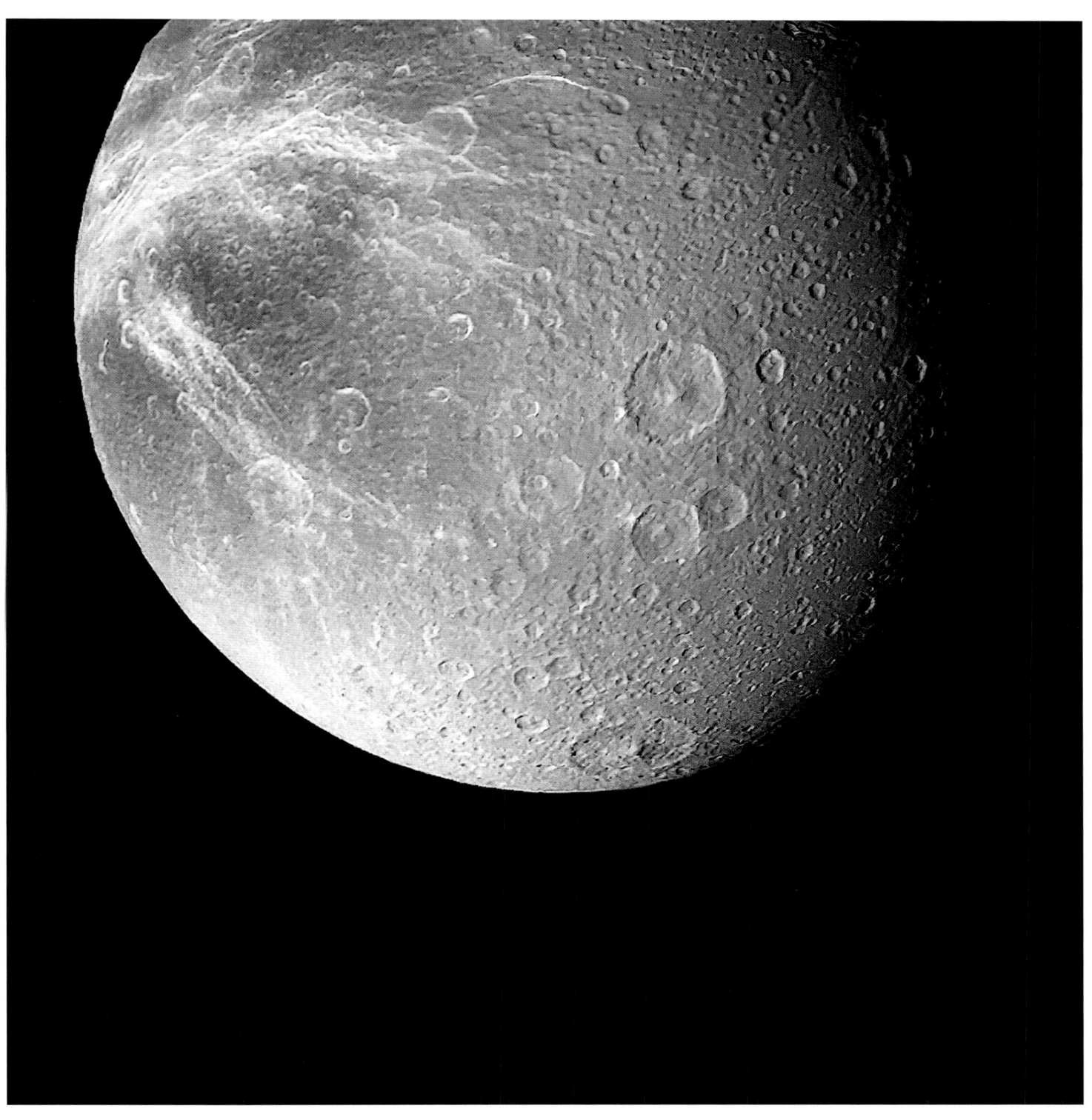

In this view of Dione we see clear bright radiating patterns. Some of
them are probably debris rays ejected by impact craters, others could
be topographic ridges and valleys. Other more irregular valleys
represent old fault troughs degraded by impacts.

DARK SIDE OF SATURN

With Voyager 1's successful flyby of Titan, and the realization that another visit by Voyager 2 to Titan would produce fruitless images of no additional value, Voyager 2 was cleared to have the close encounter with Saturn. The results of the close flyby returned brilliant images from Saturn's cloud-tops. More importantly however, it gave the spacecraft the necessary pull from Saturn's gravitational forces. By accelerating the speed and turning the craft 90°, Voyager 2 was now pointing in the right direction for its next stop at Uranus in 1986.

Ed Stone and the Voyager science team for Voyager 2's flyby of Saturn placed an important mission constraint. Trajectory planners were instructed to fly Voyager 2 behind and into the shadow of Saturn. This way the cameras and scanners could detect how the sunlight travels through Saturn's upper atmosphere and rings. Measuring the occultation of sunlight technique used in astronomy is a powerful trick that gathers and records a plethora of vital data, such as atmospheric pressures, the chemical compositions of gases in the atmosphere and the particles in the rings. A similar process was also implemented when the Voyager craft flew past behind Jupiter.

The trouble with sending the spacecraft to the dark side of Saturn was that Voyager 2 would be completely cut off from communications with Earth for 95 minutes, as the giant planet would block all incoming and outgoing radio signals. Another challenge that the imaging team had to factor in when programming the image exposures and data scans was that the craft would now be traveling faster than at any other point in the mission. As Voyager 2 would get closer to the planet, the gravity assist would accelerate the craft from 36,000 mph (58,000 km/h) to 54,000 mph (87,000 km/h). In turn, this meant that the cameras had to work far more rapidly, moving from shot to shot, almost twice as fast as during the visit to the Jovian System. The additional speed also had a big effect on the shutter exposure times in order to make sure that images of the surface, its rings, Enceladus, and Tethys returned back clear and sharp. The Voyager team expected to see amazing photographs of storm clouds, ring particles and maybe even auroras and lightning bolts if Voyager 2 got lucky.

The final program was uploaded to Voyager 2, ready for operation on the night of August 25th. At 22:26 Pacific Time, the spacecraft went behind Saturn and out of radio signal range. At that moment, Voyager 2 went silent. An eerie feeling took over the team at JPL. The connection was always there, serving as a lifeline to the mission and its continuity. The question was "Will it complete its trip in the dark and come back into range?" After 95 dreadful minutes, a few seconds after the expected time, a stream of data started to flow and Voyager 2 passed out from Saturn's shadow.

The first two images of Enceladus and Tethys returned to Earth as expected. However just moments later, the expected shots of the rings were received, revealing fully black images. As further images started to return, it became clear that the images returned were not what was planned and programmed in advance. Having it dawned on Candy Hansen and the imaging team that the targeting of the images stopped working during the trip around the dark side of Saturn, the engineers immediately went straight to work to get the camera back into full operation.

As they continued to work through different tests on the ground-based mock up, they eventually came up with the answer to the problem. The conclusion was relatively simple in the end. It was believed that they had simply overworked the scan platform during the high-speed flyby with the intense shot list, heating and drying up the lubricant and thus making the gears in the motor jam.

The simple solution was to adjust all future camera programming to run slower and fewer shots at one time. Luckily, this seemed to have solved the issue. Although a good number of important observations from the dark side of Saturn and its moons were lost, the mission was not over, and fortunately the camera was still operational and ready to go for its encounter with Uranus.

Right: the markings seen in this view of Dione are slightly brighter than those captured on Jupiter's moons, which suggests that they are surface frost deposits. Due to the patterns of the bright bands, its origin might have something to do with internal geological activity. The resolution however is insufficient to state this as a fact. Dione captured by Voyager 1 on November 12th, 1980 at 12:51:46.

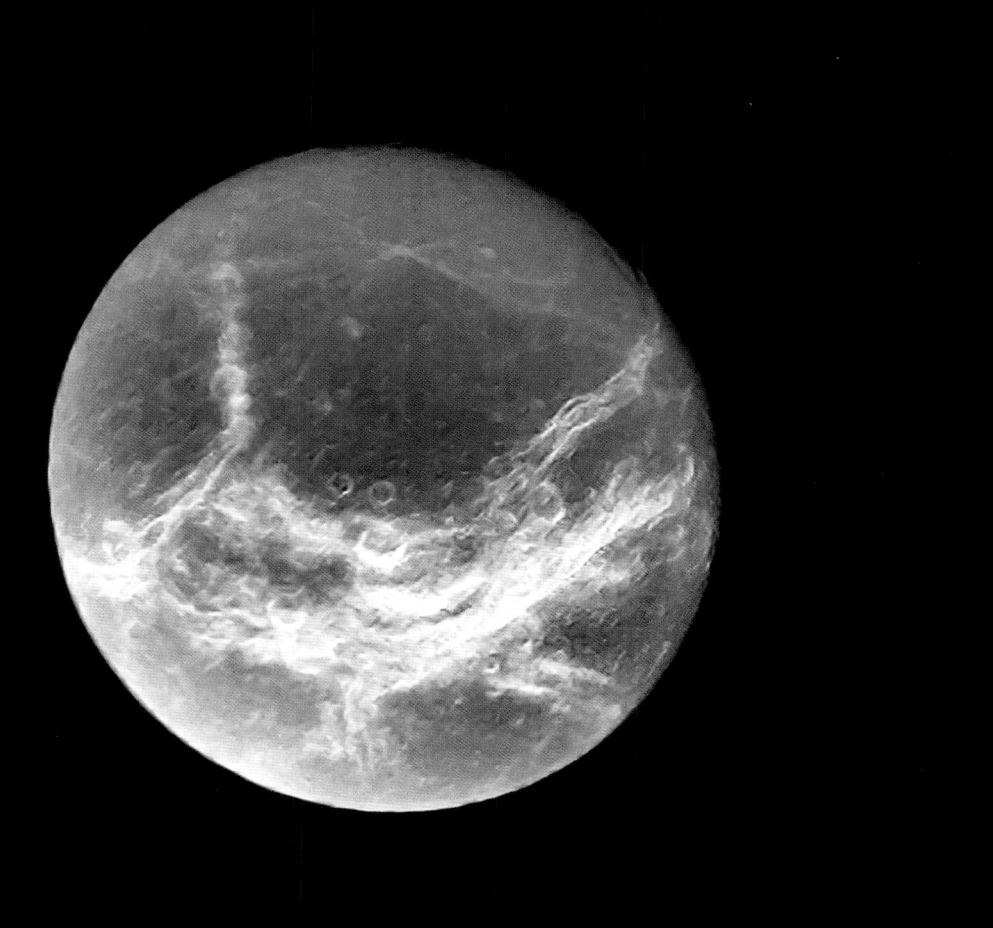

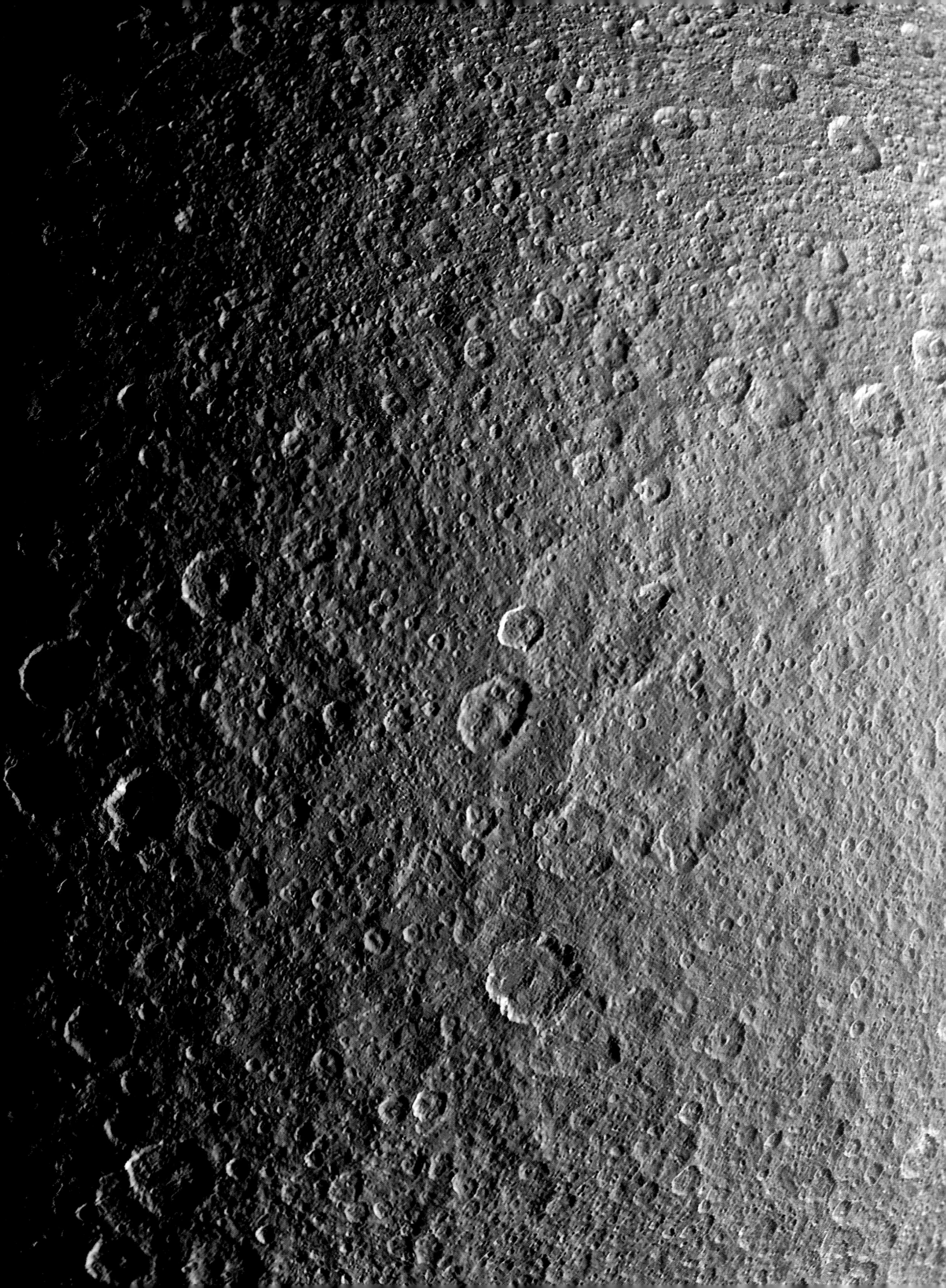

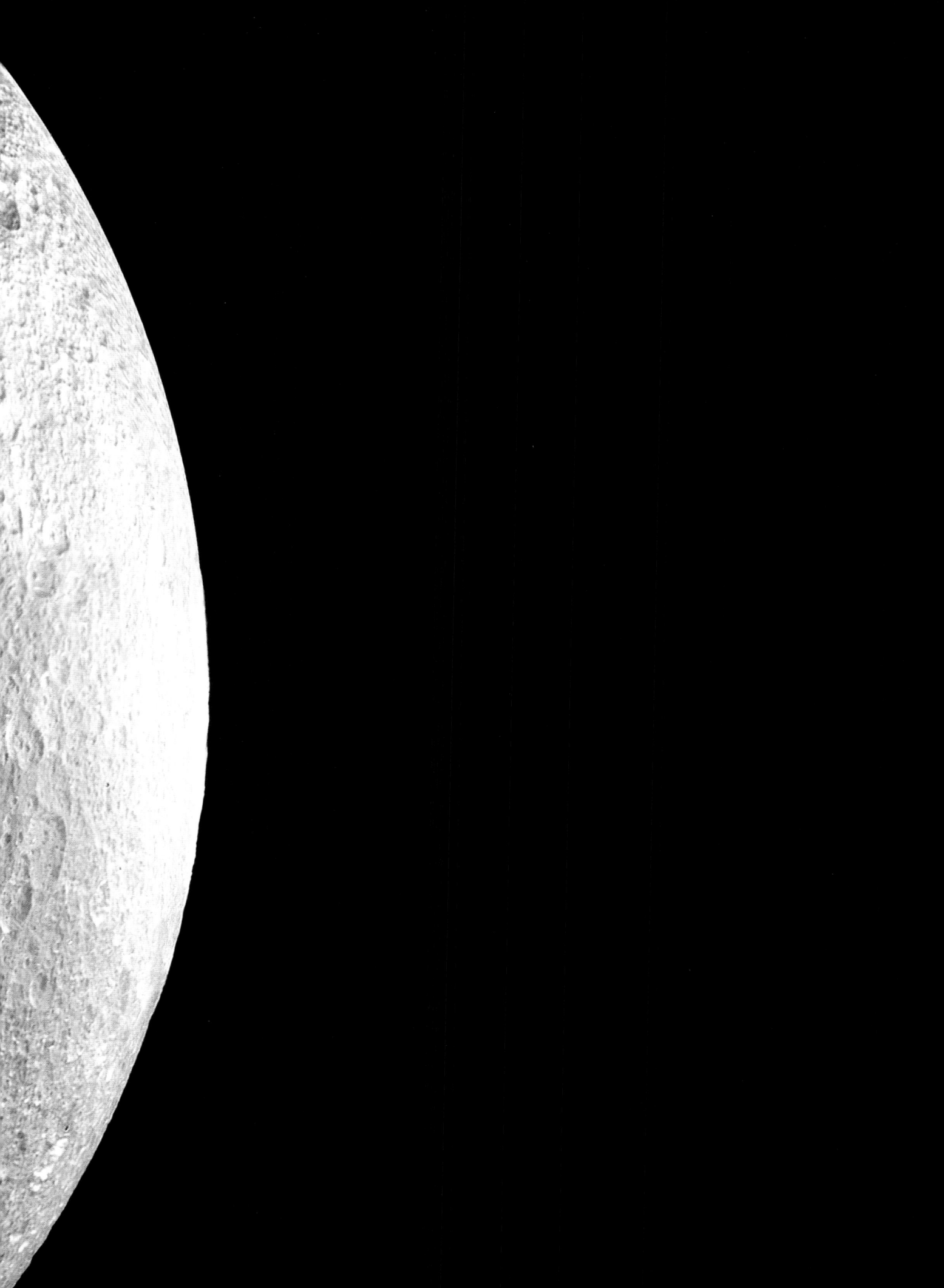

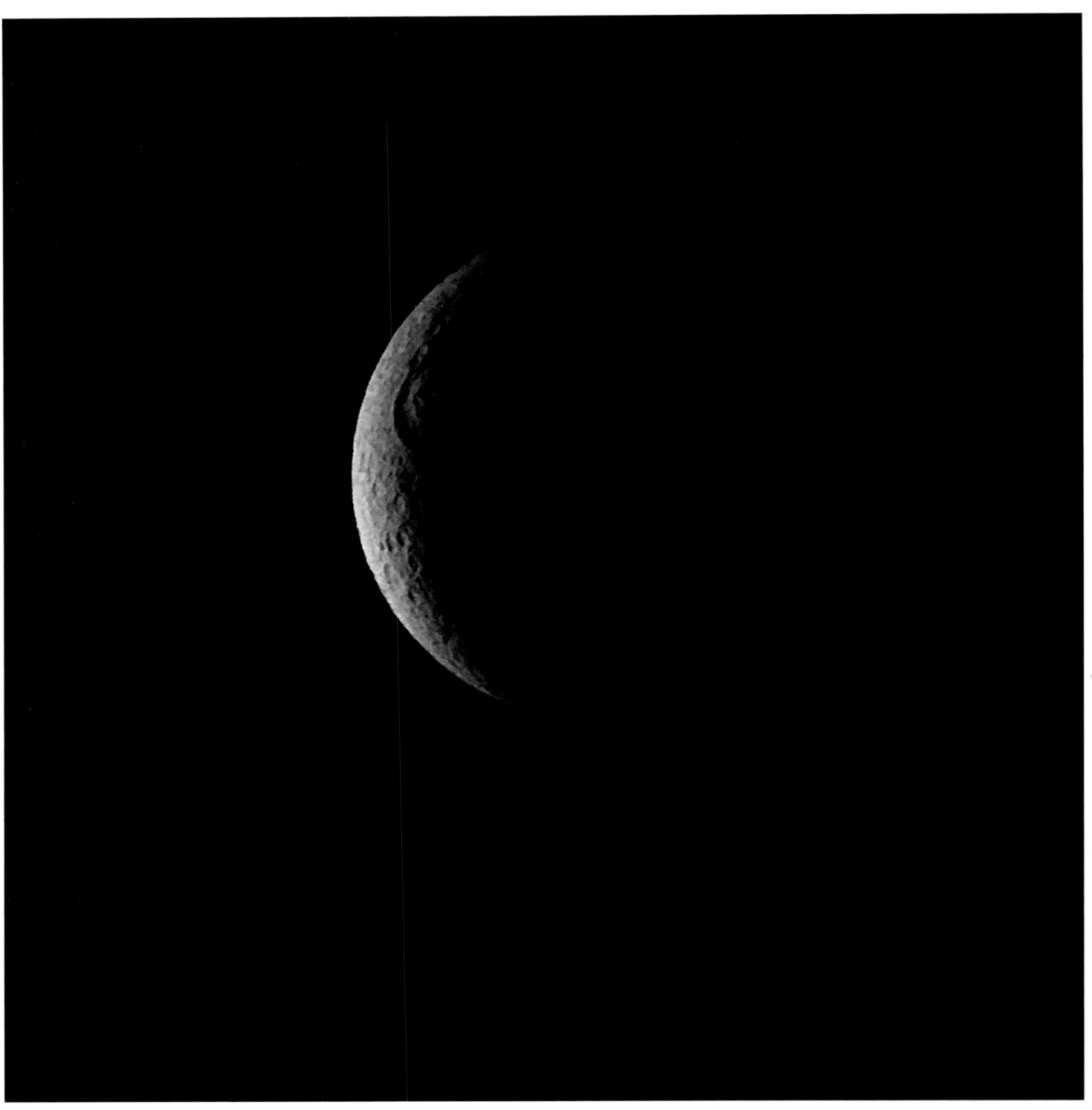

Previous page: Rhea, Voyager 1, November 13th, 1980, 06:58:09.

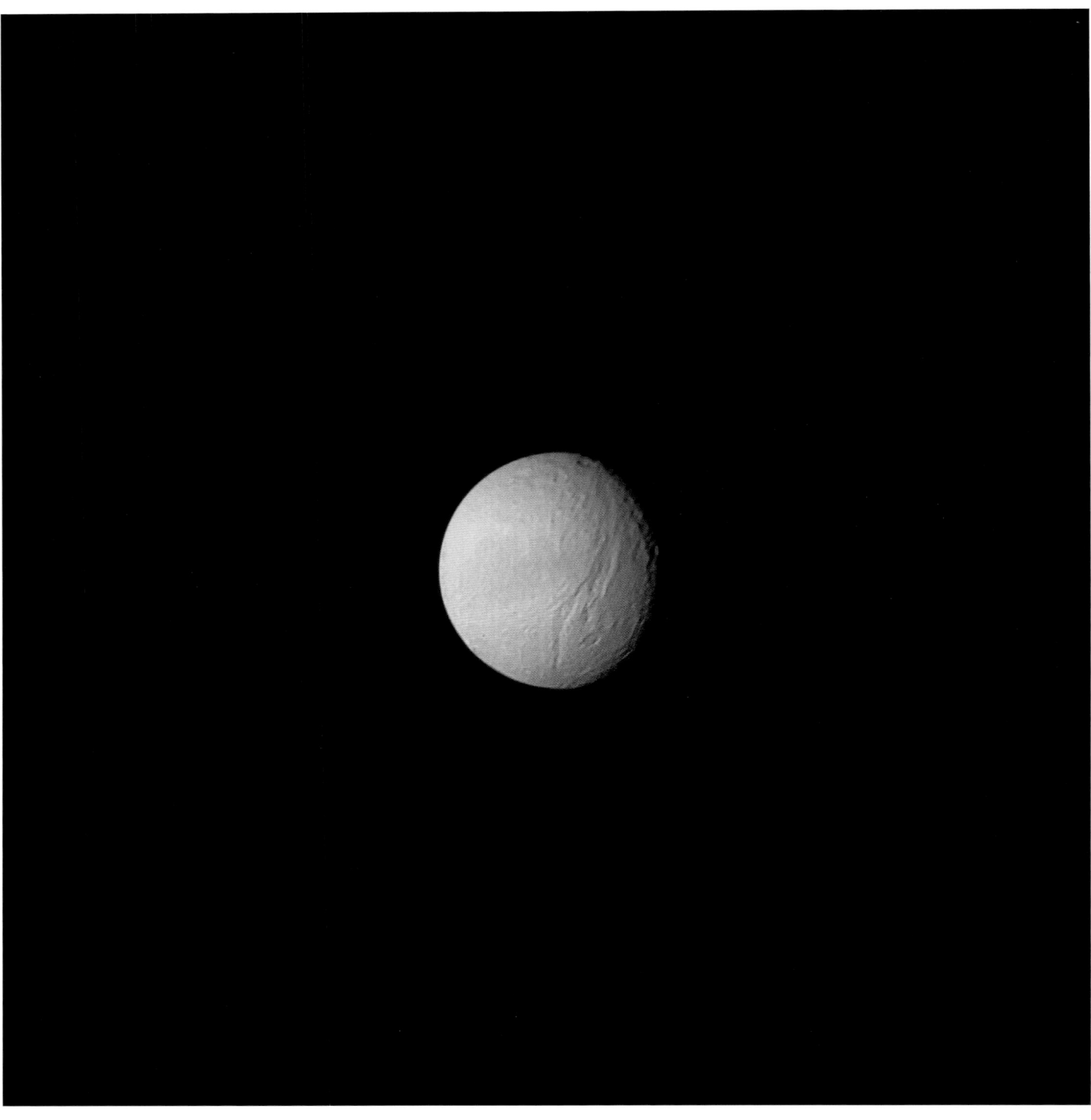

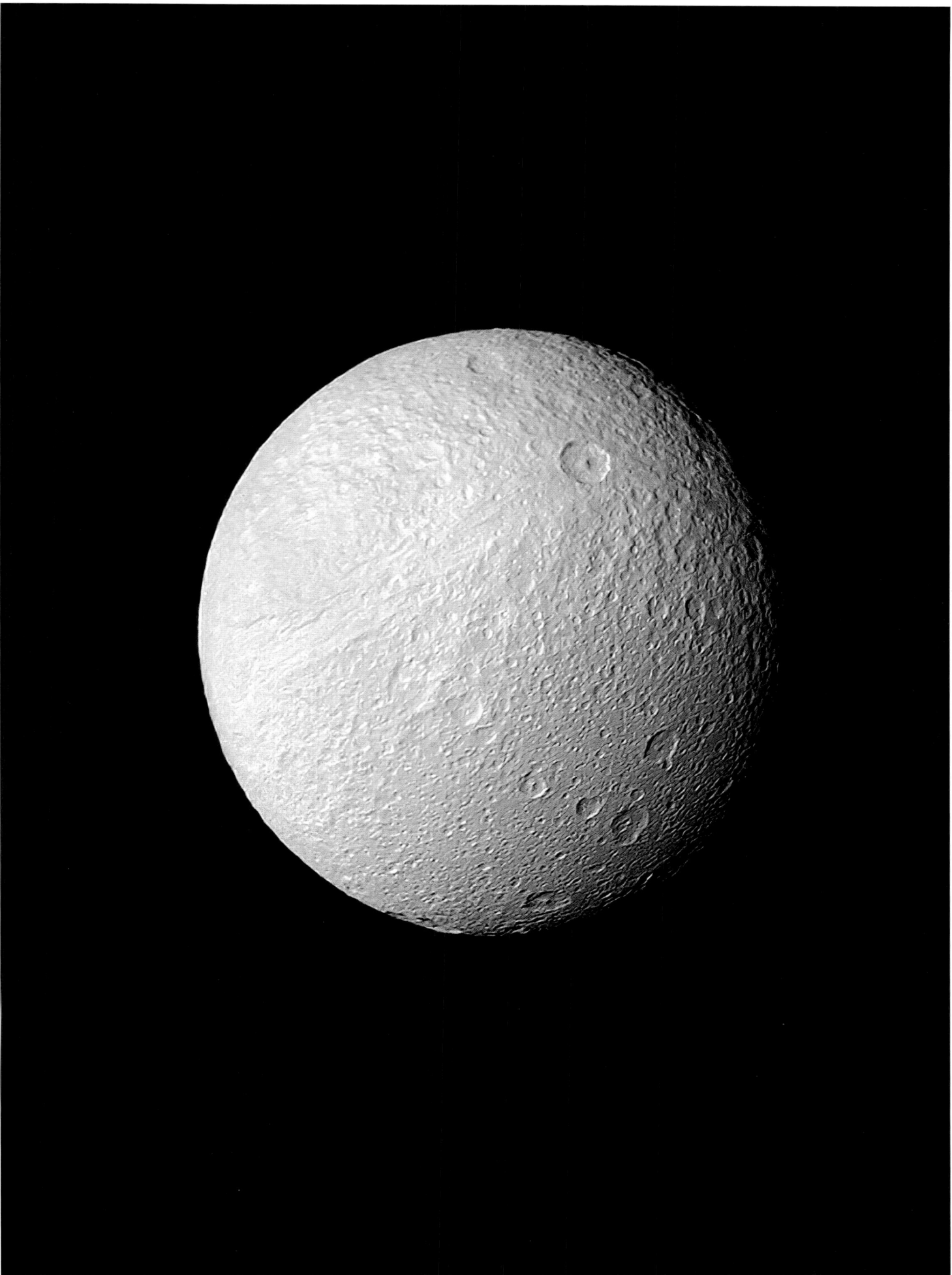

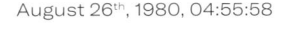

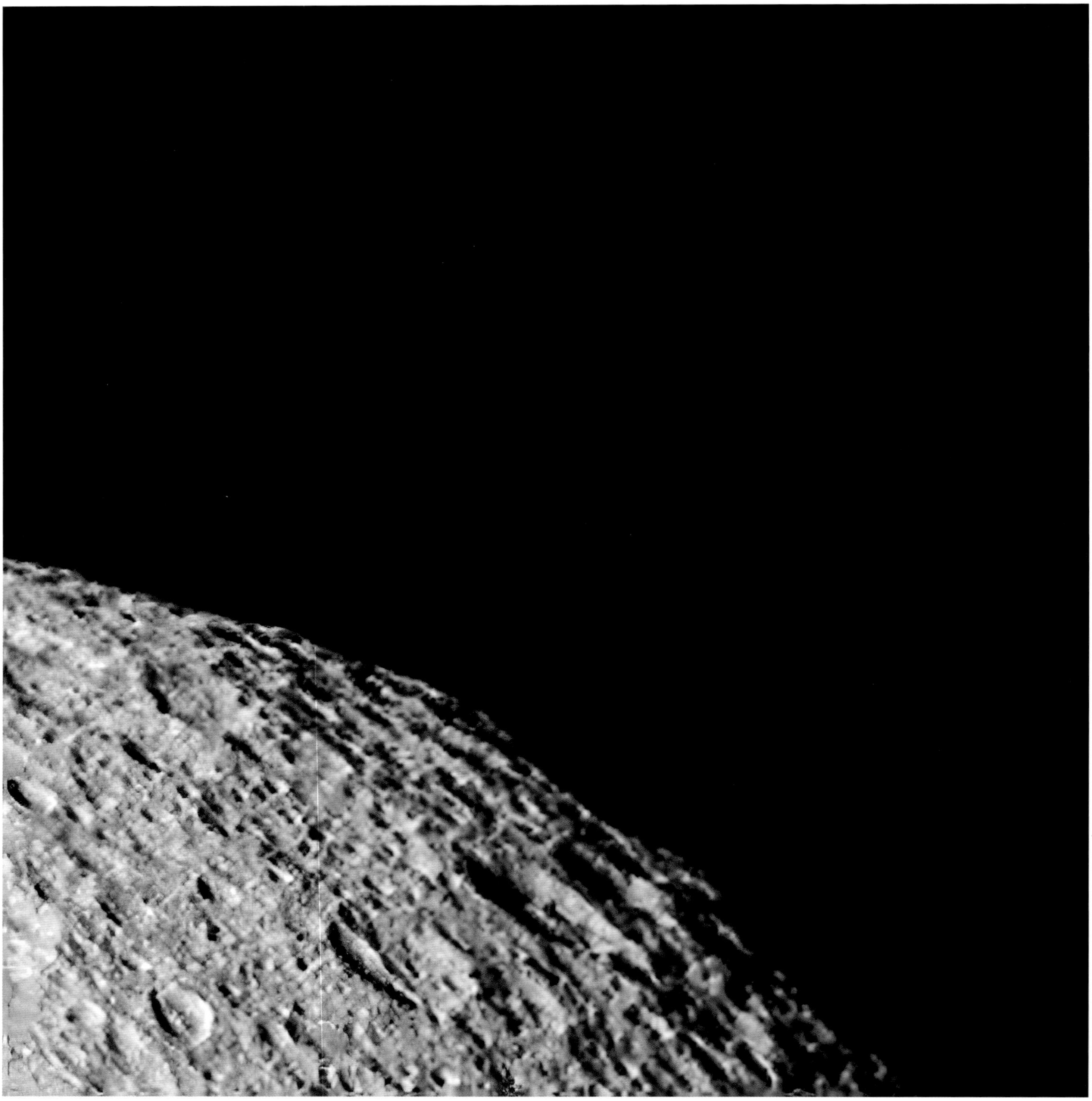

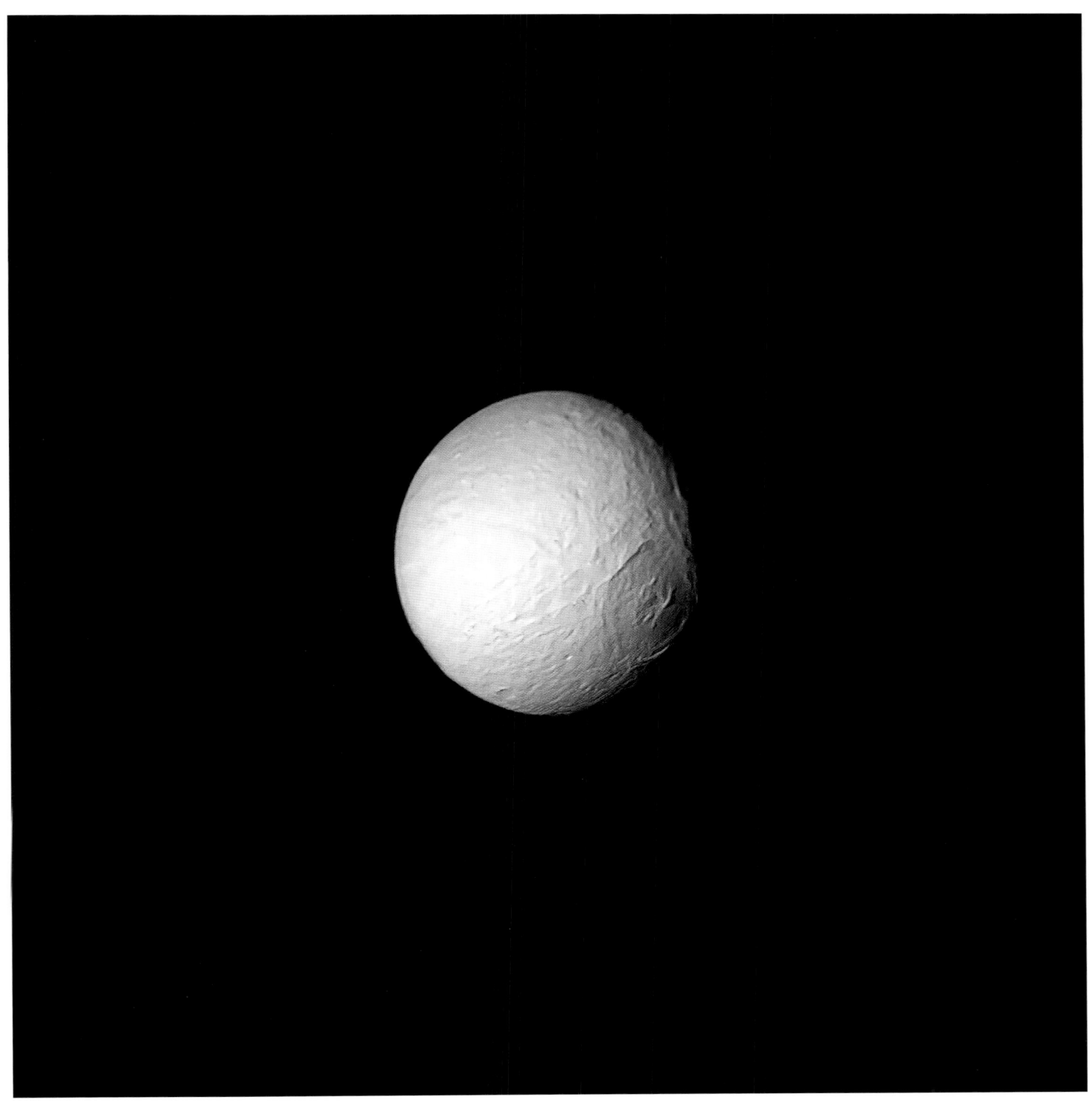

While Tethys is only 650 miles (1,050 km) in diameter, it has an enormous crater, Herschel, which is 250 miles (400 km) across and 10 miles (15 km) deep, shown here on the surface of this icy moon. Its central peak is as high as the crater is deep and is the result of rebound after the impact.

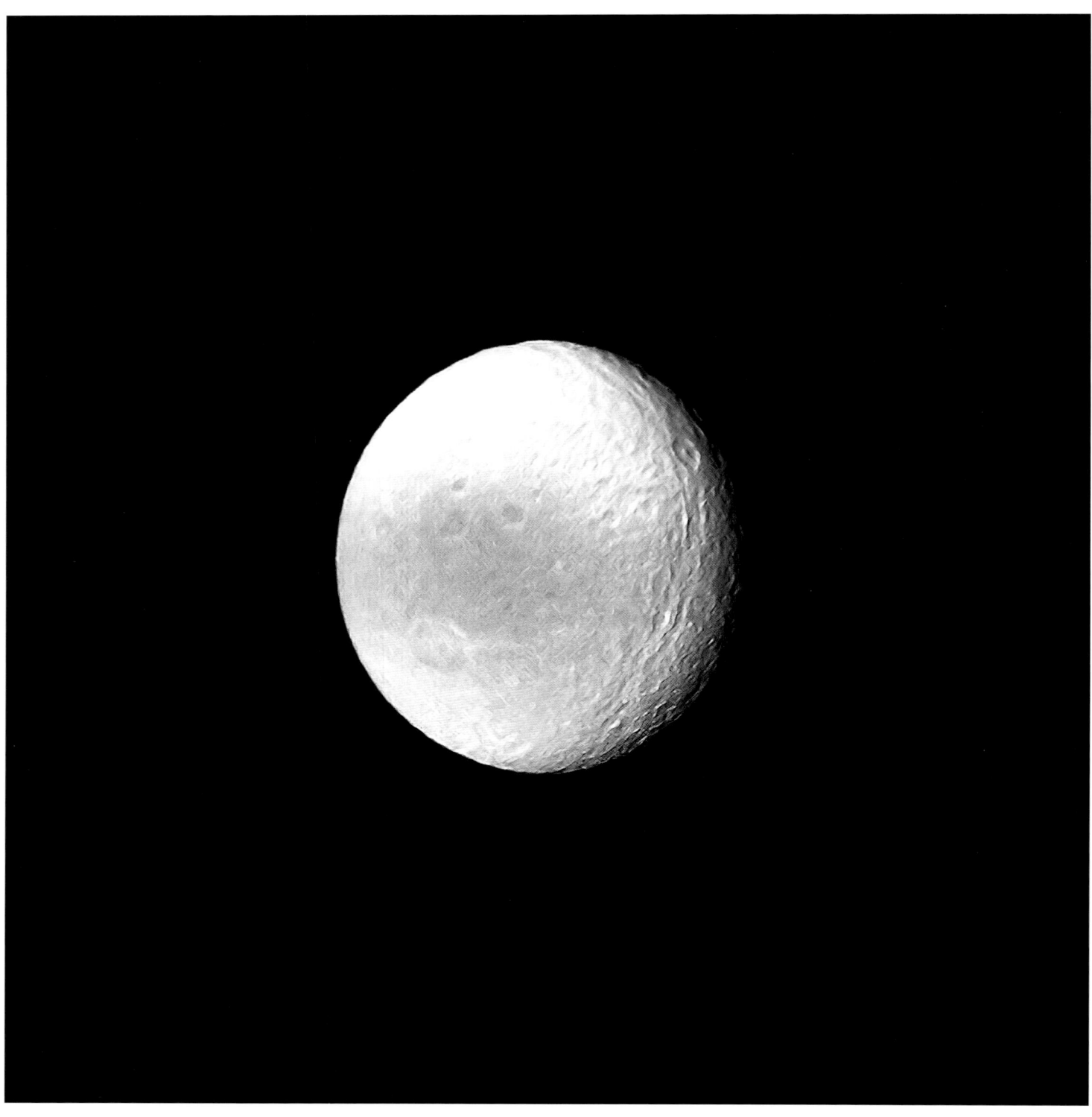

While scientists believe the densely cratered plains are part
of the ancient crust, the lightly cratered terrain is said to have
been shaped by internal processes in a later stage. The trough
that is seen running parallel to the terminator (the day-night
boundary on the right) is an extension of the huge canyon
system that extends two-thirds the distance around Tethys.

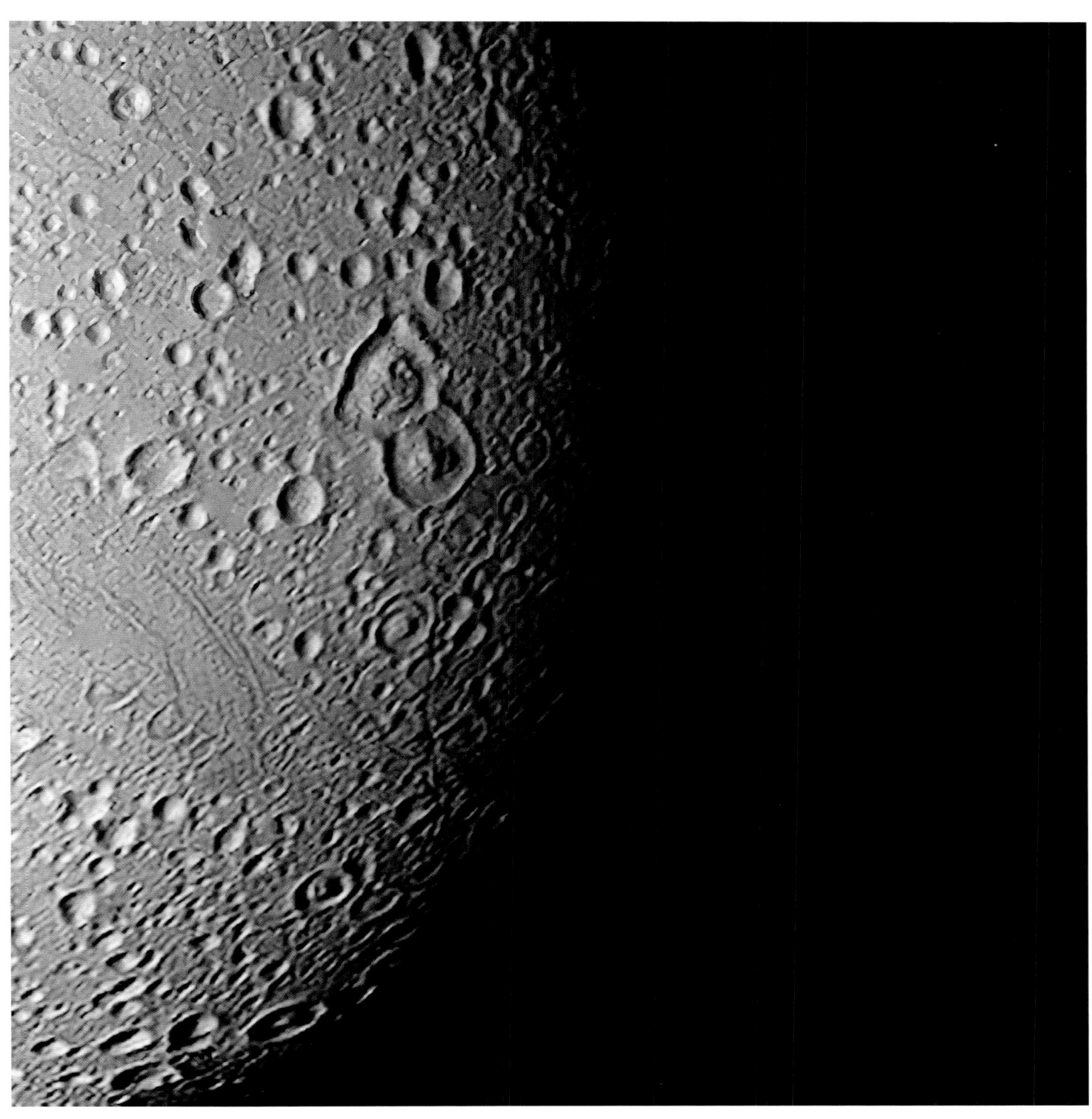

The part of Enceladus best seen by Voyager is the north polar
region, which is the oldest and most heavily cratered part. Based
on the grooves and the linear features one can deduce that the
satellite has been subjected to considerable crustal deformation.
This is the result of internal melting. The biggest crater seen here
measures approximately 20 miles (35 km) across.

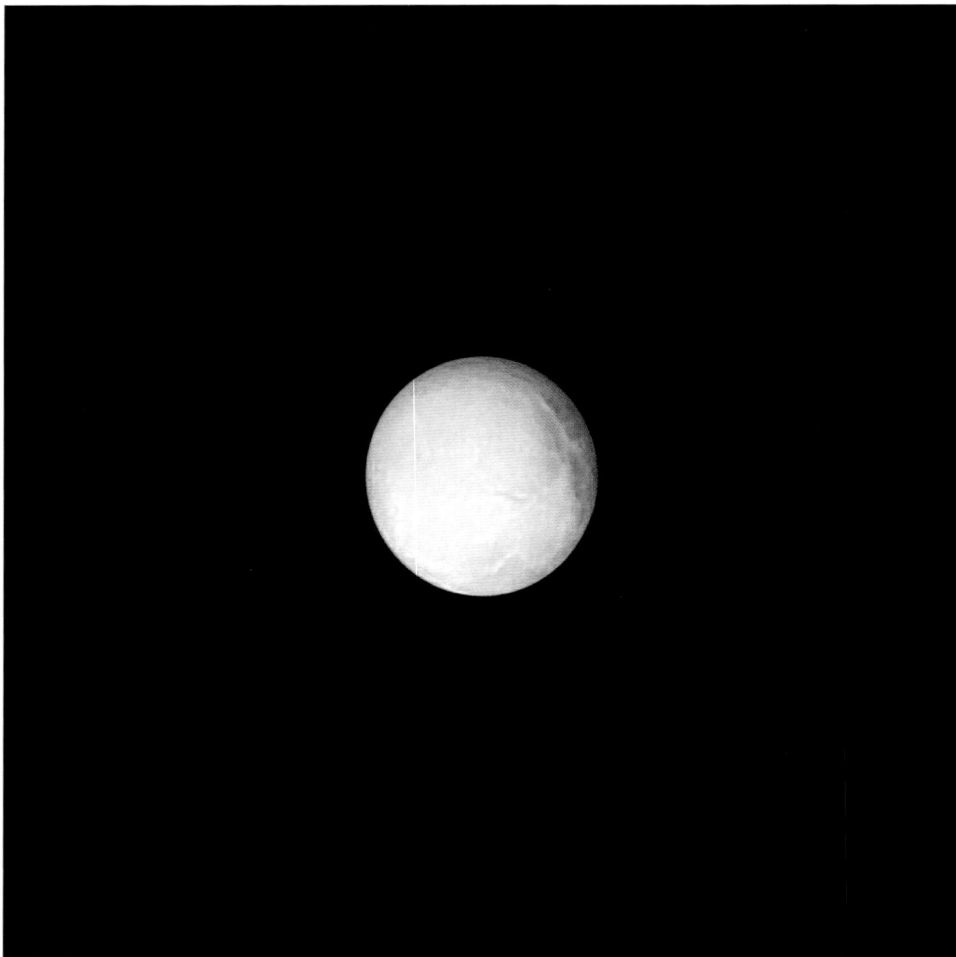

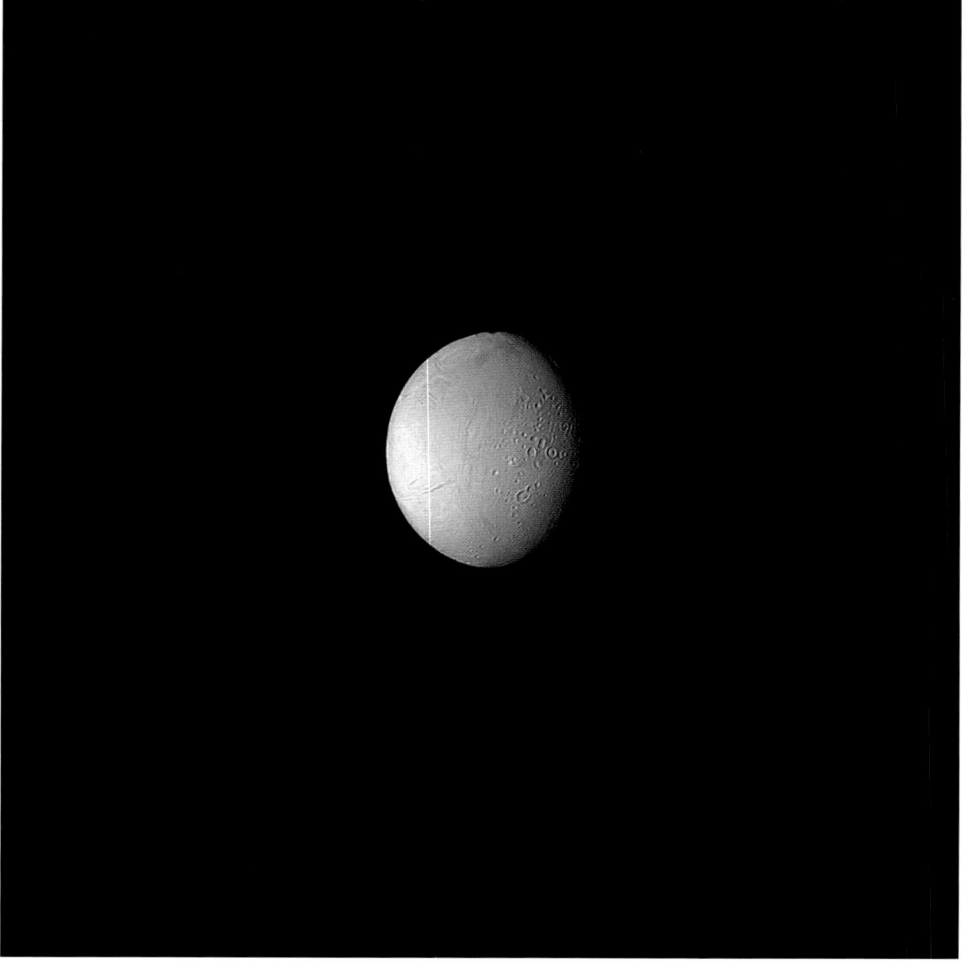

Right: Enceladus, Voyager 2,
August 26th, 1981, 02:31:58.

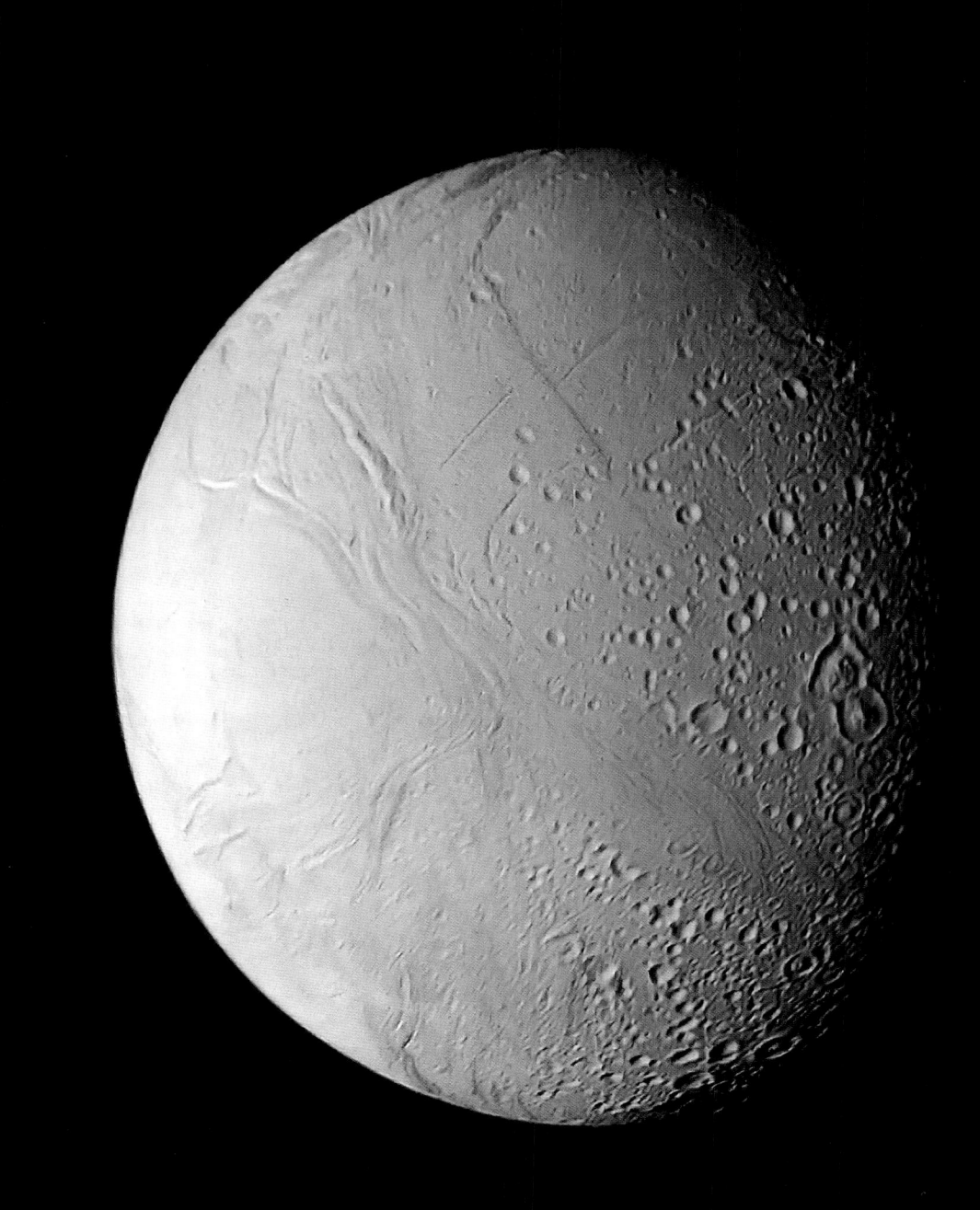

ONWARDS

The Saturn encounters enriched our knowledge and understanding of the Saturnian System greatly, providing vital close-range observations and high-resolution data that simply could not be achieved from any Earth-based observation.

To recap briefly some of the amazing findings: Voyager 1 discovered a number of new moons, in particular the remarkable shepherd moons Prometheus, Pandora, and Atlas, whose orbits keep Saturn's rings in a neat and orderly fashion. In spite of the fact that the images of Titan did not reveal as much as the Voyager team had hoped for, it did provide a new understanding of Titan's atmosphere. The faint G-ring was also discovered within the images sent from Voyager 1. The probe continued its observations, looking back at the planet for another five weeks after its closest approach, capturing beautiful images of the planet and its rings. After this point, the mission for Voyager 1 soon changed, turning its attention to the next phase of its journey, the eternal voyage towards interstellar space, to probe among the stars in the Milky Way and beyond.

Voyager 2's closest approach to Saturn provided countless new data and information, showing brilliant images of other moons and the detailed pictures of its rings, revealing the mysterious spokes and braids in the system. Moreover, while Voyager 1 was heading directly for interstellar space, Voyager 2 was only half-way through its Solar System mission. The close flyby of Saturn, with the gravitational assist, slingshot the craft into the direction of its next stop, sending the probe on a lonely journey to Uranus over the next four and a half years.

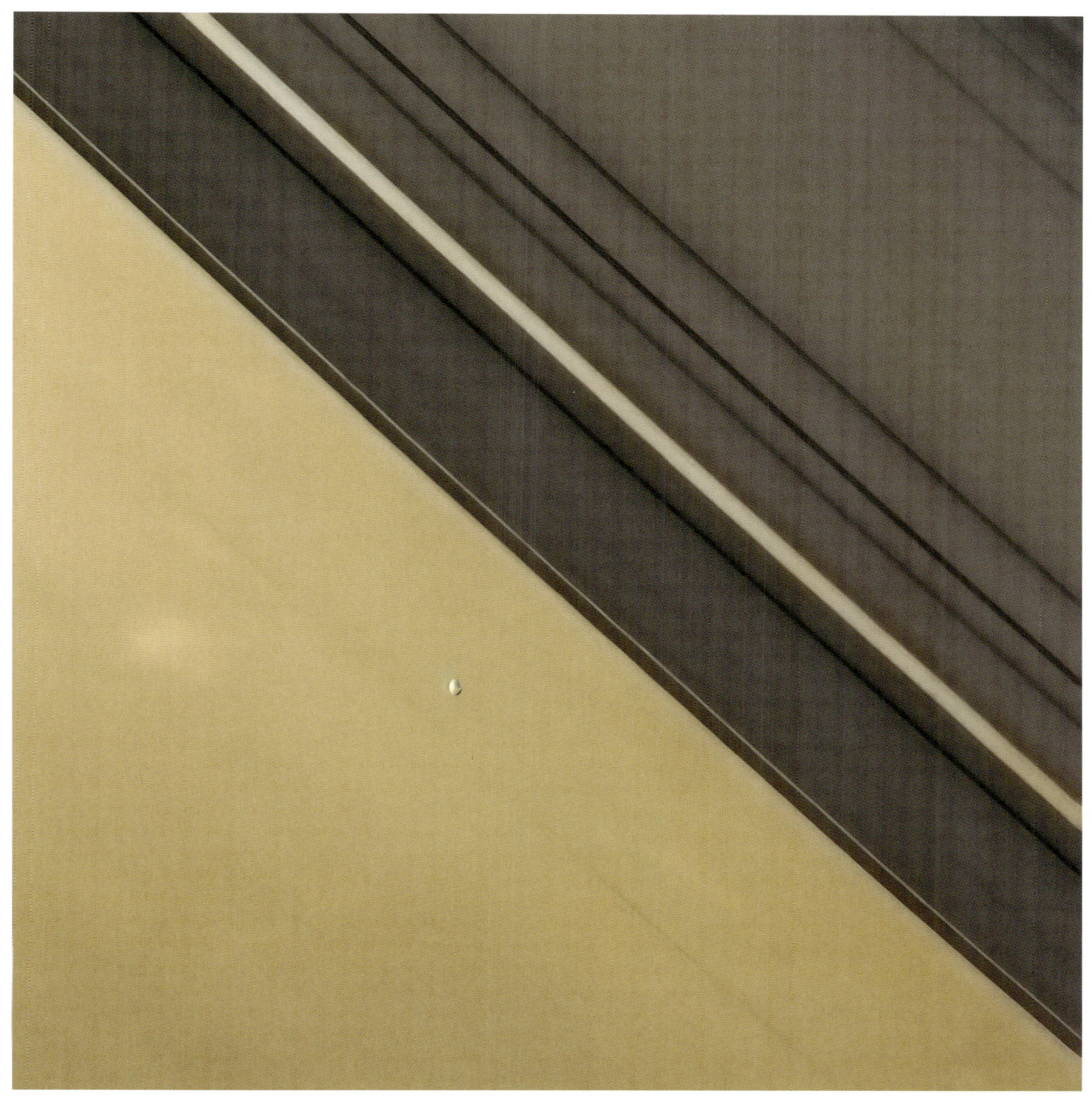

SIDEWAY SPINNER

INTRODUCTION

After leaving the Saturn system in 1981, it took Voyager 2 four and a half years to arrive at its next destination. In the meantime, our own world is spinning madly on. China announced the birth of its one-billionth citizen, the CD (Compact Disc) was introduced to the world, Michael Jordan was awarded "Rookie of the Year," and the first Apple Macintosh computer was put onto the market by brash young pioneers named Steve Jobs and Steve Wozniak.

We already know that the trip to Uranus and Neptune was not part of the original flight plan, but due to the success of both Voyager spacecraft, the now extended mission was on its way to Uranus. The five-year traveling time gave the Voyager team some well-deserved and needed downtime. It also gave many of them an opportunity to work on and develop new space exploration projects. Voyager however, was never far from their minds. No planet was as unknown and defined by mystery as this next giant. The five-year period thus became an exercise in patience and building up a sense of deeply felt anticipation. Uranus was coming up.

Right: this image is processed to bring out discreet clouds hidden beneath the Uranian haze.

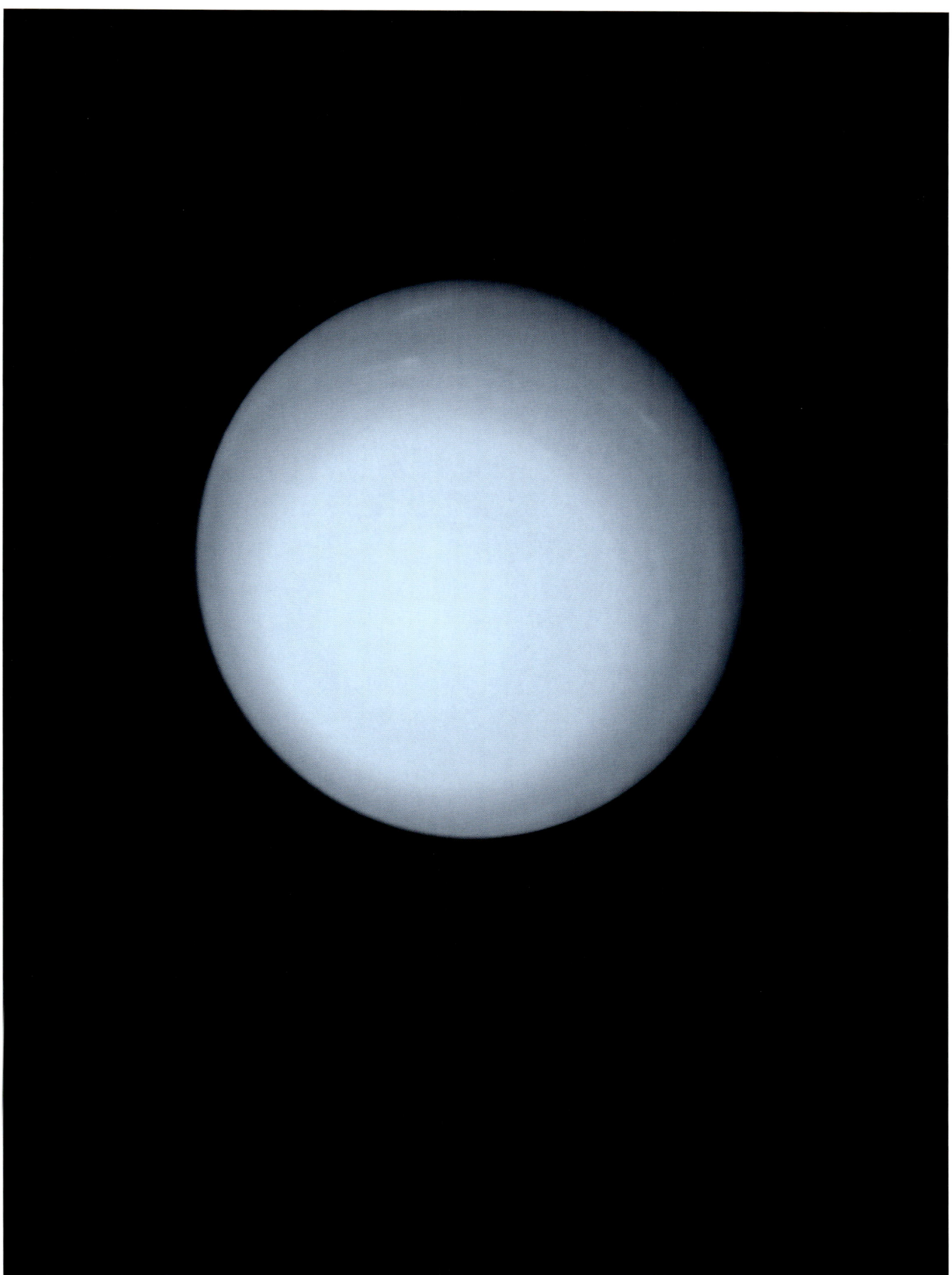

"An object is frequently not seen from not knowing how to see it, rather than from any deficit in the organ of vision. I will instruct you how to see them…"

William Herschel

AN ICE GIANT

Uranus is very far away. It does not require the services of Sherlock Holmes to figure that out. However, once you take in the figures, you realize it is actually very, very far away. As the Solar System is always in an orbit, the distance between Earth and Uranus is constantly changing. The closest the two ever get to each other is 1.6 billion miles (2.6 billion km), while the separation at its biggest is 1.98 billion miles (3.2 billion km). That's 21.3 times the distance between the Earth and the Sun. This is also the reason why it is so hard to see the planet if you try to look for it in the night sky. The planet hardly receives any sunlight.

Uranus is the third largest planet in our Solar System and one of the four gas giants. Comparing it to our little blue marble puts this into perspective. The equatorial radius of Uranus is four times that of the Earth, measuring in around 15,881 miles (25,559 km), and you can fit roughly 63 Earths inside this giant. It takes Uranus just over 84 Earth years to fully orbit around the Sun, and as revealed by Voyager 2, 17 hours and 14 minutes to rotate fully on its axis.

Voyager 2 discovered that the atmosphere around Uranus is very similar to that of Jupiter and Saturn.

While mainly composed of hydrogen, helium, and methane, it is the dominance of the latter that gives Uranus its bright green-blue color. A steamy, hot, dense fluid of "icy" materials, water, methane and ammonia sitting atop a rigid rocky core together make up the planet's mass. Although this core can heat up to 9,000 degrees Fahrenheit (4,982 degrees Celsius), it is the "icy" materials that give Uranus the name of being one of the two ice giants in our Solar System. It is no coincidence that it holds the accolade of having the coldest recorded temperature on a planet in the Solar System.

Uranus carries a few distinct attributes. Together with Venus, it is the only planet that rotates in the opposite direction compared to other planets. Also, unlike any other planet in our Solar System, Uranus rotates on its side. This has to do with a form of physical confrontation during formation. It is believed that a giant collision with another large moon or planet size body knocked it on its side early on in its life. The distance, the low visibility from Earth, and the sideways rotation made Uranus an odd and mysterious planet to astronomers. Mysteries that were turned into facts with the penetration of the Uranian System by Voyager 2 in 1986.

A NEW DISCOVERY

Our Solar System single-handedly doubled in size on March 13th, 1781. William Herschel announced to the world that he had discovered a new distant planet within our Solar System. The first planet to be discovered since ancient times and the first planet to be discovered by looking through a telescope. Herschel rose to fame not as a connoisseur of the cosmos, but as a respected musician, notably composing more than 24 symphonies and 14 concertos. Herschel's intellectual curiosity eventually led him to astronomy.

While it started out as a hobby, Herschel became serious and determined about astronomy. His ambition and drive left him frustrated with the standard telescopes and instruments available to him on the market. Based on Isaac Newton's original designs, Herschel decided to take matters into his own hands, and build telescopes himself. The result became a 7-foot long (2.1 m) telescope that he constructed, with which he started making systematic recordings of the night sky from his home Observatory in Bath, United Kingdom. As he made his regular observations, he noticed a strange "non-stellar" disk traveling slowly through the sky. Unlike a star, which appears as a point of light when viewed through a telescope, this object was round in shape. Hershel's first thought was that he was

the proud discoverer of a new comet. However, the question arose after a period of further observation by himself, his sister Caroline and some of their peers. With new calculations and orbit plotting, they could now declare the object planetary. A planet with an orbit that was much further in distance from that of Saturn.

Herschel originally named his discovery Georgium Sidus (Georgian Star), to honor King George III. His peers did not accept his modesty and commonly referred to the planet as "Herschel." After some discussion from astronomers around the world, an alternative name was chosen as the final version: Uranus, the Greek God of the sky. One year later, with a new and improved 18.5-inch (47 cm) diameter telescope, William Herschel took another close look at his discovery. Together with his sister Caroline, they were able to make out two moons orbiting around Uranus. This time around, Herschel's son John had the honor to think of the names for these new additions to our Solar System. They were named Oberon and Titania— characters from Shakespeare's A Midsummer Night's Dream. During their further observations, the strange sideways orbit rotation of both Moons was noticed traveling around Uranus. This anomaly would kick-start the discovery of a string of mysteries raised by this planet.

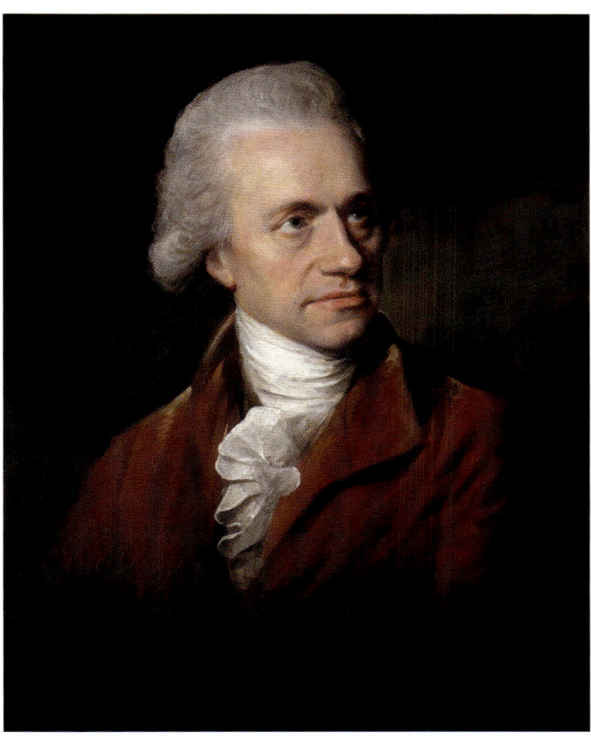

Portrait of William Herschel, painted by Lemuel Francis Abbott in 1785.

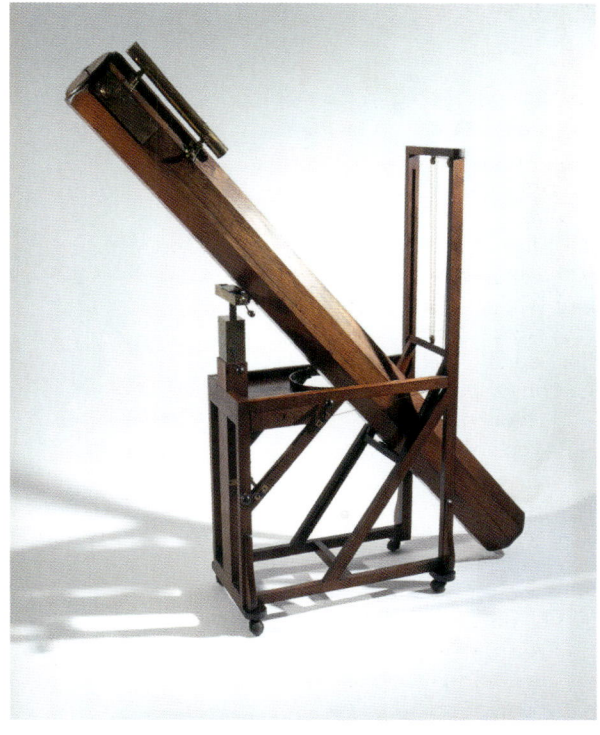

A replica telescope based on the one Herschel discovered Uranus with.

THE ONE TAKE TOUR

The Voyager 2 navigation team knew they had a challenging and high-risk assignment on their hands when laying out the flight plan for the flyby of Uranus. Unlike the longer flyby approaches of Jupiter and Saturn, this visit would see Voyager 2 darting straight through the Uranian system at 51,000 mph (82,000 km/h). This meant that the imaging team was given a mere ten hours to take all of the close-up photographs and other data observations of Uranus, its rings and moons. There was simply no room for error as the probe hurtled through the system. It was truly a "one take-only" kind of operation.

The short amount of time for this close-up encounter was not the only problem. With its moons orbiting around the planet in a different direction to that of moons found on other planets, the mission planners had to ensure that the planned trajectory would allow a safe distance for Voyager 2 to pass through the Uranian system without a potential collision. To add further challenges to the trajectory, the planners also had to approach Uranus at a very specific angle in order to make use of the gravitational field. This was required to give the spacecraft the needed gravity assist to slingshot on to its last stop: Neptune. This meant that the flyby would only allow a close-up pass of just one of the five known moons at the time, Miranda.

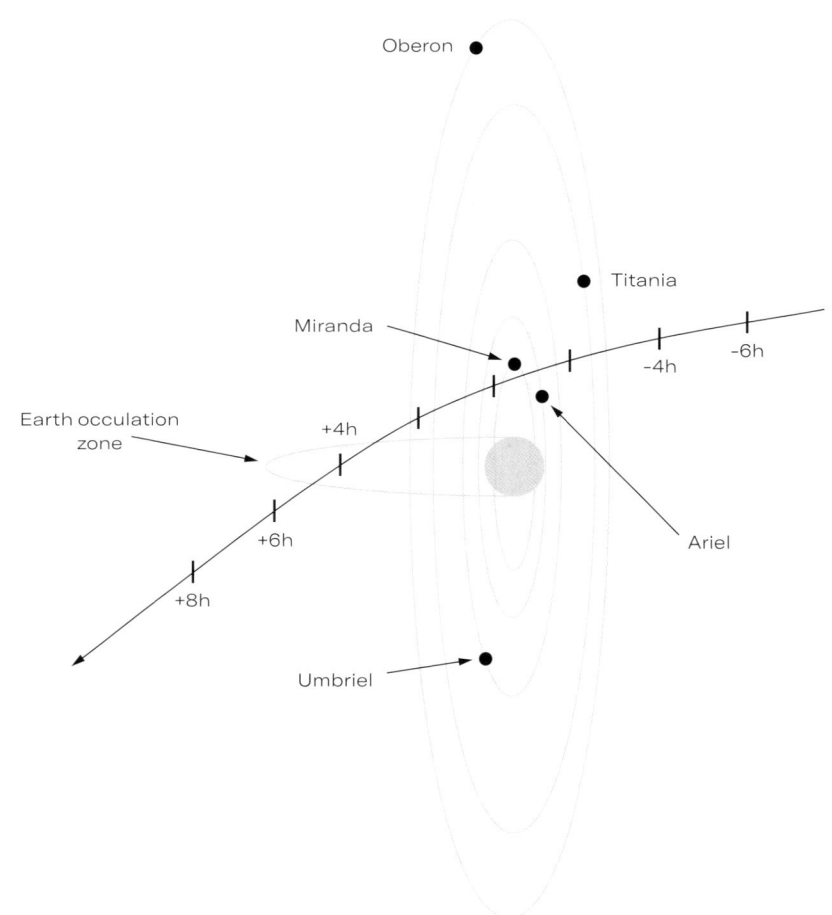

The trajectory and timings of Voyager 2's flyby through the Uranian System.

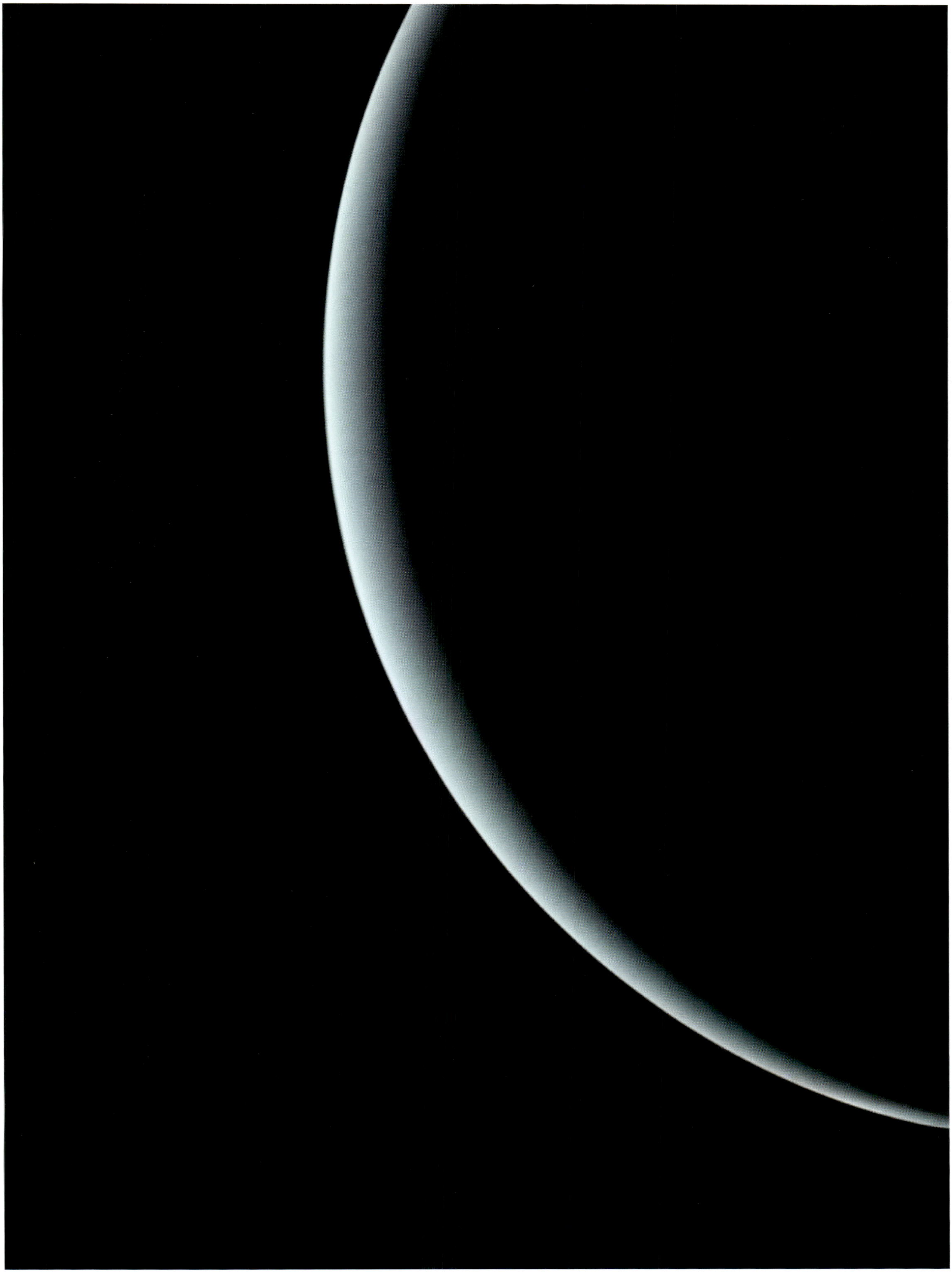

FIRST GLIMPSE

Voyager started recording its first distant observations of the Uranian system in November 1985, at around 6 million miles (9.7 million km) away from the planet. Image analysts scoured images as they were being broadcast back to Earth. With every little bit closer to the planet, they learned a little bit more. Towards the end of December, just a month before Voyager would pass Uranus and its closest point, a new moon was discovered. Originally called "1985UI," it was later renamed officially as "Puck," following in the tradition of Uranian moons being called after characters from Shakespeare's A Midsummer Night's Dream. As Voyager 2 continued its approach, it dawned on the mission team that with a slight change to the pre-planned schedule, images could be taken of Puck. The schedule was adjusted and the probe reprogrammed. During the Miranda encounter, the camera rotated and got a faraway yet crystal-clear glimpse of the newly discovered moon.

A little over two months later, on January 24th, 1986, Voyager 2 had arrived at its closest crossing point with Uranus. After much suspense and anticipation, the imaging camera could finally start to spring into action. The probe was passing Uranus at 50,600 miles (81,500 km) away. This one-take tour meant that the cameras and sensors only had ten hours to shoot and record the data before blasting out of the Uranian System and heading forward to its next stop.

During these brief ten hours, an invaluable amount of data was collected and some of the most beautiful space photography was captured. After further analysis, these images revealed ten new moons and discovered several further rings around the planet. Some of the most remarkable and interesting findings were how varied and unique the Uranian moons were, with Oberon and Umbriel showing vast craters scars that they have picked up during their 4-billion-year-old lifespan.

Titania and Ariel both reveal a turbulent and active past. Huge surface fractures that were up to 3 miles (4.8 km) deep and large rifts, all forming tell-tale signs that these moons (once upon a time) had very active cores, internal heating and tectonic shifting. However, the most remarkable images came from Miranda, the smallest of the big five moons and the one that Voyager 2 would pass closest. The images would show Miranda to be one of the most scientifically interesting celestial objects in our Solar System.

Right: Uranus, Voyager 2, February 1st, 1986, 06:54:15.

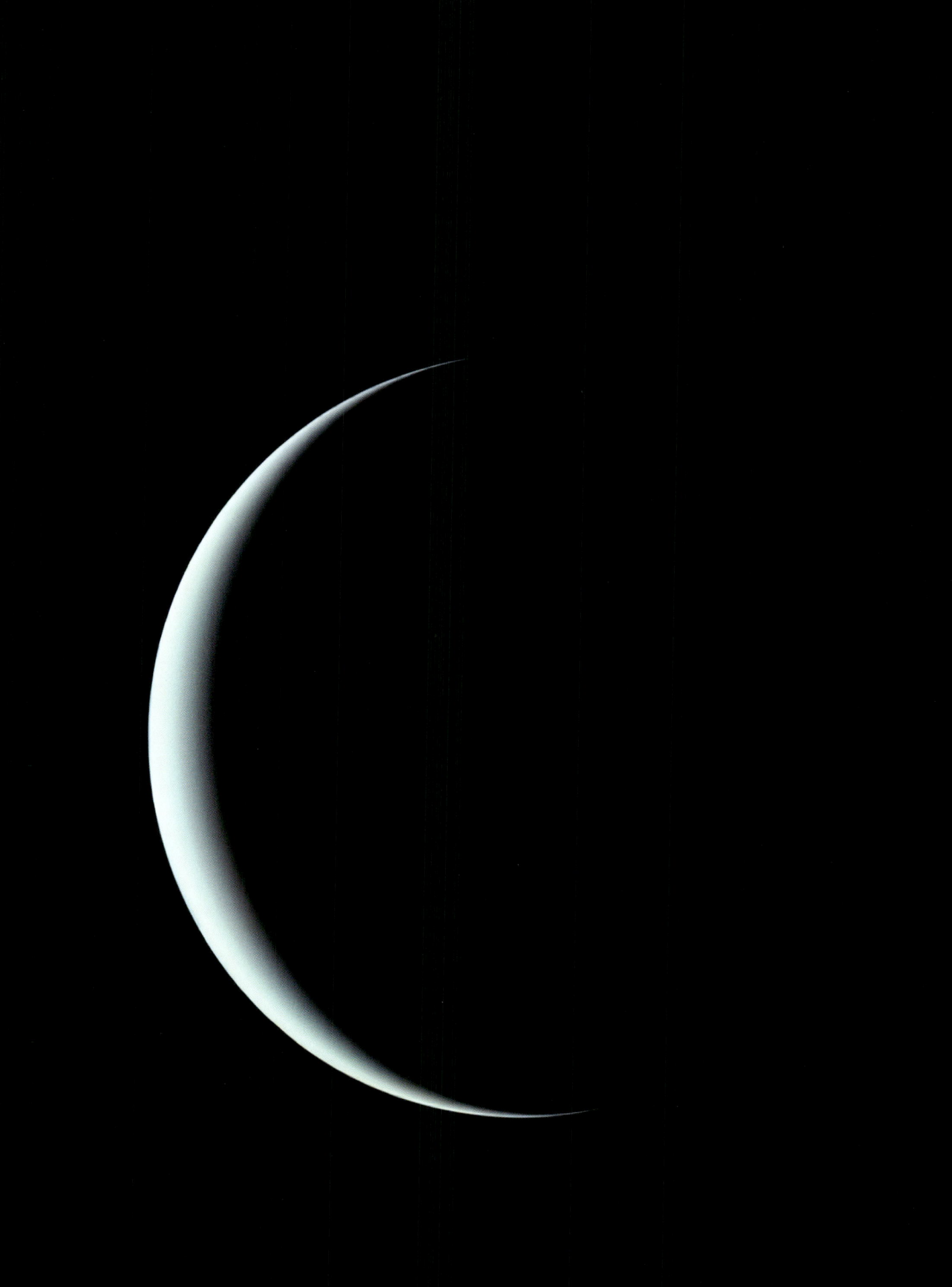

SUMMER SOLSTICE

Voyager 2 showed the world a series of stunning images that show a softly colored blue-green gradient sphere that is Uranus, capturing unrivaled images in the history of space photography. For more specially trained eyes however, these images lacked a certain scientific wow factor.

Uranus turned out to be somewhat of a dull boring planet with an atmosphere that had no remarkable features. Yes, astronomers and the Voyager team already knew what to expect with the close-up shots of Uranus. There weren't going to be amazing cloud bands of varying patterns and colors like found on Jupiter, nor a prominent and defined ring system like that seen orbiting around Saturn. Even so, it was claimed that the Voyager team was still surprised at how bland the planet was.

It turned out that Voyager 2 was passing by the planet during the peak of the southern hemisphere's summer, each season lasting for 21 Earth years. This meant that all the remarkable cloud features around the pole were hidden by a haze of methane gas in the upper atmosphere. Unfortunately, there was not the equipment on board that could penetrate through the haze to show any of the hidden planet below.

The planet is tilted at an extreme angle and has a nearly sideways angle with respect to its orbit around the Sun. This means that during the winter-summer season, the pole facing away from the Sun is in darkness for 21 years, and the other pole has full daylight for 21 years. Spring and Autumn time on Uranus is a completely different story. The planet, which would have been either in total darkness or in total light for decades on end, would now experience both nighttime and daytime during the seventeen-hour day. These extreme seasons have a dramatic effect on the cloud-top gases being either heated or cooled. At the time of writing in 2020, Uranus was well into its northern spring. If Voyager 2 were to fly by today, the probe would see a much more varied atmosphere — with bands encircling the planet and vivid dark and light spots of clouds.

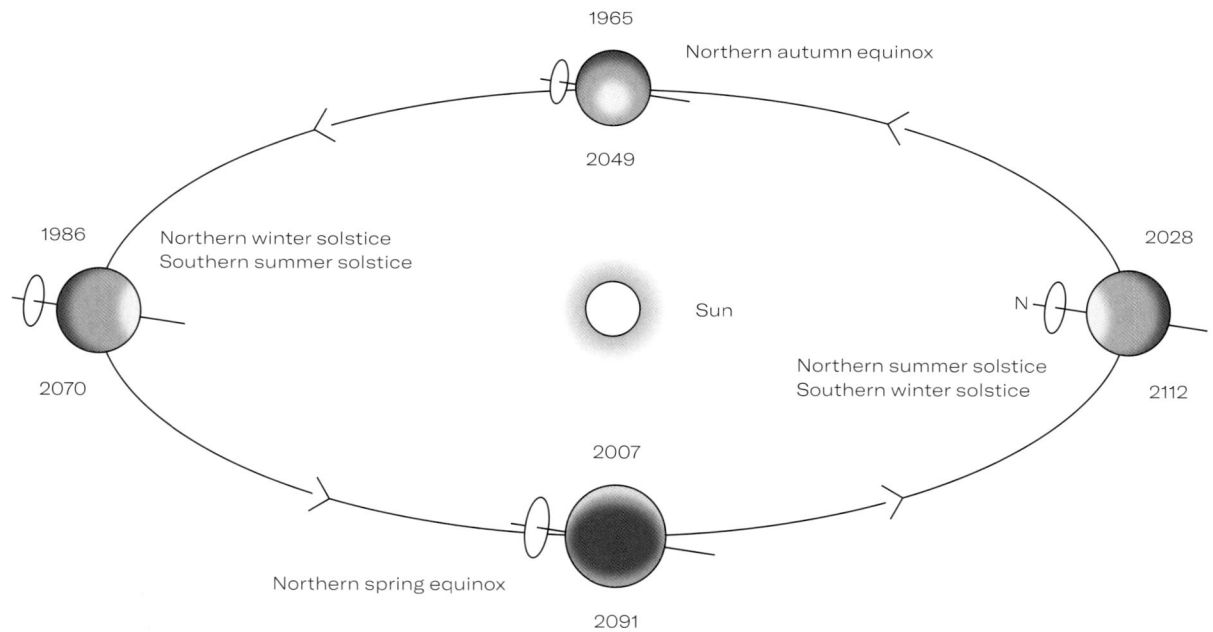

1965
Northern autumn equinox

2049

1986
Northern winter solstice
Southern summer solstice

2028

Sun

N

2070

Northern summer solstice
Southern winter solstice

2112

2007

Northern spring equinox

2091

This diagram shows how the 98° axial tilt places its equator almost perpendicular to the ecliptic. This results in the planet experiencing the most extreme seasonal variations in the Solar System. The equatorial regions have two warm seasons (around the equinox periods) and two cold seasons (solstices) each Uranian year, resulting in the poles being alternately kept in complete darkness for 42 years at a time.

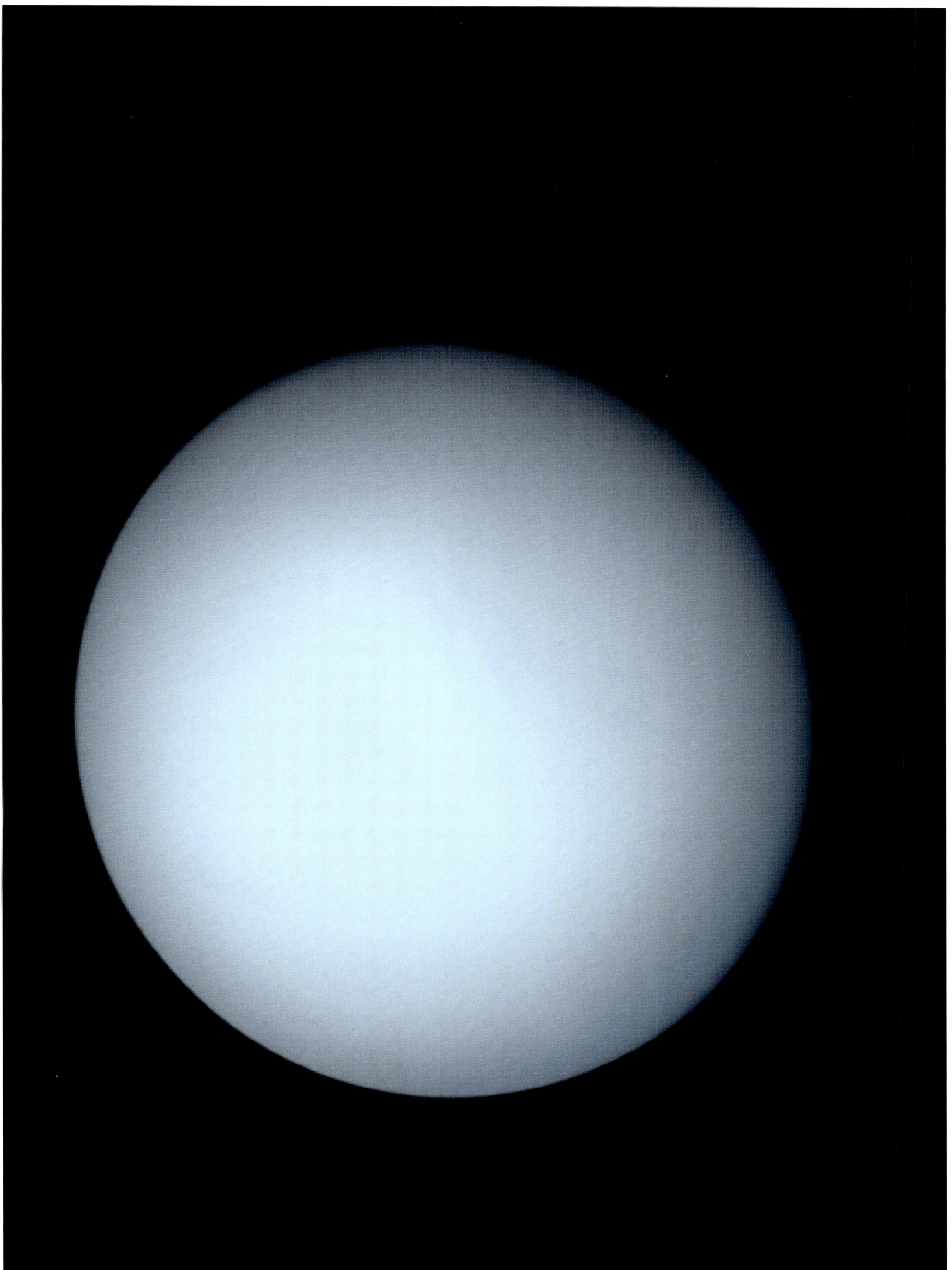

RINGS OF URANUS

On March 10th, 1977, the rings of Uranus were discovered by James Elliot, Edward Dunham and Jessica Mink. Taking into account the entire history of astronomy, it's quite astonishing that this was only a short nine years earlier than when Voyager 2 was to pass by the planet. The team of astronomers led by Elliot planned an observation experiment on board the NASA Kuiper Airborne Observatory, a modified C-141A Jet. The plane was able to fly high into the stratosphere, above the clouds and water vapor, giving clearer and uninterrupted views of the night sky.

Their plan was to observe the occultation—this is when one object is hidden, in this case the distant star SAO 158687, by another object— Uranus—which passes between that star and us. This would allow Elliot and his team the opportunity to study the planet's atmosphere. When their observations were analyzed, it showed that the star disappeared briefly from view five times, both before and after it was eclipsed by the planet. From this, it was concluded that a system of five narrow rings was orbiting around Uranus.

When Voyager 2 visited the planet, it gave a much more informative and detailed picture of these rings. The first pictures show the rings to be very dark, and composed of varying sized objects, ranging from a couple of inches (5 cm) to 10 feet (3 m) in size. The pieces of ice-carbon materials were seen to be further darkened by the solar wind radiation and that of the planet's own magnetic field.

Voyager 2 was able to capture the clearest and scientifically most interesting images of the ring system after it traveled past the planet. Mission planners had programmed the camera to turn around, looking back towards the Sun, similar to how the camera was programmed to observe the Jovian rings. This maneuver paid off, with images and stellar occultation data collected showing a further two thick rings and a thin dark sheet of material in between the dark bands. This increased the total number of rings encircling Uranus to 11.

The images indicated that the rings were young and evolving, as ring debris had not yet settled down into an ordered and even distribution. Much of the rings could be seen to vary in width and circumference, and in some sections of the ring, there is very little material, which you can see in the breaks and gaps. It was concluded that the material found in the ring was probably from a moon or celestial object that got too close to the planet and was then ripped apart by its gravitational forces, scattering the material into an orbit around the planet. These rings, just like those of Saturn, will probably be only a temporary feature, which in a few million years will dissipate and vanish into the surrounding space.

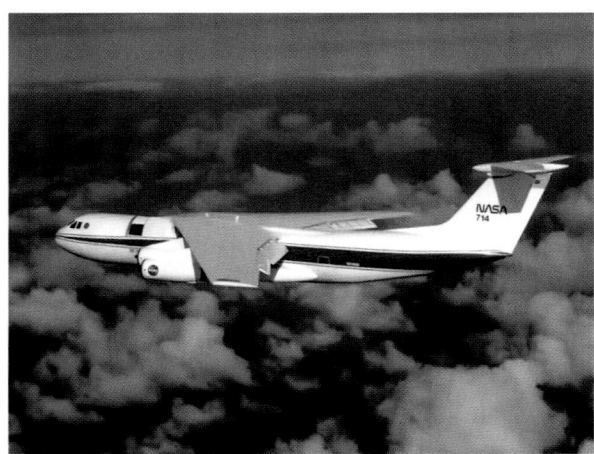

Gerard P. Kuiper Airborne Observatory.

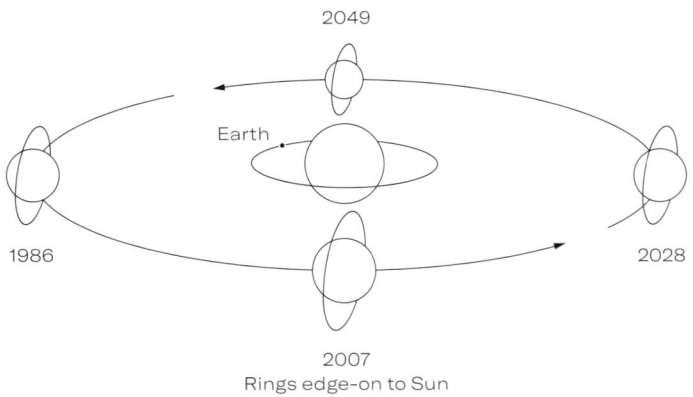

The position of the ring system around Uranus between 1986 to 2049.

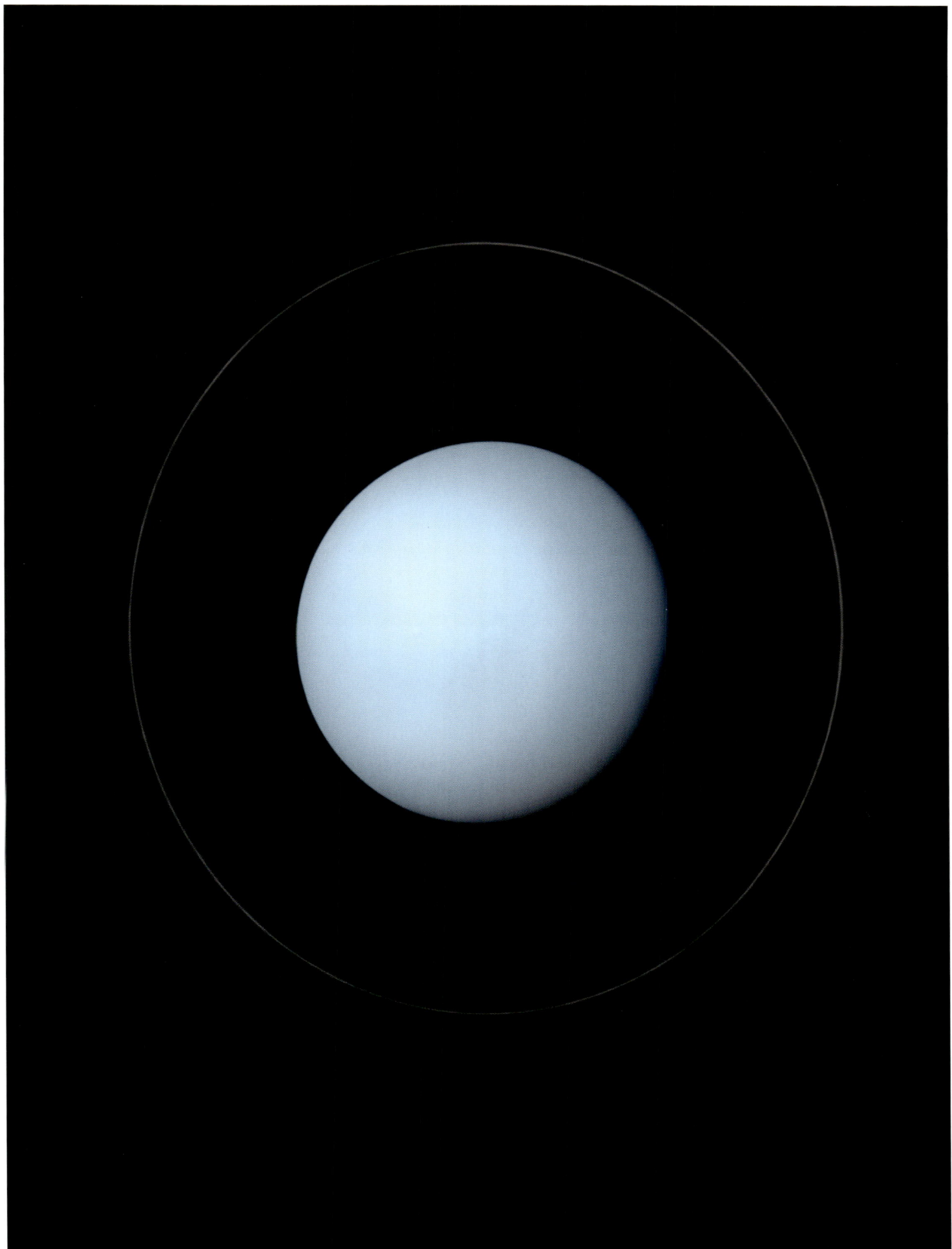

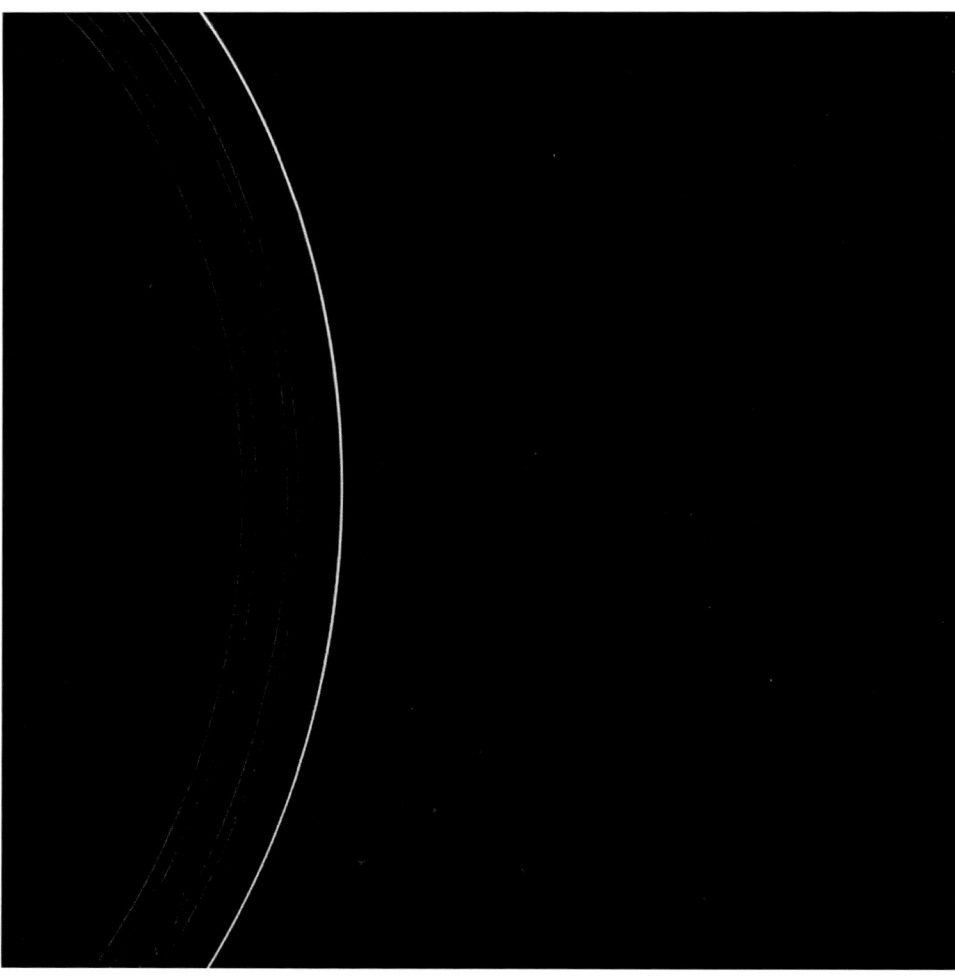

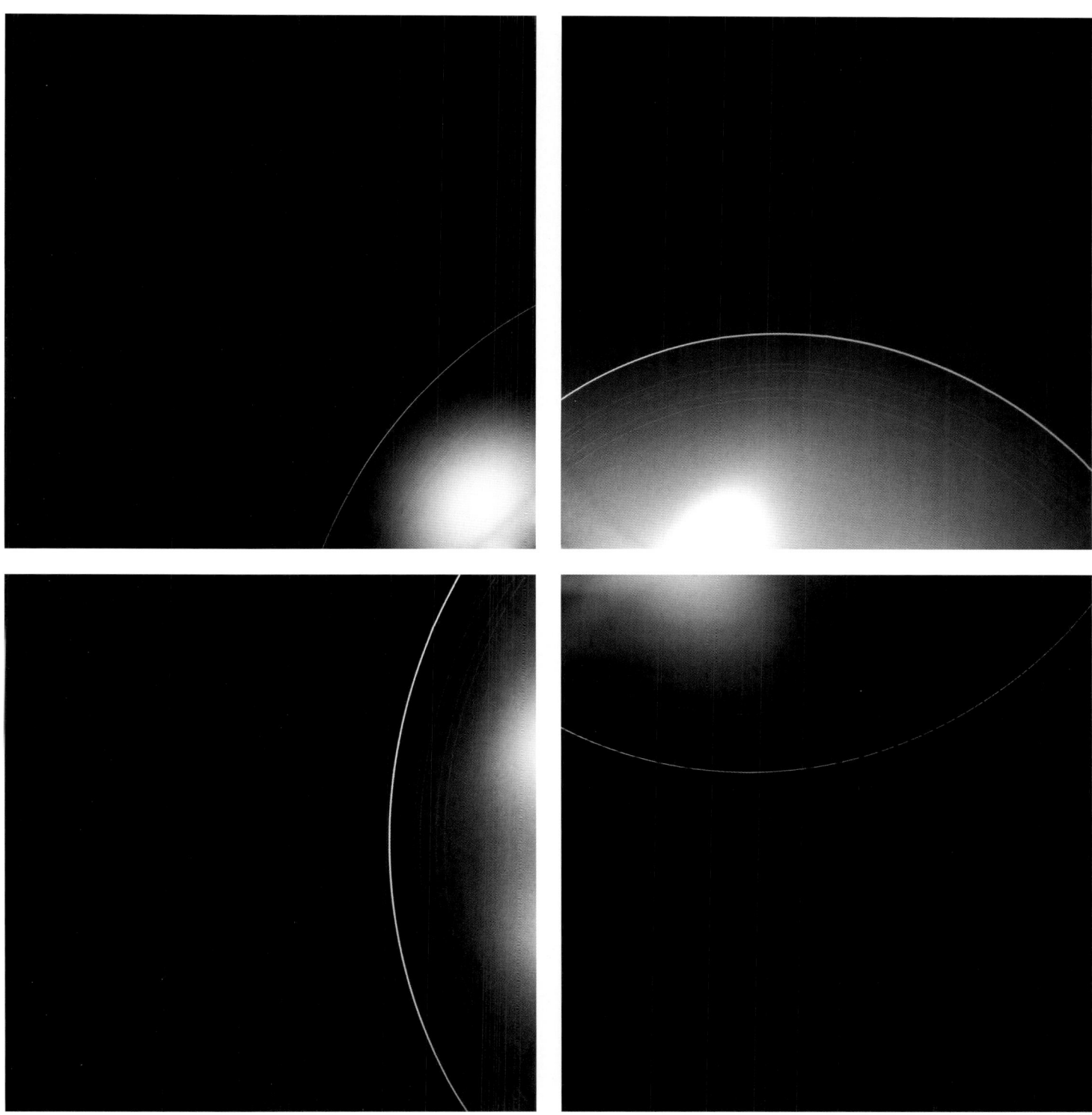

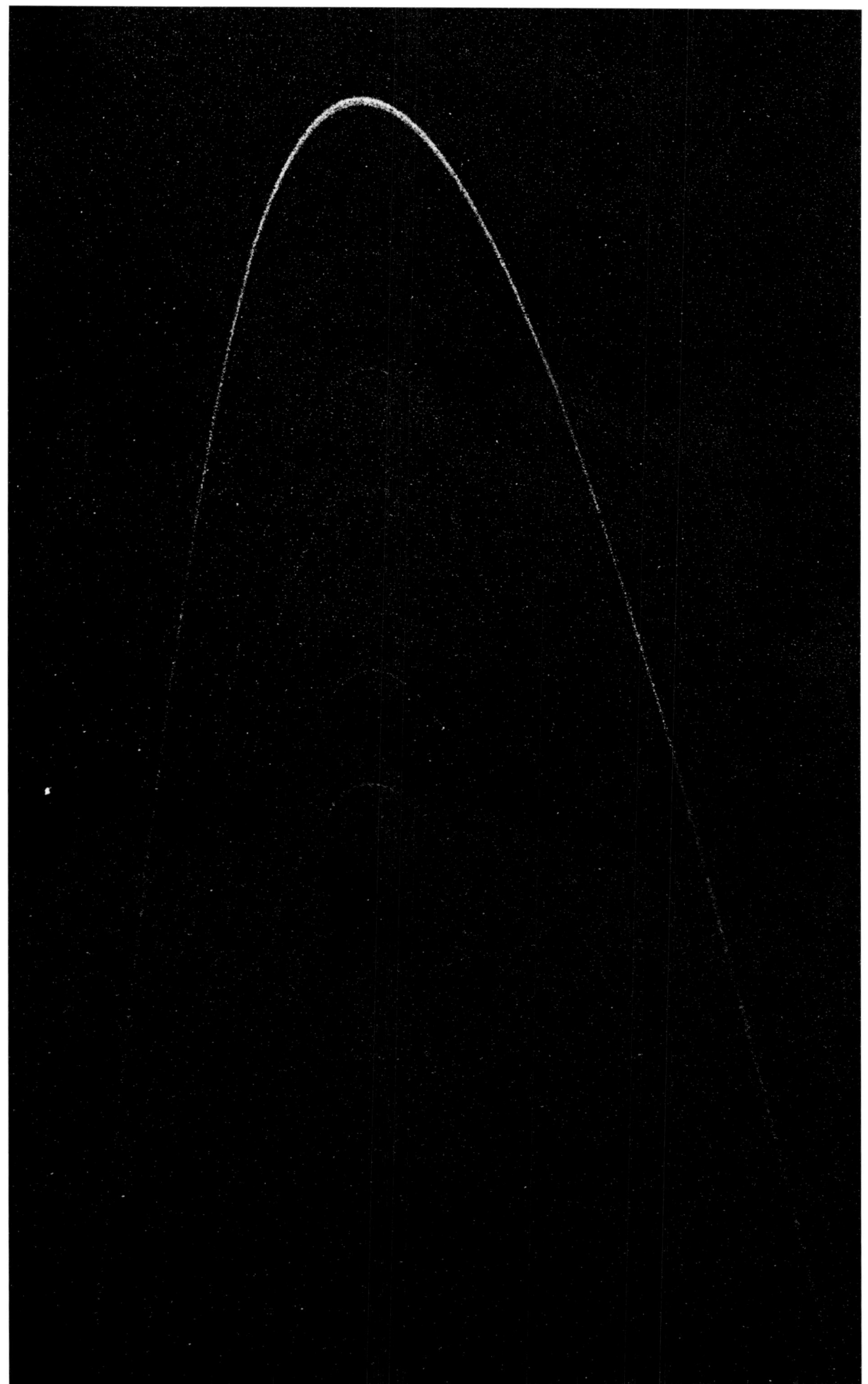

The rings of Uranus presenting
themselves in front of the
cloud tops.

Due to the highest phase angle we get a glimpse of the lanes of fine
dust. Despite the fact that all the previous known rings are present here,
some of the bright dust lanes were not previously seen. A spectacular
observation made possible by the combination on a unique geometry
and a long, 96 second exposure.

MANY MOONS

In spite of the flyby being a brief ten hours, the images of the Uranian moons captured did not disappoint the Voyager team back on Earth. Although it took some time to receive them on Earth, the radio data signals took 2 hours, 45 minutes to travel back to Earth – this did not spoil the fun. Imaging team member Larry Soderblom was quoted to be "completely astounded by the diversity of the Uranian satellites" when each of the moons were seen for the first time.

As these tiny circular discs started coming into view, the images first showed five large, icy moons and later a further ten new, smaller moons, which were hidden deep inside the images. These ten new moons were all named after Shakespearean characters, with Belinda being the exception, whose name comes from Alexander Pope's poem "The Rape of the Lock".

Images of Oberon and Umbriel showed countless crater marks, indicating that little has changed to the surface features of the moons over their four billion plus lifespan. The dark color seen on Umbriel was different to that of the other moons, and to this day, the reason for this color anomaly is still unknown. The other moons all showed other varying geological features. Titania illustrated a history of an active interior core, evident from the 3 mile (4.8 km) scars, fractures, and cliff faces visible in the photographs. Similar evidence on Ariel also showed a past active life, with large rifts on its surface and what looked like icy, cryovolcanic flows. All indicators of past tectonic movement and radiating heat out from the moon's core.

It was the tiny, 300 mile (482 km) wide Miranda however, that harbored the most surprising and notable images. The tiny moon made from ice and rock, to which Voyager would pass closest during its flyby, was seen to have one of the most extreme and varied landscapes of any object in the Solar System. It was the opinion of the scientific community that small celestial objects such as Miranda were too small to have had active interior heating, and what heat it may have had would have radiated out rapidly into the space. Miranda quashed this theory, its surface being unlike any other object in the Solar System.

Thought to be made of a composition of ice and rocky material and having an average temperature of -335 °F (-187 °C), Miranda is a real mix-match of different terrains that all look like they have been glued together. Heavily cratered areas, flat terrains and large patches of contrasting colored curving ridges and flows, all alternate each other. However tiny the moon is, these were all clear indicators of a landscape that has been altered, twisted, and distorted over its lifetime through internal heating, cryovolcanic activity, and meteor strikes. Some of the most spectacular landscapes were discovered, towering giant cliff faces and canyons, some as deep as 6 to 10 miles (9.6 to 16 km), around 12 times deeper than the Grand Canyon. These images led astronomers to believe that Miranda had once been smashed into pieces, and somehow managed to "glue" itself back together again over its long history. More recent theories however suggest that an upwelling from the moon's core of partially melted ices may have created the unique and peculiar surface features.

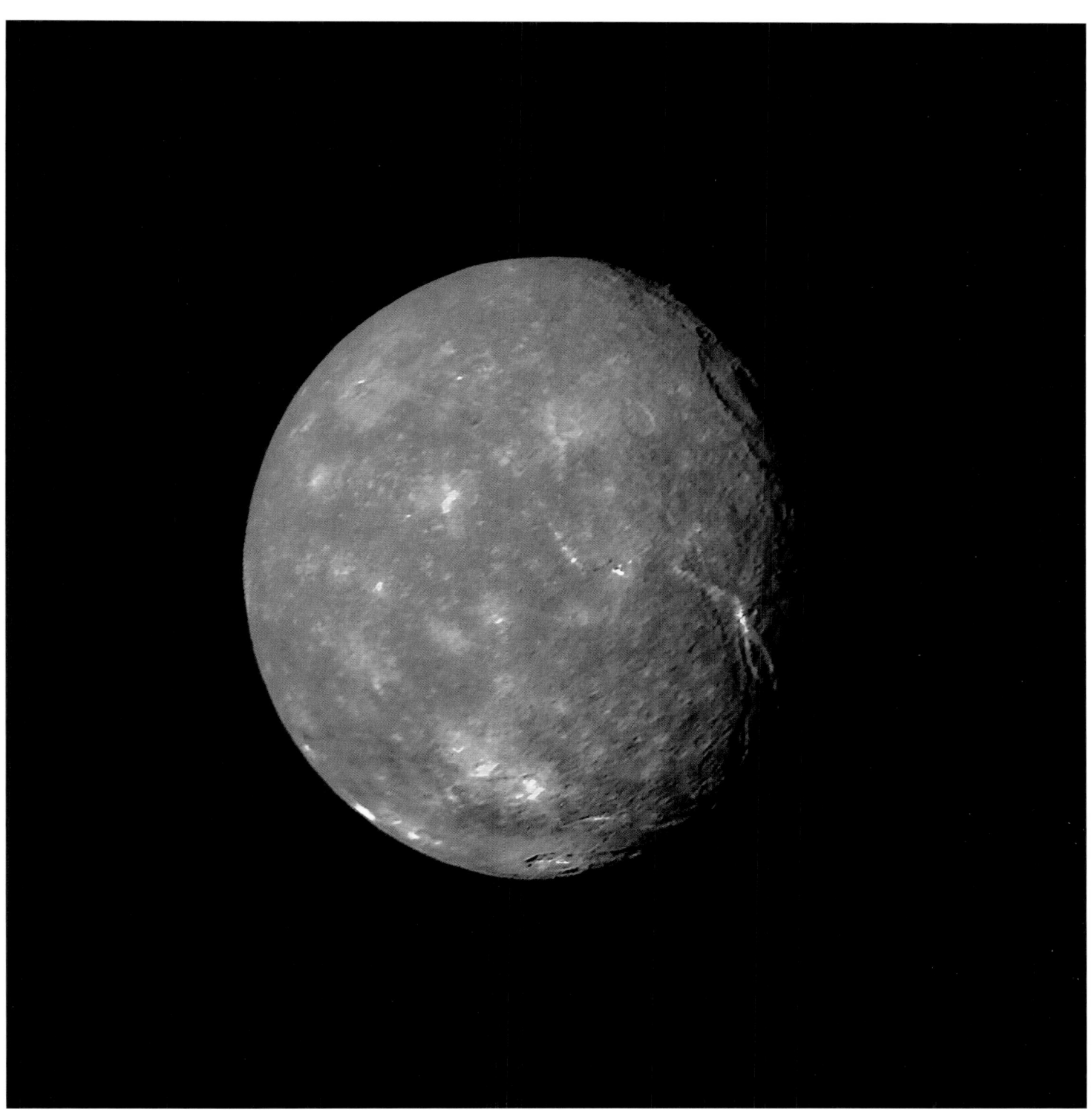

ALL JUST A BLUR

How do you take a photograph of a planet with a camera mounted to a space probe that is traveling at over 51,000 mph (82,000 km/h)? How do you make sure that your image comes out crystal-clear, and not blurry, like when you try to take a photograph from within a traveling car? Moreover, how do you know the lighting will facilitate proper viewing without over- or under-exposure? This was a big challenge for the image engineers working on the Voyager project, who had already noticed a small amount of blurring, or more correctly termed "image smear," on some of the low-light images taken during the Jupiter and Saturn flybys. Nevertheless, that would be child's play compared to the challenge Uranus posed, since it had four times less sunlight than Saturn.

The low-light meant the camera would have to expose the image sensor for longer in order to capture the necessary pictures. This required leaving the shutter open for four times as long, whilst traveling 18,000 mph (29,000 km/h) faster than when passing Jupiter. These factors massively increase the chances of image smear. Luckily, the team had five years to work out a solution whilst Voyager 2 was traveling from Saturn to Uranus. The Voyager team re-configured and programmed the probe during its journey to Uranus. The team had previously noticed that some image smearing had occurred when the mechanical moving parts of the data tape recorders whirred into action. This caused the spinning tapes to jolt and shake the probe while in the process of taking a long exposure photograph.

The team came up with three ingenious technical solutions to solve this problem. First, they reprogrammed the flight software to move the probe to slew, changing its attitude and orientation in a slower and smoother way. This kept the spacecraft steadier and more stable as the photographs were being taken. An Image Compensation Program was installed on Voyager's flight computers, which would anticipate and compensate against the movement caused by the data tape-recorders. The spacecraft would gently spin in an opposite motion relative to the subject when taking a photograph. This opposite movement in effect made the image subject to pass by the lens more slowly, further reducing the risk of image smear.

The second major alteration the team rolled out was to reprogram the computers to make the camera save the images directly to the backup computer. This way the main computer did not have to process and compress them, preventing an overload due to extra image data while it also had to handle all the other flight data recordings. This allowed the camera on Voyager 2 to take a picture 70% faster.

Thirdly, a new data encoding scheme using the Solomon Data Encoding Box, that was fitted onto the probe but which had not been put into service during the Jupiter and Saturn flybys, was activated and used for the first time. The new encoding scheme was more efficient and broadcasted data with a much lower risk of data corruption. This was an important change as the vast distance between Uranus and Earth meant that there was a higher risk that the data packages being sent across the Solar System could be transmitted poorly due to the weaker radio signals.

Overall, it was an amazing technical feat to program a spacecraft – with very rudimentary technology by today's standards – that was over 1.6 billion miles (2.6 billion km) away. These smart and intelligent reprogramming adjustments allowed Voyager 2 to capture the remarkable and never seen before images of the Uranian System in a clear and vivid way.

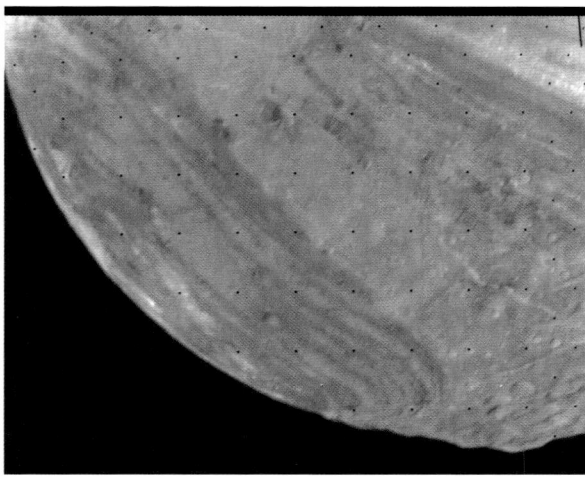

The raw image, with original reseau marks. Taken by Voyager 2, image smear can be clearly seen in the soft, blurry areas around the bottom left section of Miranda.

Right: Miranda, Voyager 2, January 24th, 1986, 16:36:43.

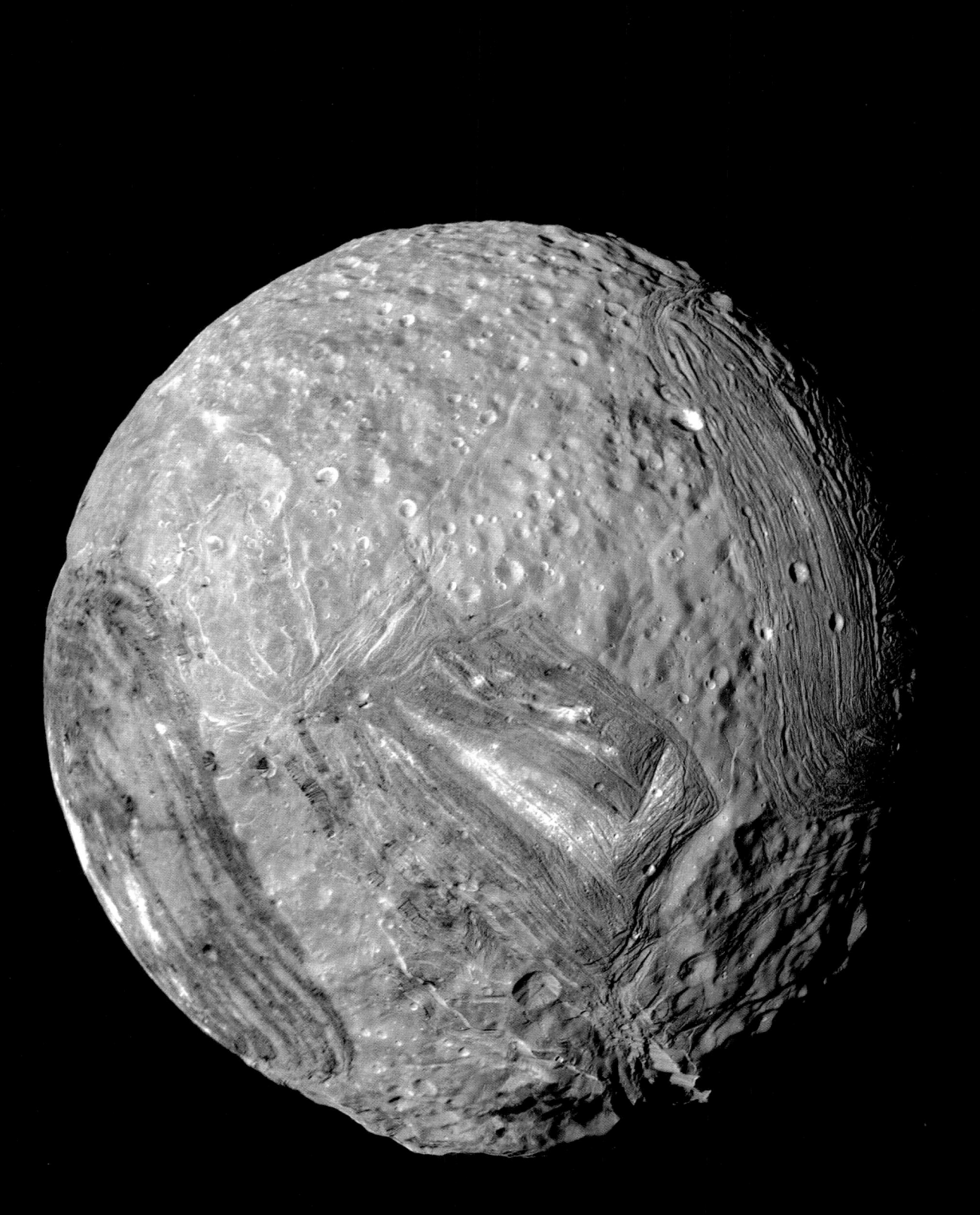

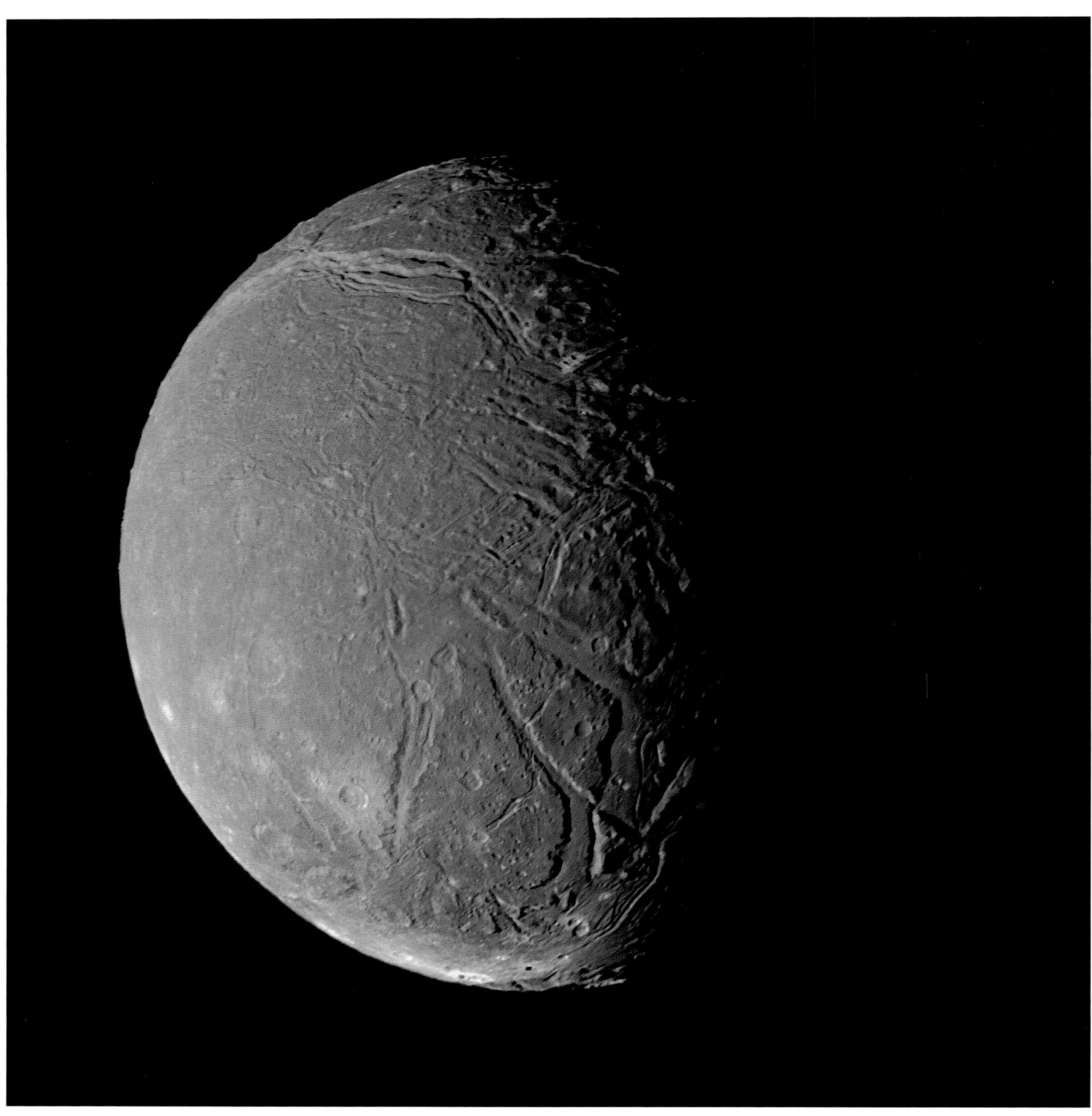

Some details surprised the scientists. The faults that bounded the linear valleys being invisible where they transect other valleys, was one of those details. These valleys are believed to be filled with deposits after being formed by tectonic processes. This left them flat and smooth. Sinuous rilles (trenches) were later formed, most likely due to some form of flow process. It is speculated that some type of fluid may have been the driving force of their evolution.

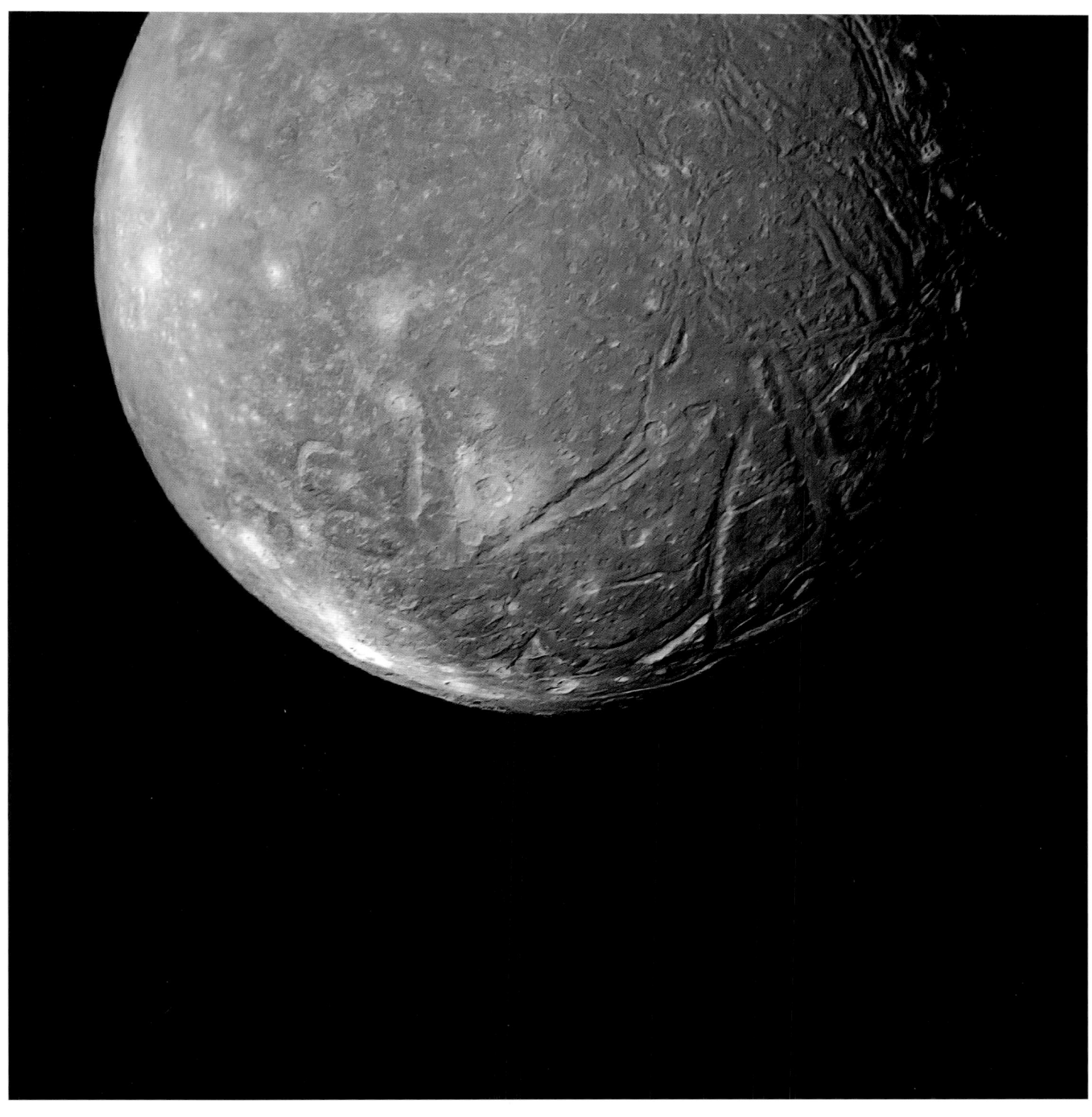

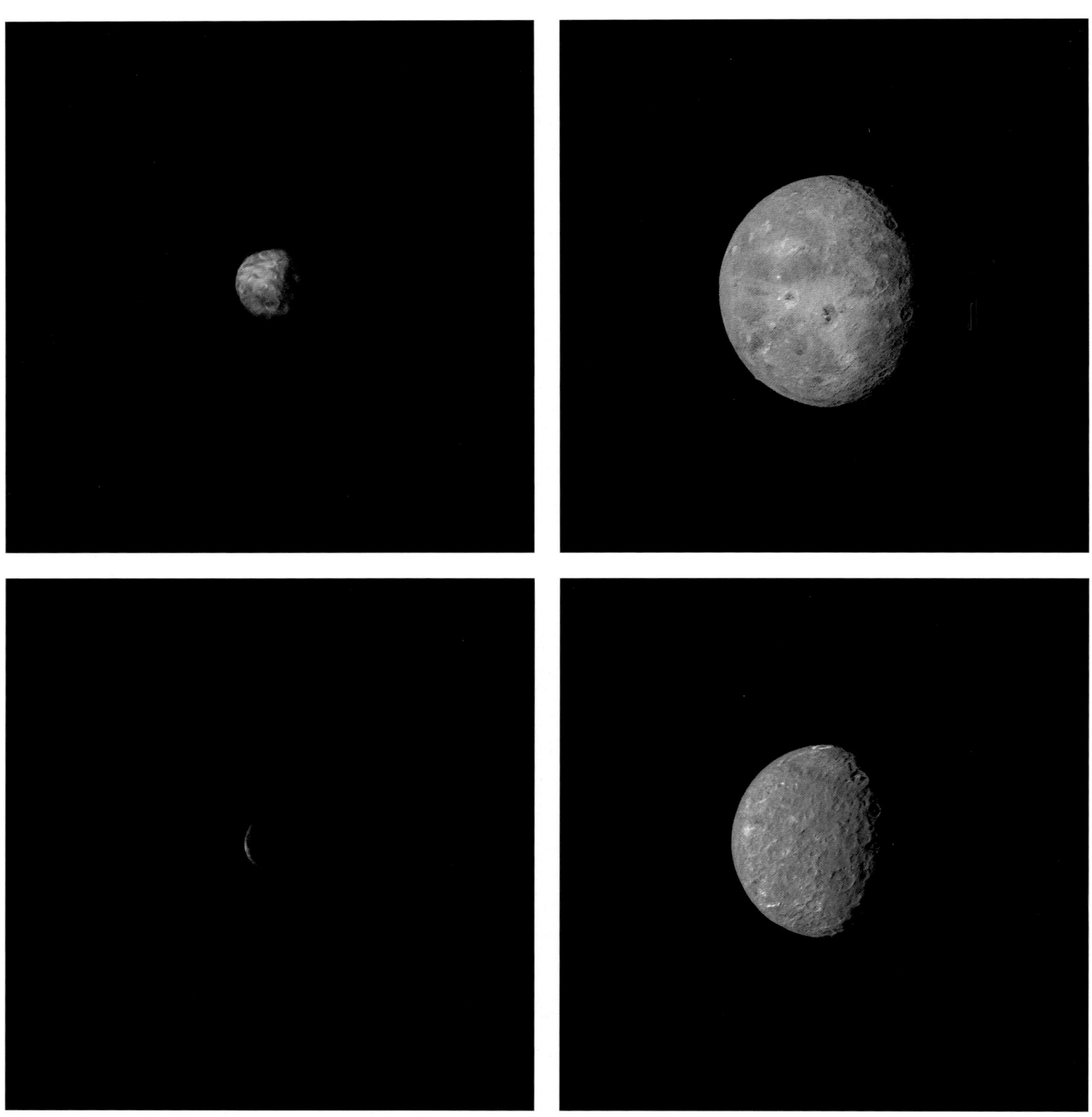

Top right: bright ray craters, similar to the ones that can be seen on Pluto's moon Charon, surround several large impact craters in Oberon's icy surface.
Bottom right: Umbriel is unusually dark except for the bright ring at the top, "Wunda".

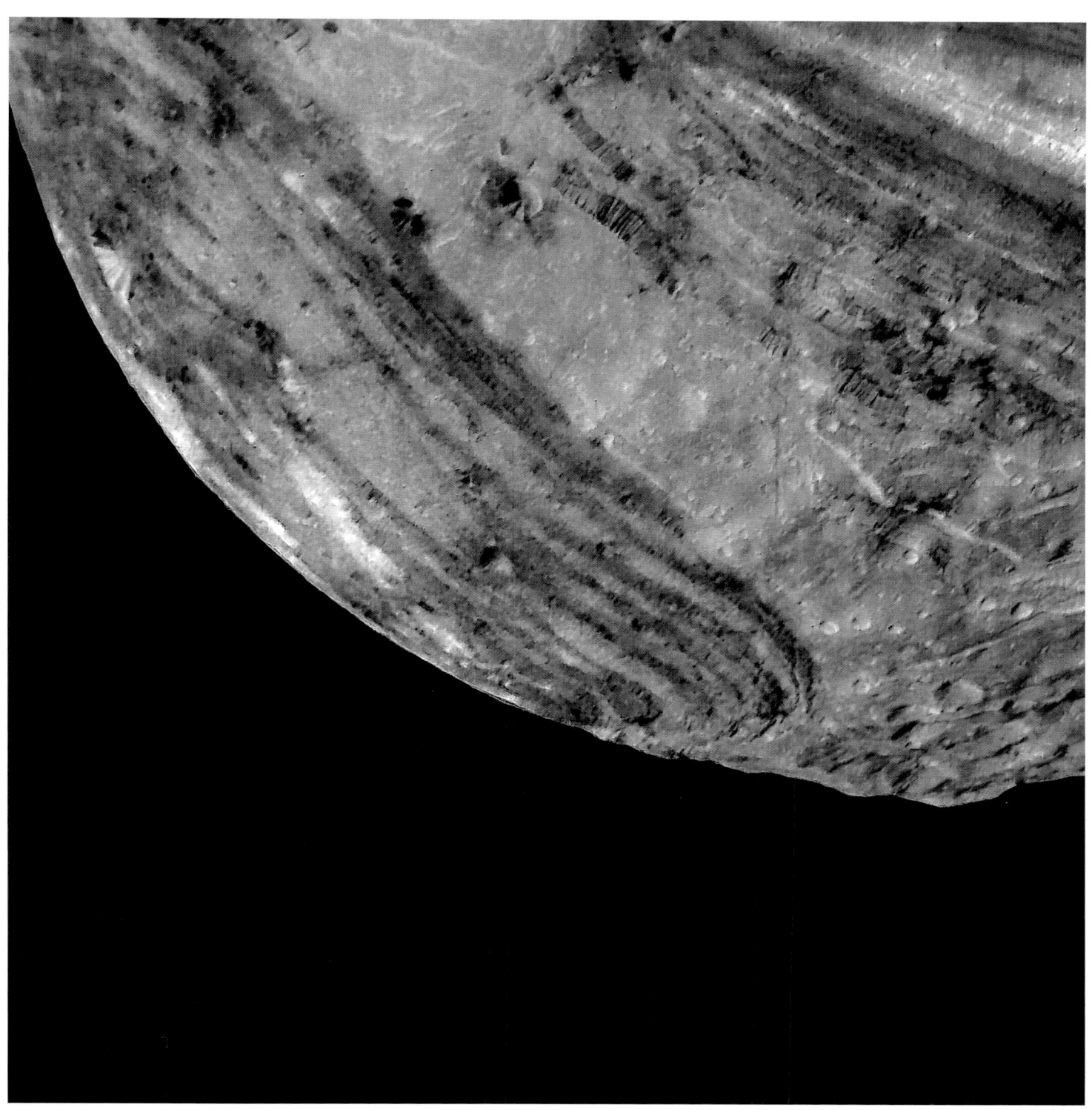

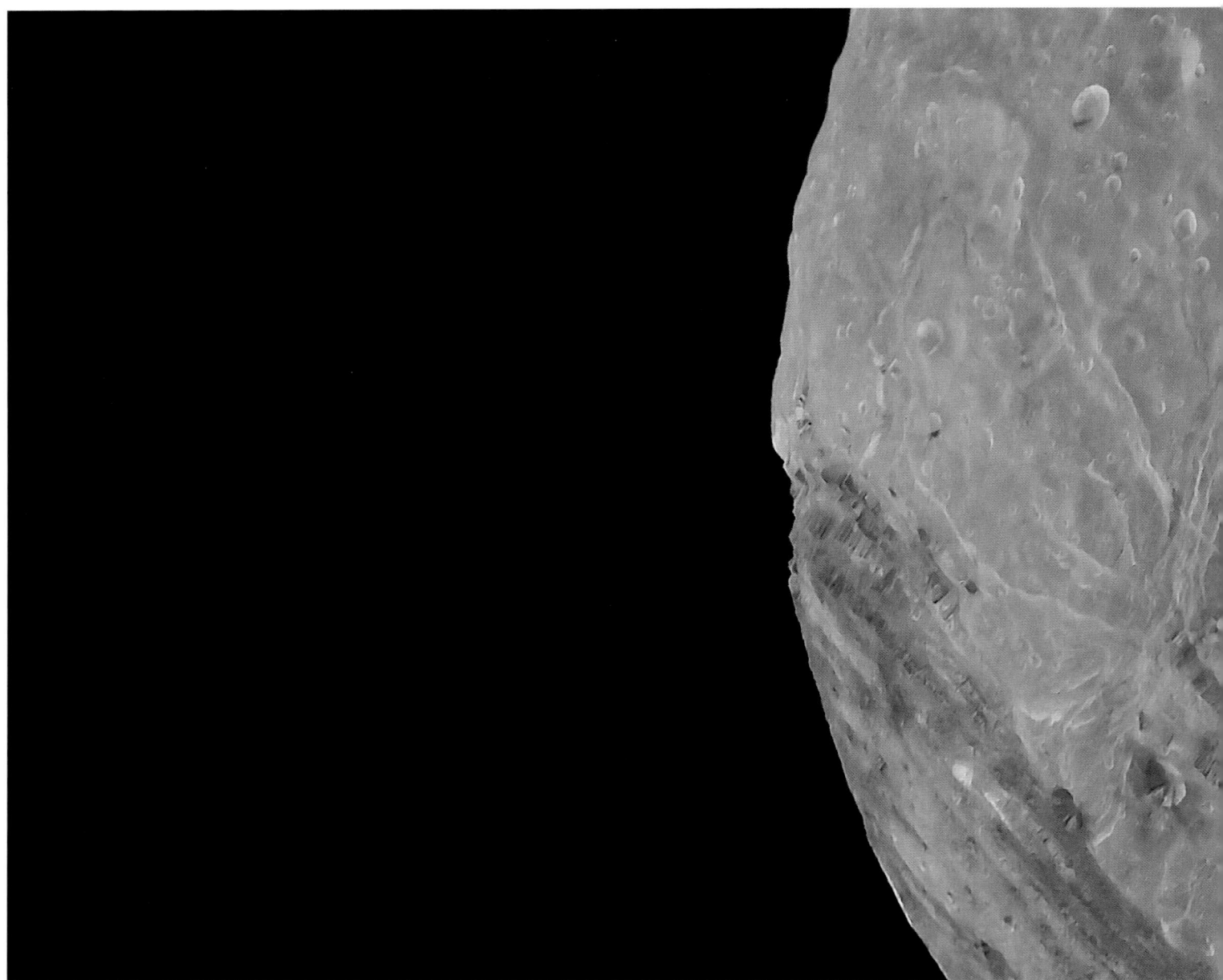

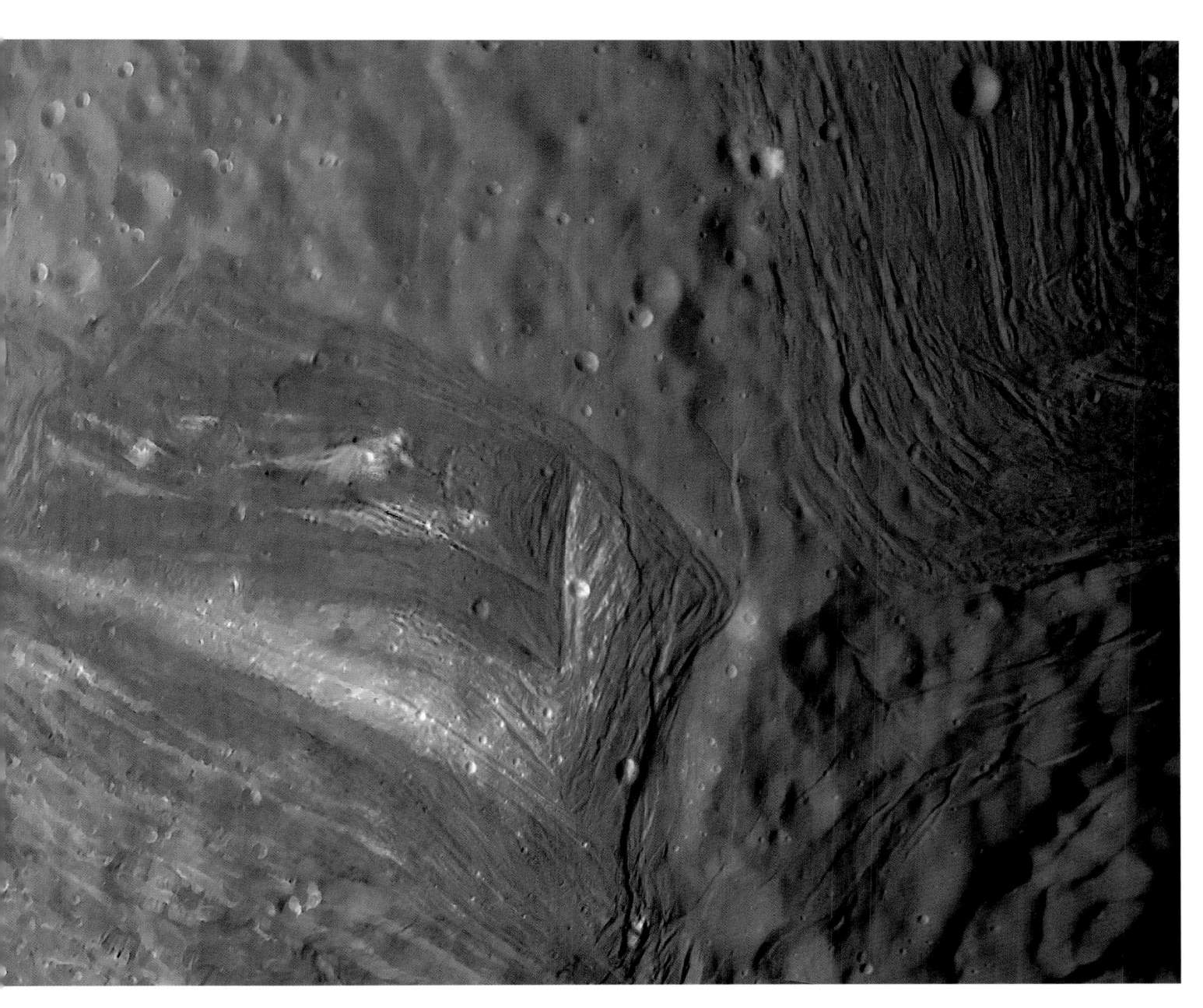

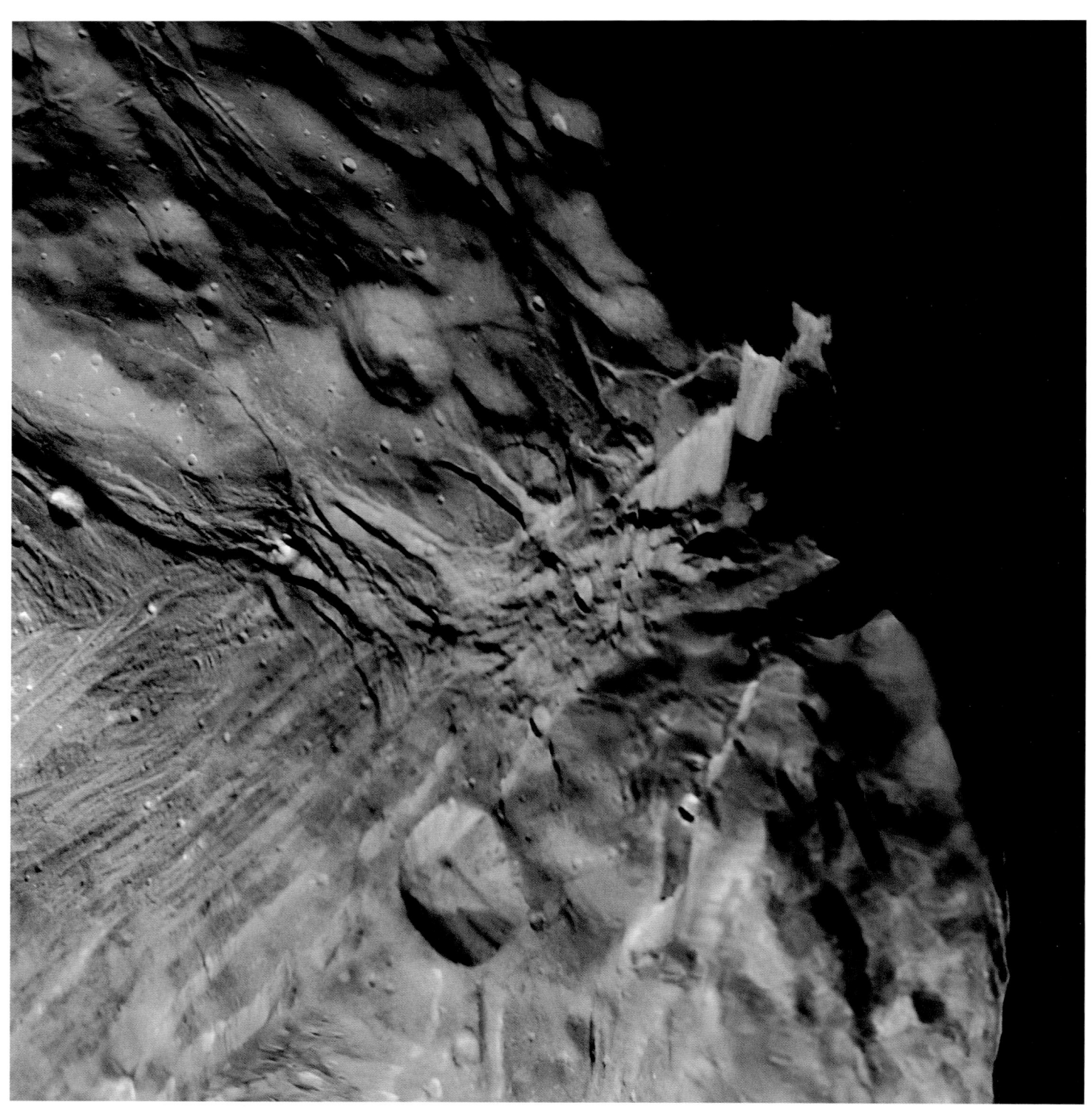

Verona Rupes, the large cliff, is 12.4 miles (20 km) high.

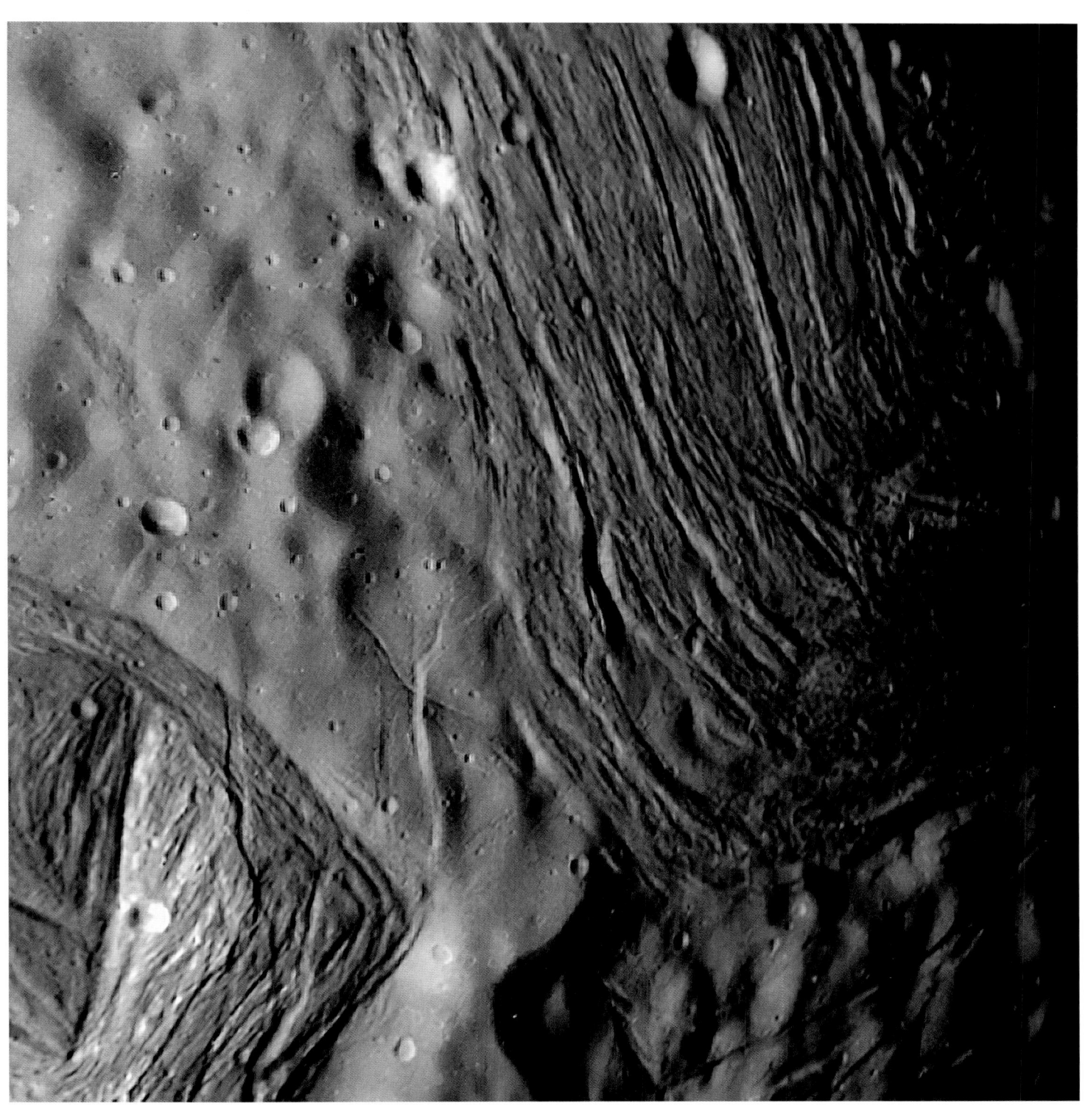

Notice the two distinct terrain types in this image; one is a lower striated terrain (left), while the other is higher–elevated and more rugged terrain. The numerous craters on the higher terrain reveal that it's the older of the two.

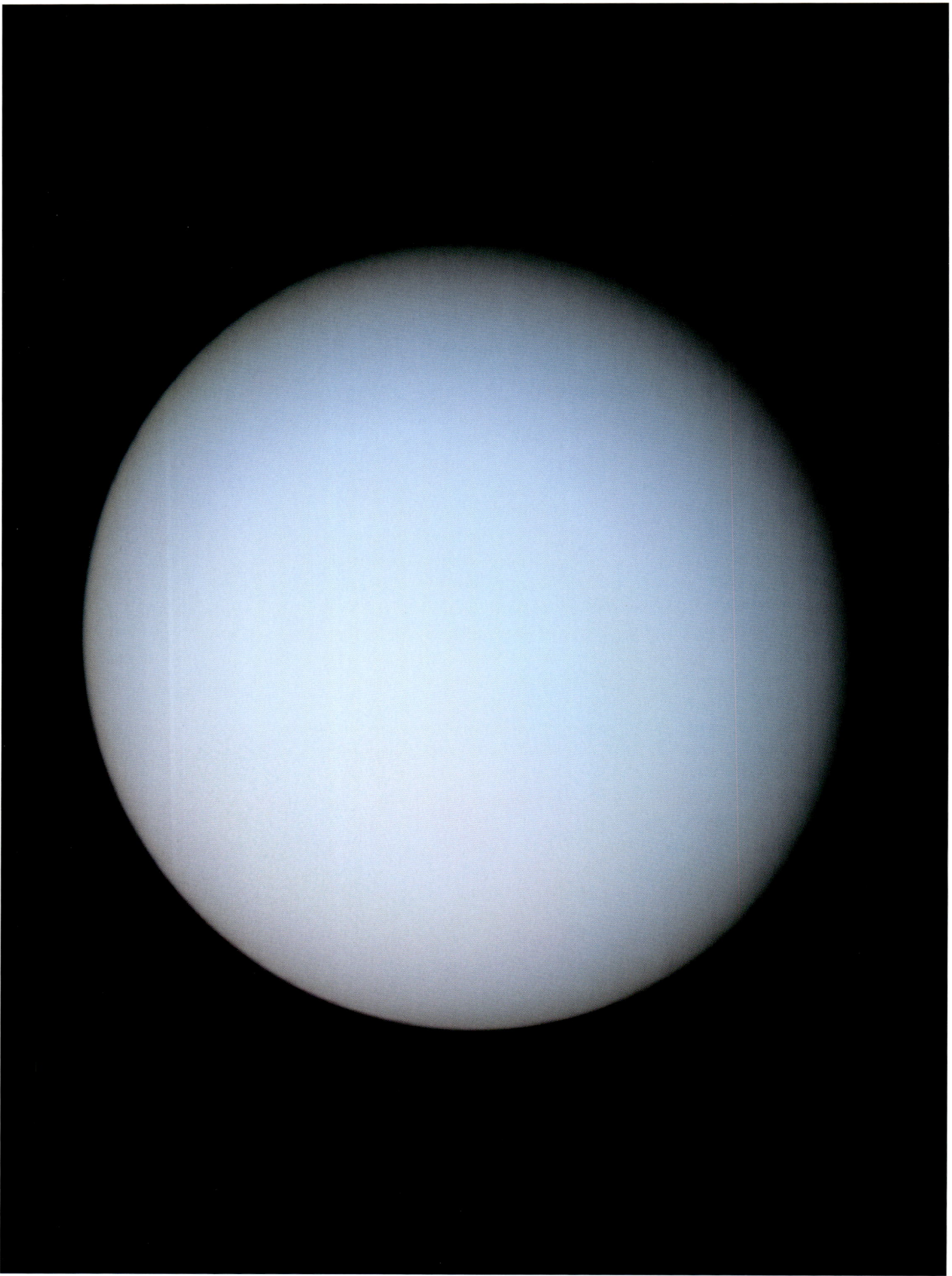

ONWARDS

Within the time frame of a mere ten hours, Voyager 2 revealed to the world wonderful and varied landscapes within the Uranian System. An invaluable amount of data was recorded which immeasurably increased our knowledge and understanding of this planet and its moons.

In summary, the probe discovered numerous weird and peculiar attributes on the planet, such as an extremely low surface temperature. An almost fully tilted axis that creates the extreme seasonal differences. The newly discovered bizarre off-centered magnetic field around the planet, which strangely spirals at the tail-end of the planet in a corkscrew fashion. This spiraling magnetic field was probably caused by the planet's tilted spin and further pushed outwards by the solar winds. More recent analysis and re-evaluation of old data in 2020 indicated that a giant plasma bubble full of energized hydrogen was pulled out from the planet's atmosphere by this messed up spiraling magnetotail, releasing a 127,000-mile-long (205,000 km) plasmoid into space.

IT'S A WILD WORLD

Uranus was a small bluish green dot in the telescope and Neptune was an even smaller bluish dot. And that's all.

Brad Smith - Imaging Science Team Leader

INTRODUCTION

Out of the entire Solar System family, Neptune is that distant cousin that lives far away abroad, whom you only know from speculation and stories. Voyager 2 became the first, and still the only, space probe to visit this beautiful blue planet and end some of that speculation. After both NASA's approval of the budget to continue funding the Voyager mission, as well as the successful Uranus flyby, Voyager 2's mission was extended to continue onward to Neptune. The trip had an estimated time of arrival of three years. With Neptune being the eighth and farthest planet from the Sun, this would become the last milestone of this great planetary journey.

Neptune is the fourth largest planet in the Solar System and is the smallest out of the four gas giants. The planet's diameter measures 34,503 miles (55,528 km) and you could fit just over 57 Earths within Neptune's immense volume. Neptune is the third most massive and the densest of all of the gas giant planets. Although it is smaller than Uranus, the higher density is due to its greater mass, creating more gravitational compression of its atmosphere. The planet is 2.8 billion miles (4.5 billion km) away from the Sun, orbiting the star once every 164.8 Earth years. The planet is so far away, it would

take the spacecraft over four hours to send pictures and data back to Earth, even though the radio signals were transmitted at the speed of light. A day on Neptune is 16 hours, 6 minutes, and 36 seconds.

The vast distance between the Sun and Neptune fed into different perspectives among astronomers. Sunlight on Neptune has only about 3% of the brightness compared to what is seen at Jupiter. Many predicted that the weak solar energy that normally powers the intense storms seen on Jupiter and Saturn would result in Voyager finding a relatively bland and boring atmosphere at Neptune, like what they found at Uranus. Other scientists disagreed with that notion, speculating that Neptune could have substantial reserves of internal energy that would heat and power a lot of atmospheric activity.

Scientists back on Earth have held a special amount of curiosity for Triton, Neptune's largest moon since its discovery. Triton has a highly unusual retrograde orbit, rotating around the planet in the opposite direction to the way Neptune spins on its axis. One of the main goals of Voyager 2's visit to Neptune would be to go hunt for clues to help explain this unusual retrograde orbit.

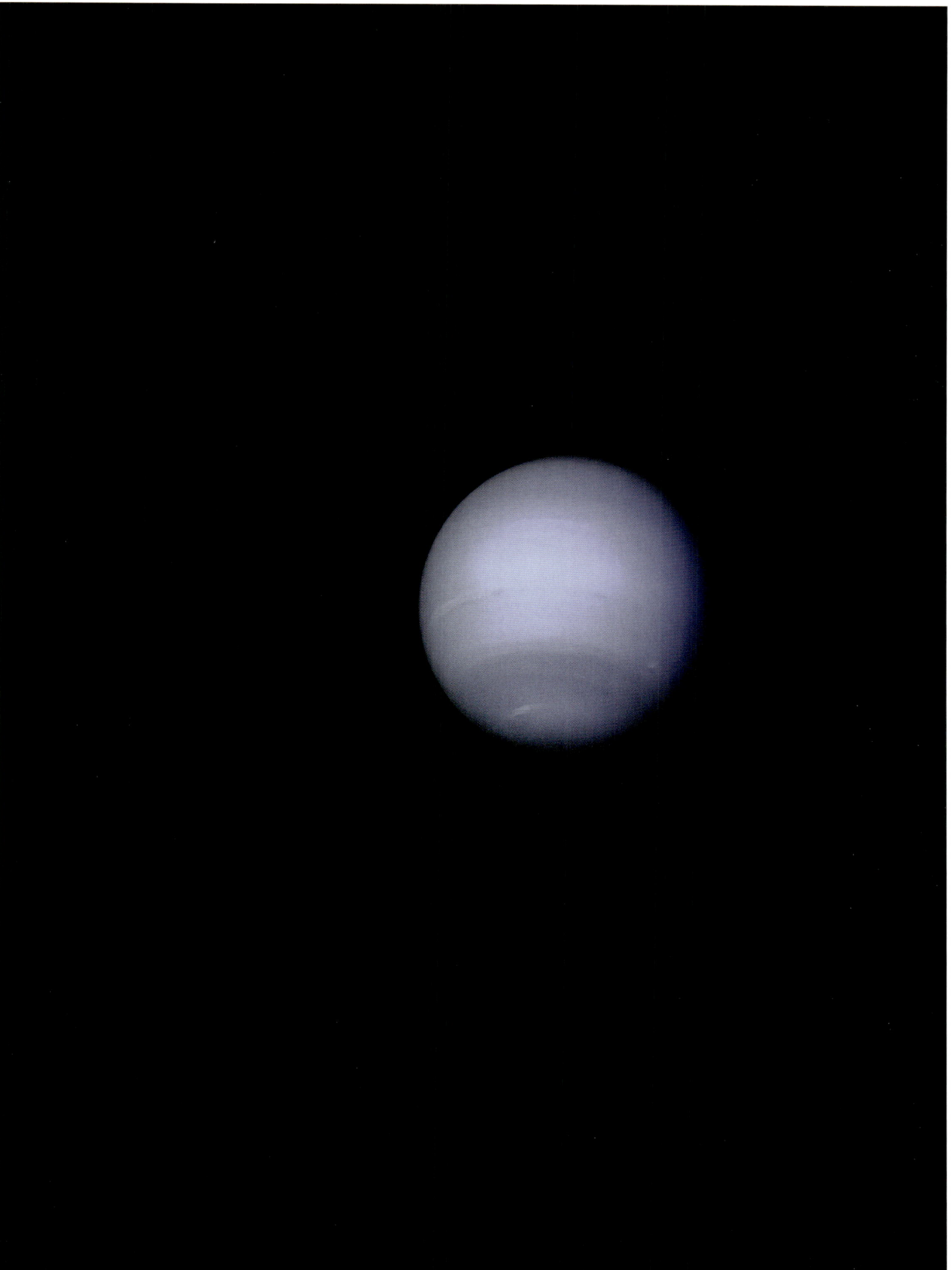

A TALE OF TWO MATHEMATICIANS

The discovery of Uranus in 1781 was monumental news that travelled around the world, making William and Caroline Herschel major celebrities. The newfound fame and status that discovering a new planet brought to the siblings encouraged a revolution among the European scientists and their wealthy benefactors. Who will be the next astronomer to discover another new planet in our Solar System? Advancements in telescope technology rapidly increased, allowing humankind to see further and clearer into the night sky than ever before.

Astronomers however, were not the only ones who were working to discover a new planet. Those with the power to calculate were stepping into the game as well. With the more powerful telescopes, it was possible to accurately track the movement of Uranus and the other planets within the Solar System. Any tiny deviations of changes from any planets orbit – calculated to predict the future positions of the objects in the night's sky based on the motion, orbital, and gravitational laws written by Kepler and Newton – caused by gravitational influence from another object, could now be more easily detected. Hence, mathematicians were also joining the cause.

During the 65 years following the discovery of Uranus, astronomers spent countless hours tracking the orbit of Uranus. Over the decades, they started to detect a series of irregularities in its path that could not be explained or calculated away by Newton's Law of Universal Gravitation. These irregularities led astronomers and mathematicians to believe that there must be a farther, undiscovered planet, whose gravity was pulling Uranus off from its calculated orbit. Finding this distant planet however seemed almost impossible. Even with the advancements in telescopic technology, and without knowing roughly where to look in the night sky, it was the proverbial, "looking for a needle in a haystack."

Two mathematicians from France and the United Kingdom, unbeknownst to each other, independently started to work on the "Uranus problem" in 1845. In Paris it was Urbain Le Verrier, while in Cambridge John Couch Adams sank his teeth into the same challenge. Their goal was to calculate and determine the position and nature of the hidden planet that was creating the disturbance of Uranus's orbit. Using just mental arithmetic, (as mechanical calculators and primitive computers were still decades away from being invented), both men calculated and plotted their predicted position of the hypothetical distant planet using charts of Uranus's orbital progress.

Urbain Le Verrier managed to get there first. He roughly predicted the location of Neptune, and outlined it to astronomer Johann Gottfried Galle at the Berlin Observatory in a written letter that was received on September 23rd, 1846. Inside the letter were instructions and prediction charts on where to look. Galle started his search

that evening. It was a little after midnight, and less than an hour of searching, when Galle found Neptune. Remarkably, Le Verrier calculated the position to less than one degree of its exact position. A further two nights of observations of Neptune were made to track its movement and position. Galle wrote back to Le Verrier, "the planet whose place you have calculated really exists," confirming its position. This identified the eighth and most distant planet in the Solar System, and also solved the problem of Uranus's unpredictable orbit at the same time. It is claimed that John Couch Adams had also correctly predicted the location of Neptune some time a little earlier. Adams sent a letter to the Cambridge observatory with instructions, but it either consisted of the wrong location or the university did not have the correct maps. Thus, Le Verrier claimed the prize for the discovery.

Overall, it was a monumental moment for 19th Century science. The confirmation of the planet did not just increase the size of the Solar System by another billion miles or so, but Neptune's existence was predicted using math, proving that Newton's theories were correct and that they now became laws. Fellow French astronomer aptly described that Le Verrier had discovered a planet "with the point of his pen." It took a little time for the scientific community to agree on a name for this new planet. Janus, Oceanus and Le Verrier were all considered, but Le Verrier's choice of Neptune, the Roman God of the sea, was the one that was collectively accepted.

It is worth mentioning however, Le Verrier was not the first person to have recorded the presence of Neptune in the night sky. If we look back two centuries ago, it is once again the astronomer from Tuscany, Galileo Galilei, who can lay claim to being the first to record seeing Neptune. On December 28th, 1612, and again shortly afterwards on January 27th, 1613, Galileo depicted Neptune in his observational drawings. He mistook the planet for a fixed "blue star" that appeared to sit close to Jupiter in the night sky, and so is not credited with the planet's discovery.

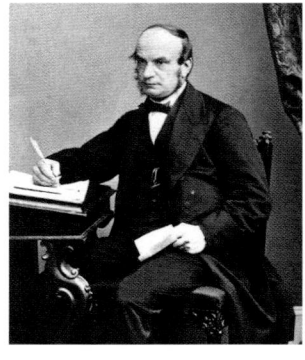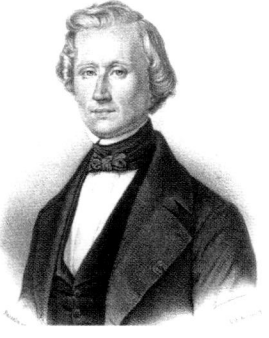

John Couch Adams (left), and Urbain Le Verrier (Right).

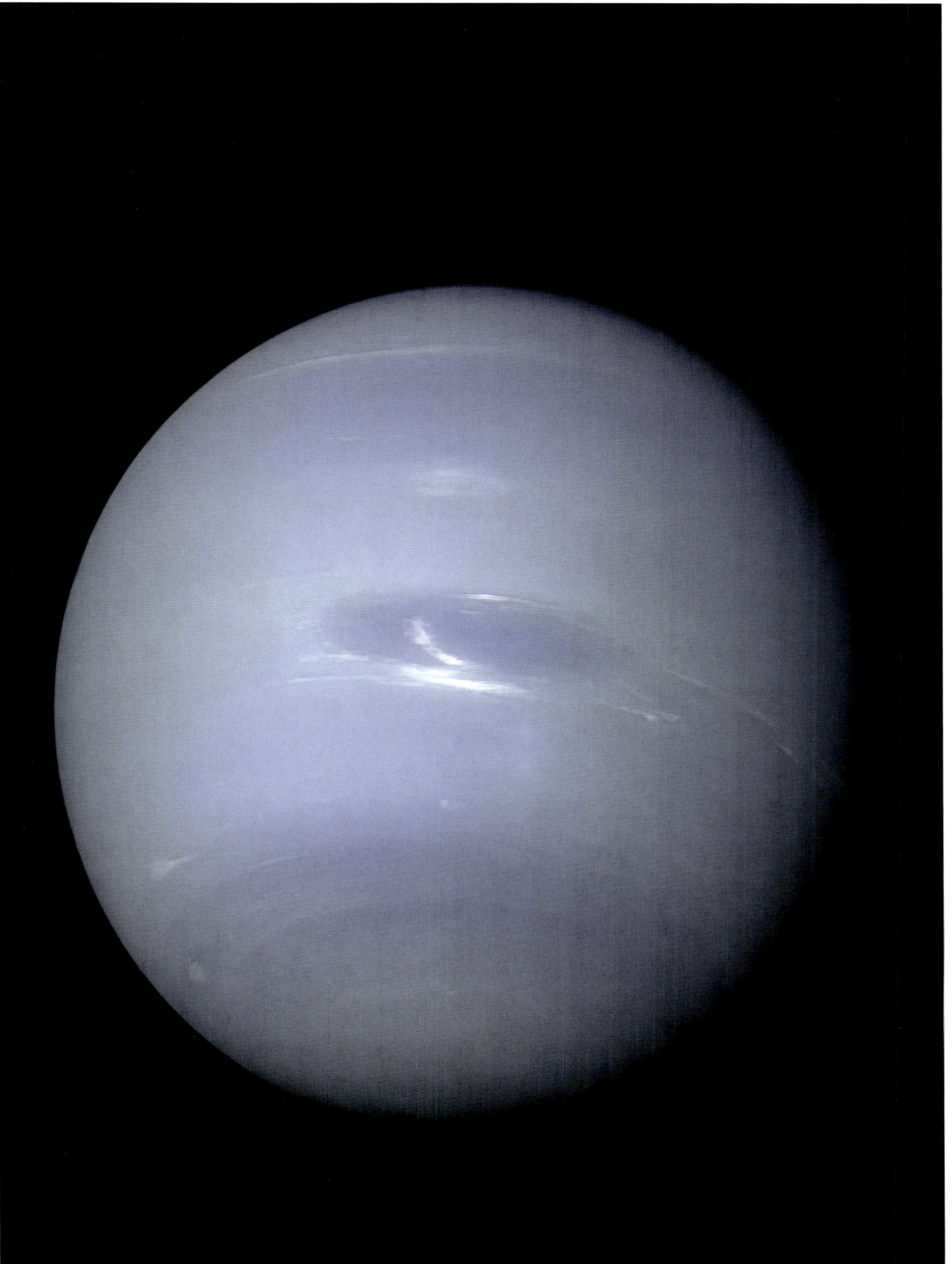

THREE FAMILIAR CHALLENGES

When Voyager 2 left the Uranian System in February 1986, it would take the next three years and six months to traverse the billion miles (1.6 billion km) to get to Neptune. Besides a mid-course correction in 1986, the Voyager team could focus on fixing up the aging craft to make sure it was up to the job and capable of handling all the new challenges it would face during the flyby of Neptune. Similar to the preparations made with the approach to Uranus, the Voyager team had to work out solutions and upload improvements to the probe to solve the same three issues that Voyager 2 would encounter: low-light, higher speeds, and the vast distance between the blue planet and Earth.

Again, we do not have to ask Sherlock Holmes to find out that Neptune is extremely far away from the Sun. At a distance of a whopping 2.8 billion miles (4.5 billion km), Neptune receives less than half of the light that is shining on Uranus. The incredibly low-light conditions would challenge the cameras on Voyager 2, pushing the imaging systems capabilities farther than to what they had been originally designed. To add another challenge to the mix, the spacecraft would be reaching speeds of around 60,000 mph (97,000 km/h). The long exposure time required for the low-light conditions at this speed would result in blurry, smeared images.

Over the years, the Voyager team ran a number of simulations to help reduce the risk of image smear. They uploaded a program that controlled the craft's thrusters, gently firing them during the close approach. This would in turn slowly rotate and slew the spacecraft to keep the camera focused on its targets without affecting Voyager 2's speed or course.

Similar to the program uploaded during the Uranian flyby a few years earlier, another program called the Nodding Image Motion Compensation (NIMC) was uploaded. The software allowed the spacecraft to remain virtually Earth-pointed whilst making a small 'nodding' movement. This movement would,

in turn shutter a frame, after the image was taken, the spacecraft would 'nod' back to make the spacecraft position to face Earth-point again. With the antenna pointing back to Earth, it meant all the images could be transmitted back to Earth immediately, rather than having to store the images on the digital tape recorder. The new program gave the craft the capability to perform higher scan platform rates with a lower risk of image smear, which would come in useful during its closest approach to Neptune.

Now that the light and speed issues had been tackled, the next problem to resolve was distance. Radio signals would be making a round trip of 2.7 billion miles (4.3 billion km) from the Neptunian System to Earth and back. The signals traveling at the speed of light would still take more than eight hours to return to Earth. The vast distance meant that the low-powered radio signals would also be extremely faint. Improvements to NASA'S Deep Space Network were made to better detect the faint signals. Larger antennas spanning nearly 230 foot (70 m) across were built and new receiving stations were constructed in New Mexico and Japan.

Due to the vast distance, the signals had to travel, the chance of data corruption increased greatly. To reduce this risk, two new techniques were implemented. Firstly, some of the editing of the data was done directly. Voyager 2 would delete some of the image pixels. This resulted in a smaller file size, but also lowered the resolution of the image, thereby only returning a part of the 800 x 800 pixel image at full resolution. Secondly, a new data compression process was taught to Voyager's onboard computers. The brightness of the first pixel of each line was sent back, followed by the difference in brightness of each adjacent pixel, rather than the full pixel brightness information. Transmitting the image data in this way reduces the amount of information to be sent by 60 to 70%, but keeps the full image size and resolution.

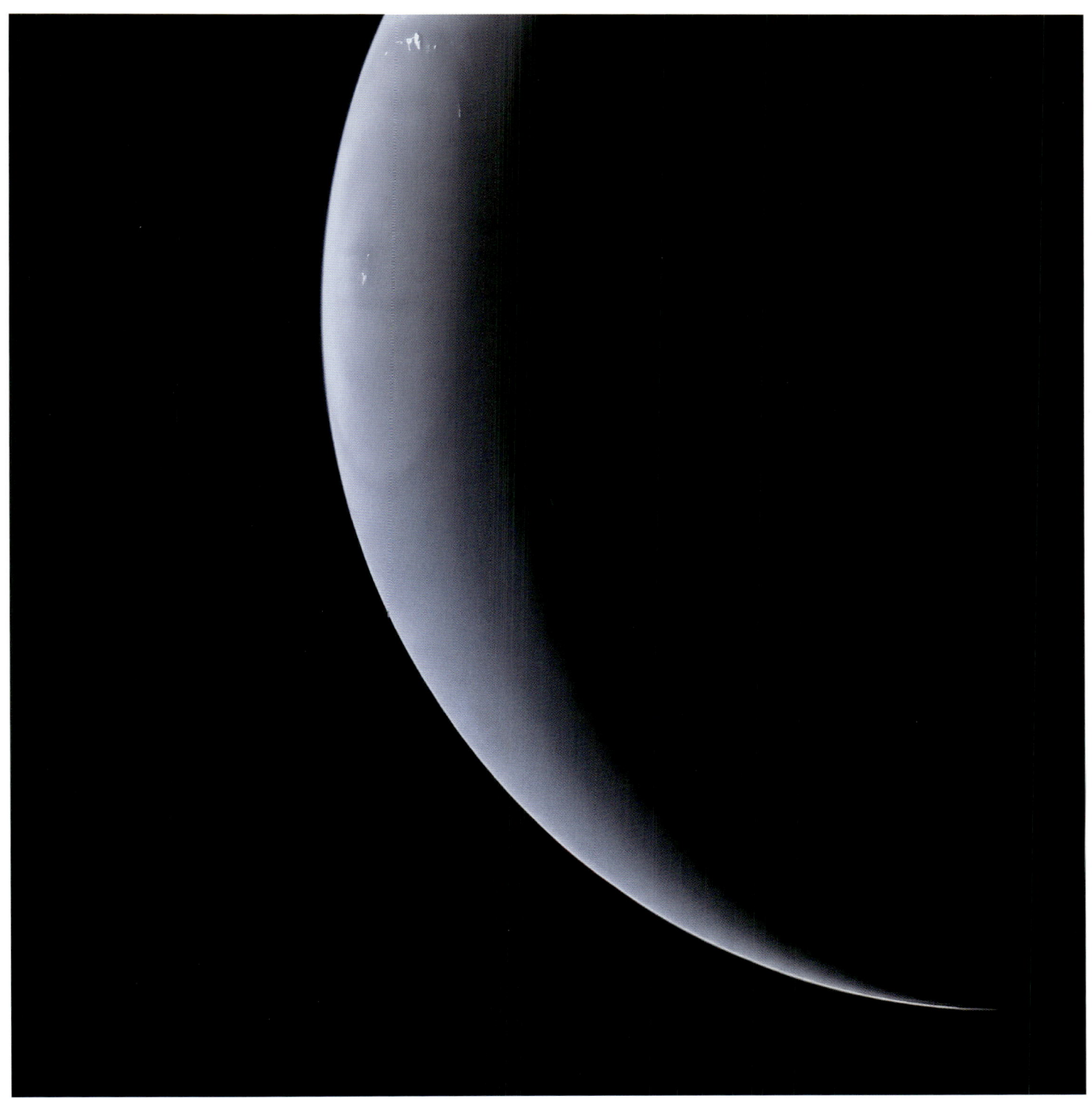

CLOSE CALL

The original idea for the flyby of Neptune was to get extremely close to the planet. This trajectory would see Voyager 2 skimming just 810 miles (1,300 km) over Neptune and then continuing its journey onward to Triton for another close encounter, passing by the largest moon at a distance of 5,100 miles (8,200 km).

Voyager started making its first navigation images of the Neptunian System in May 1988. As more data and observations of the Neptunian System were received and given to the Voyager team for analysis, it became clear that the original trajectory needed rethinking. Particles were discovered by stellar occultations along the flight path from the recently discovered ring system (only known to us five years before the Voyager flyby). This resulted in the trajectory being adjusted for a more distant flyby of the planet and its moons to avoid the potential mission ending in a violent collision with the ring system.

To make a new safer route, as well as achieve a close encounter with Triton, Voyager 2 was required to fly over the top of Neptune's North Pole, passing above it at 3,000 miles (4,828 km). This was Voyager 2's closest approach to any object in space since leaving Earth back in 1977. Passing by the North Pole at this distance would give the craft the needed gravity assist for a "handbrake" turn, pulling the craft

onto a downward heading out of the ecliptic plane towards Triton. Even with the more distant encounter with Neptune, it still wasn't without risk. Although the Voyager team was aware of Neptune's thick atmosphere, very little was known about its depth. It was not clear if a 3000-mile clearance would be safe enough. This, however, did not concern the trajectory planners too much, as with no future planetary encounters lined up, there was little consequence if the spacecraft were to be damaged.

After Voyager 2 would pass by Neptune's ring system, it would first capture images of Proteus (discovered only two months earlier by Voyager 2 from distant observations) and Nereid. Then, five hours later after passing Neptune, it would zoom past the largest moon in the Neptunian System, Triton, from 25,000 miles (40,000 km). The encounter with Triton would be the last encounter Voyager 2 would have before joining Voyager 1's journey to interstellar space. Voyager 2 started its full observation phase on June 5th, 1989, and had its closest encounter with the planet and moons on August 25th. It continued collecting images and data from the Neptunian System until October 2nd. After that point, Voyager's planetary mission stopped and the craft started to prepare for its interstellar journey.

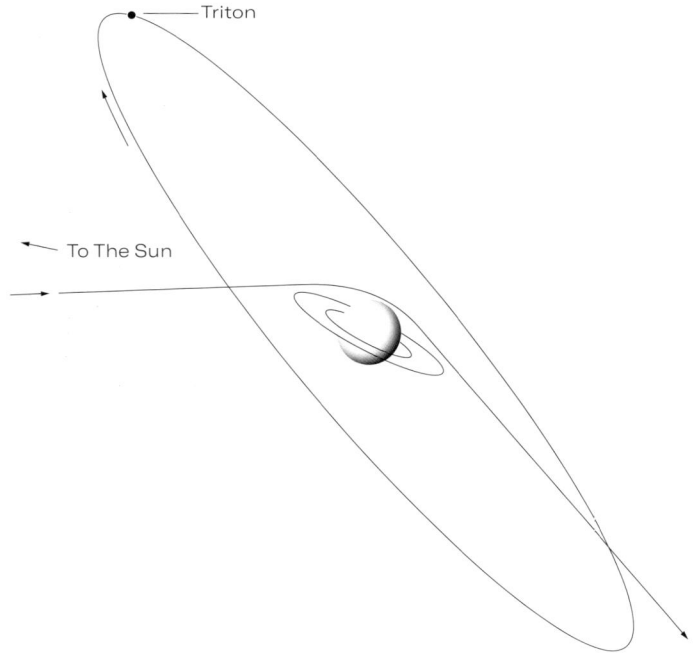

Voyager 2's close trajectory through the Neptunian system

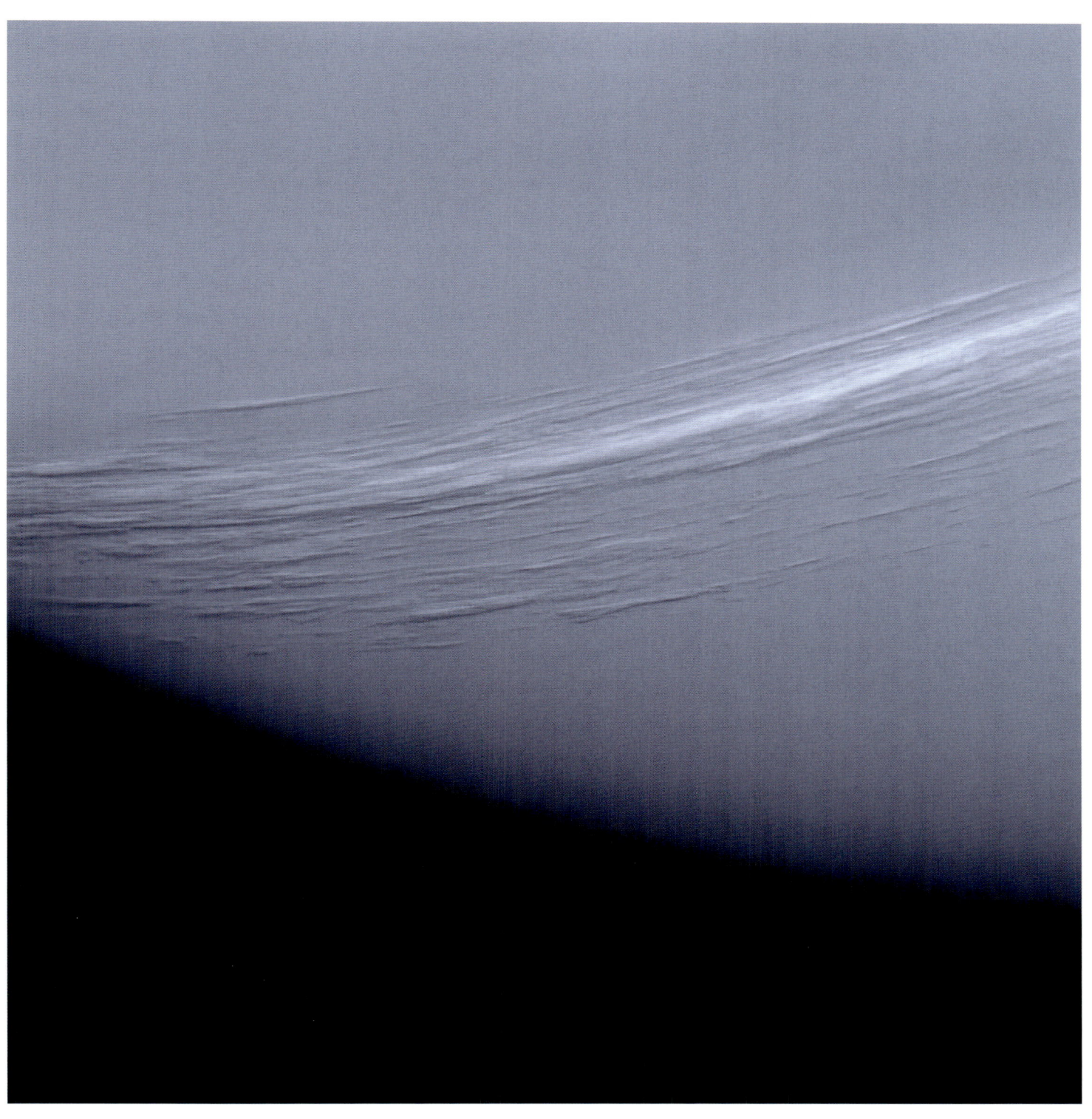

The linear cloud forms are stretched along lines
of constant latitude. Being more directly exposed
to the sun, the sides of the cloud are brighter than
the surrounding cloud deck. On the opposite side
of the Sun, shadows can be seen.

DEFYING EXPECTATIONS

With the first distant observations of the planet, it became apparent that those that thought Neptune would have a bland and boring atmosphere would get the short end of the stick. Despite its vast distance from the Sun, Neptune seemed a hotbed of activity, with Voyager 2 recording Neptune radiating about 2.6 times as much energy as it absorbed from the Sun. This already indicated to the Voyager team that there was a huge amount of energy being generated from within the planet's core. With this internal heating, similar to the planets from the inner Solar System, this creates extreme weather conditions, which are easily noticeable when observing the planet's atmosphere.

With Neptune being so far away from the Sun, very little solar energy reaches Neptune's outer atmosphere. This makes it one of the coldest places in the Solar System. Voyager 2 recorded temperatures of about −360 °F (-218 °C) at the tops of the atmosphere. However, incredibly high temperatures of around 9,300 °F (5,150 °C) were found at its core. The atmosphere is similar to Uranus, with the planet's aquamarine color as the result of methane being present in the atmosphere, Neptune's deeper blue color, also influenced by methane, is thought to be due to a thinner haze layer or additional unknown atmospheric factors, though the exact cause remains under investigation. This gives it a darker, less green appearance than Uranus.

While Voyager 2 was still about 60 million miles (96 million km) away, and two months away before its flyby with the planet, the images already returned prominent and strong cloud features within Neptune's atmosphere. This was in line with what the team had predicted, given the internal heating measurements and extreme contrasts in temperatures found in the upper atmosphere and core.

One of the many observations the Voyager craft took during its journey through the Solar System was to measure the amount of solar energy going into each planet, and the total amount of thermal energy given off. Both Voyager craft measured almost twice as much thermal energy radiating out from Jupiter and Saturn compared to the solar energy absorbed. It was clear that for these two gas giants, there was an additional heat source at its core to contribute to the extra energy. It was speculated that this extra heat could be the result of gravitational energy being stored at high pressure and temperatures at the deep interiors of the planets, or possibly from radioactive elements decaying within their rocky cores.

A pattern that Voyager observed was that the further the space probe travelled away from the Sun and the strength of the solar energy decreased, the less pronounced cloud belts, storms and other extreme weather phenomena appeared. This observation gave grounds for planetary scientists to believe that it was solar energy that was mostly powering the weather on these worlds. Moreover, with the Uranus calculation, the amount of energy going in and coming out was roughly the same. A completely different story to what had been seen at the previous two gas giants. It led scientists to believe that the smaller gas giants were not big enough to have their own internal heat sources. Given this theory, the Voyager team expected Neptune to be even less active than Uranus, as the blue planet gets 40% less solar heating than Uranus.

Neptune threw this theory out of the window. When Voyager 2 recorded the data at this planet, a similar gas giant in size and composition, it showed that Neptune was found to emit about 2.6 times the energy it received from the Sun. Still today, it remains a mystery as to why Neptune has strong internal heating and Uranus has not.

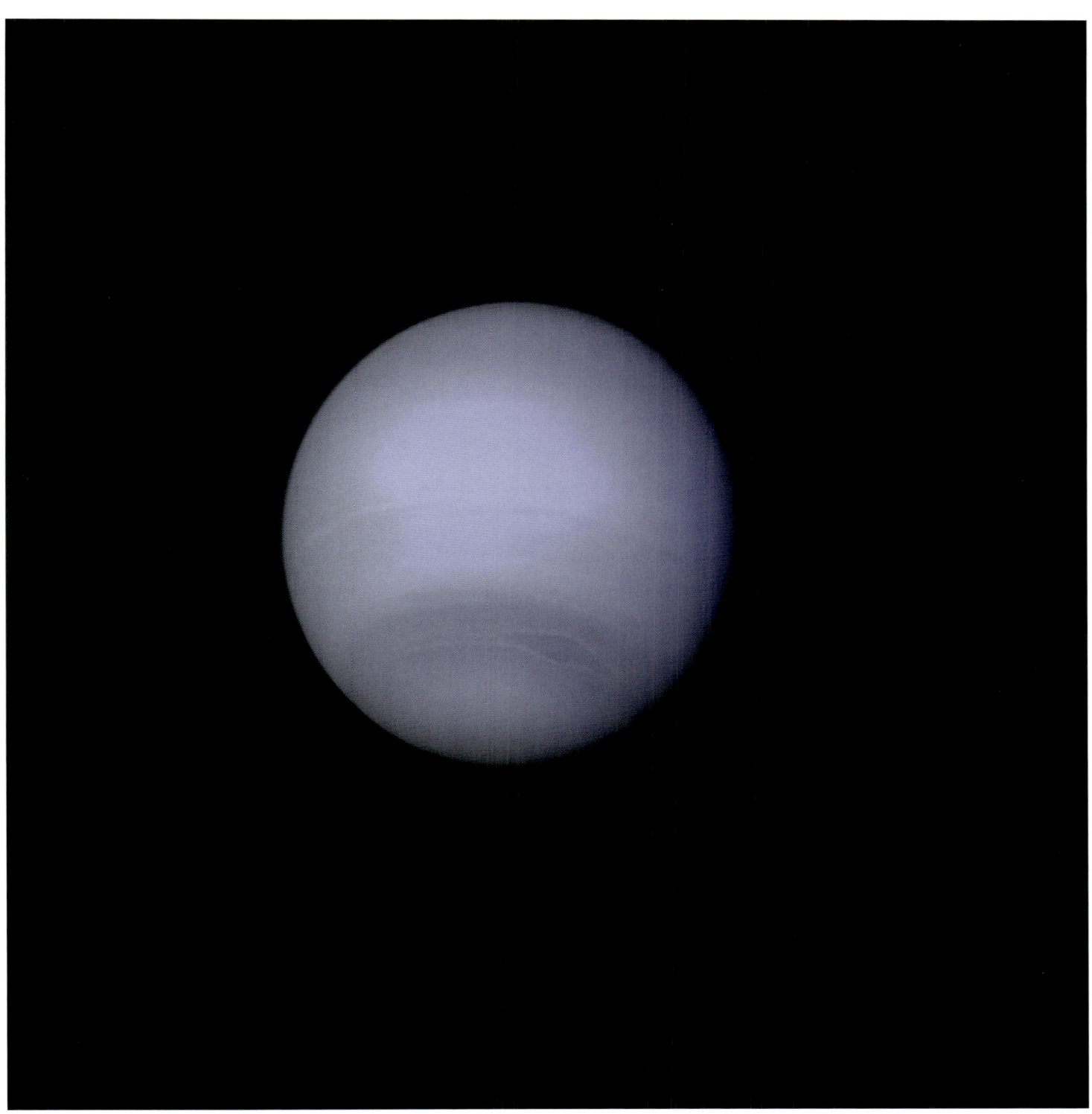

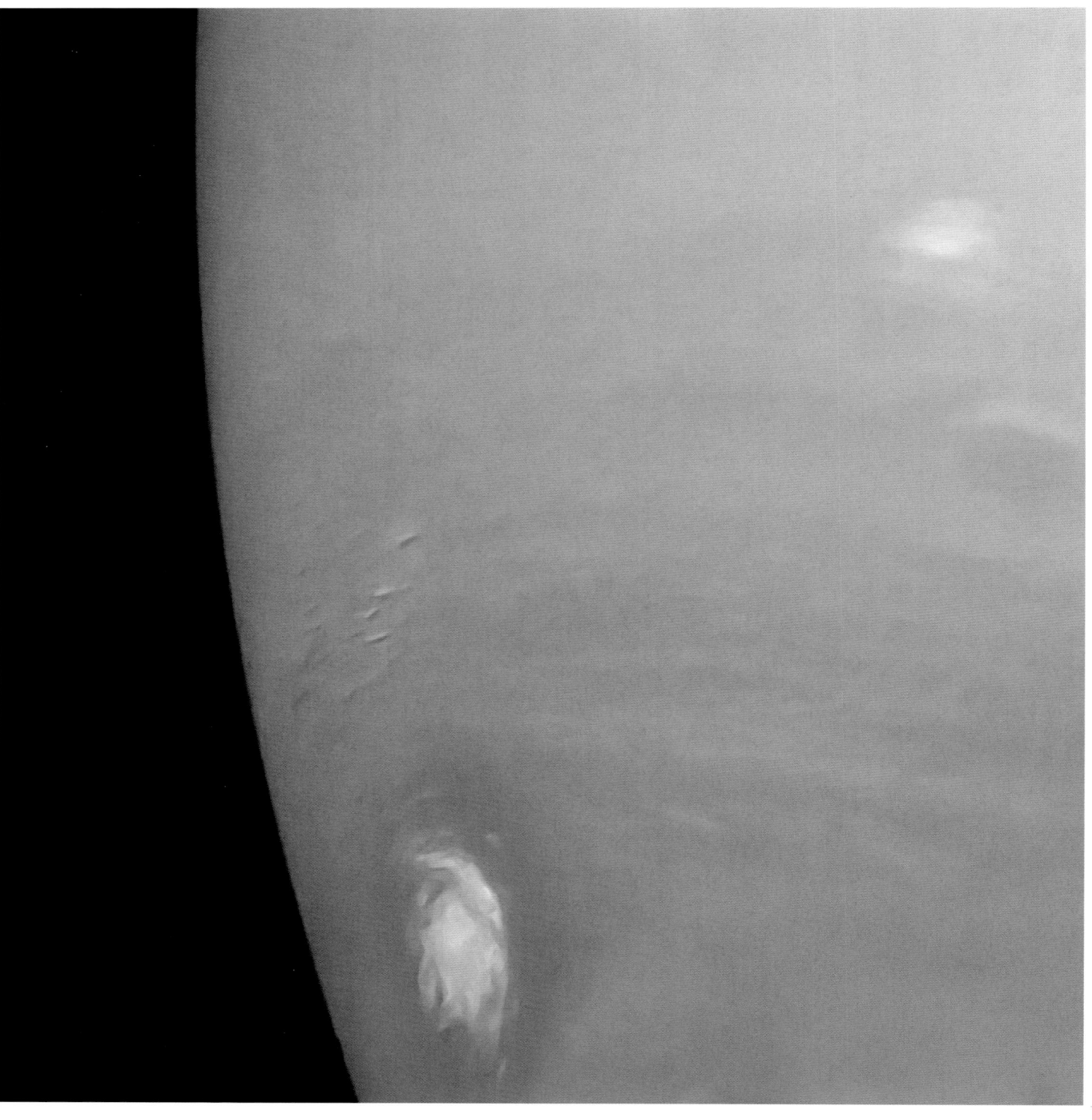

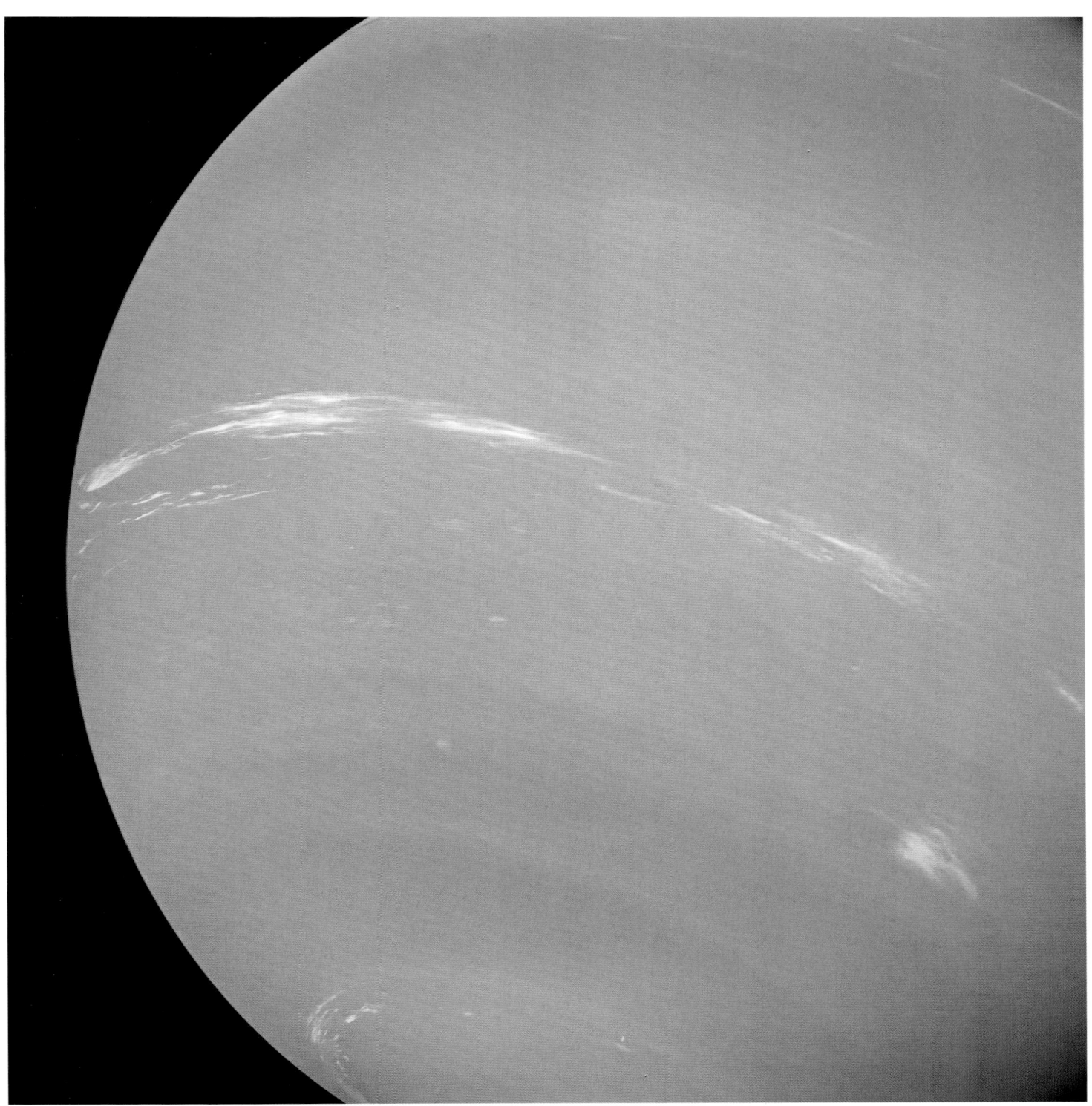

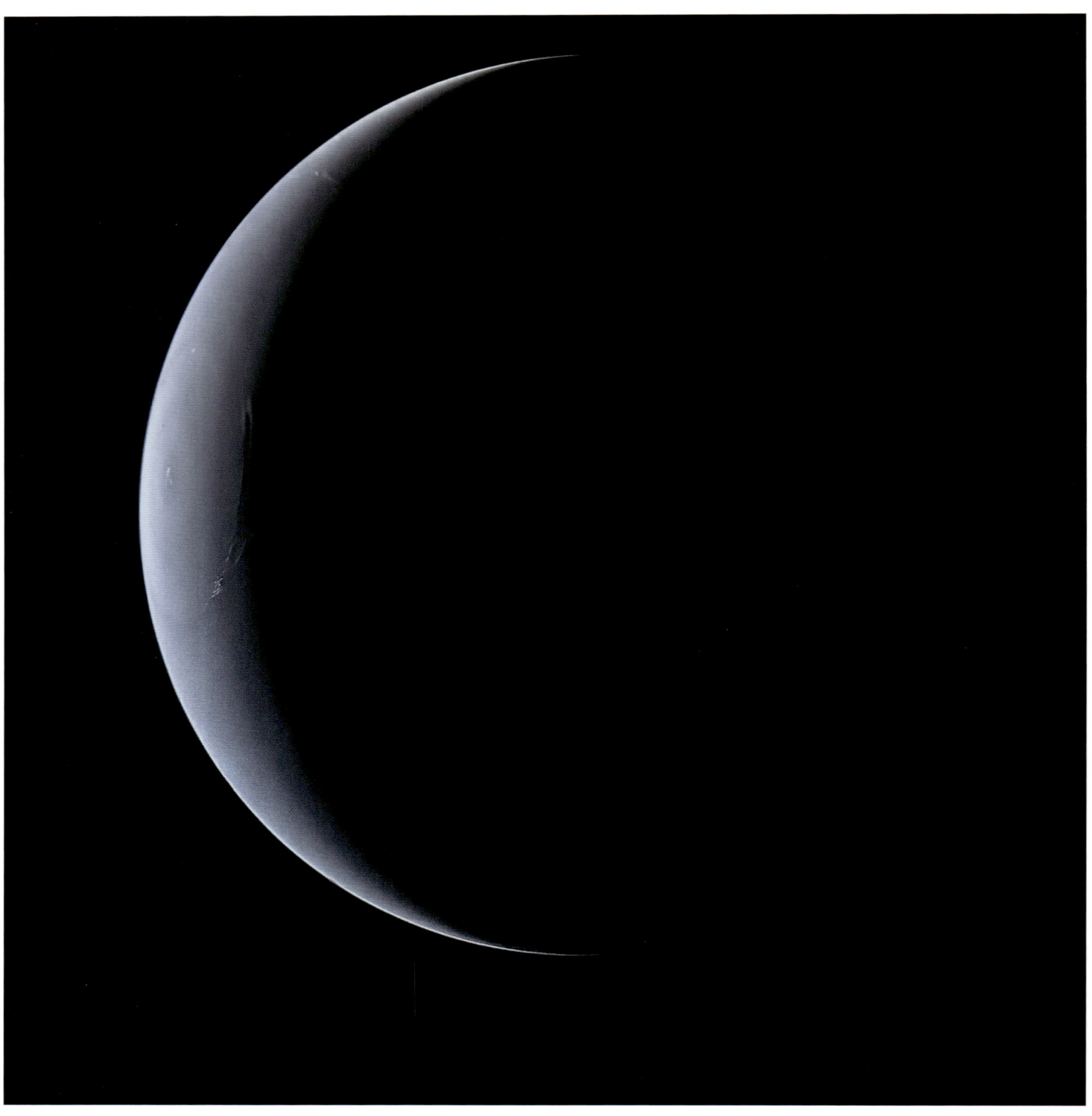

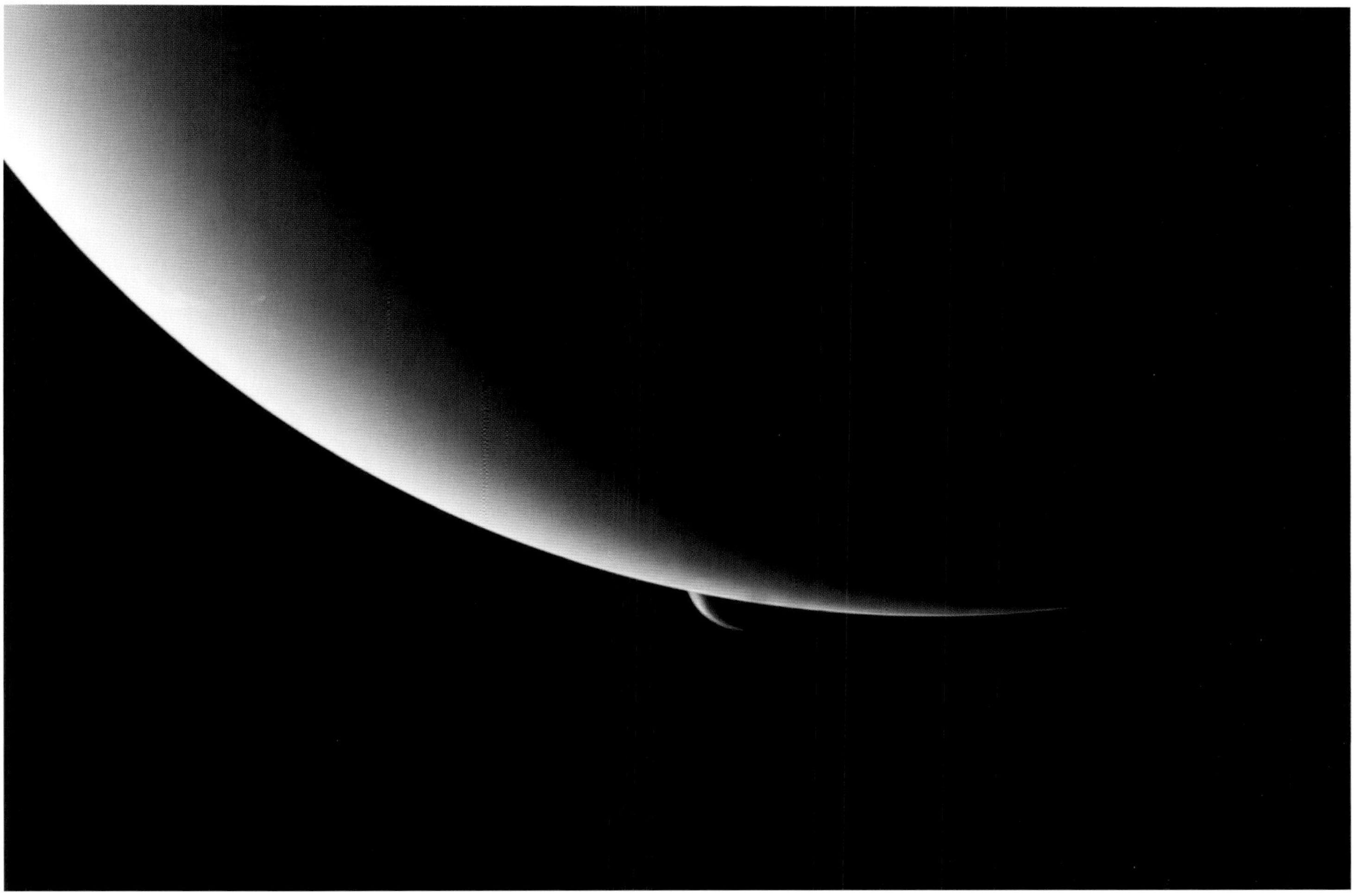

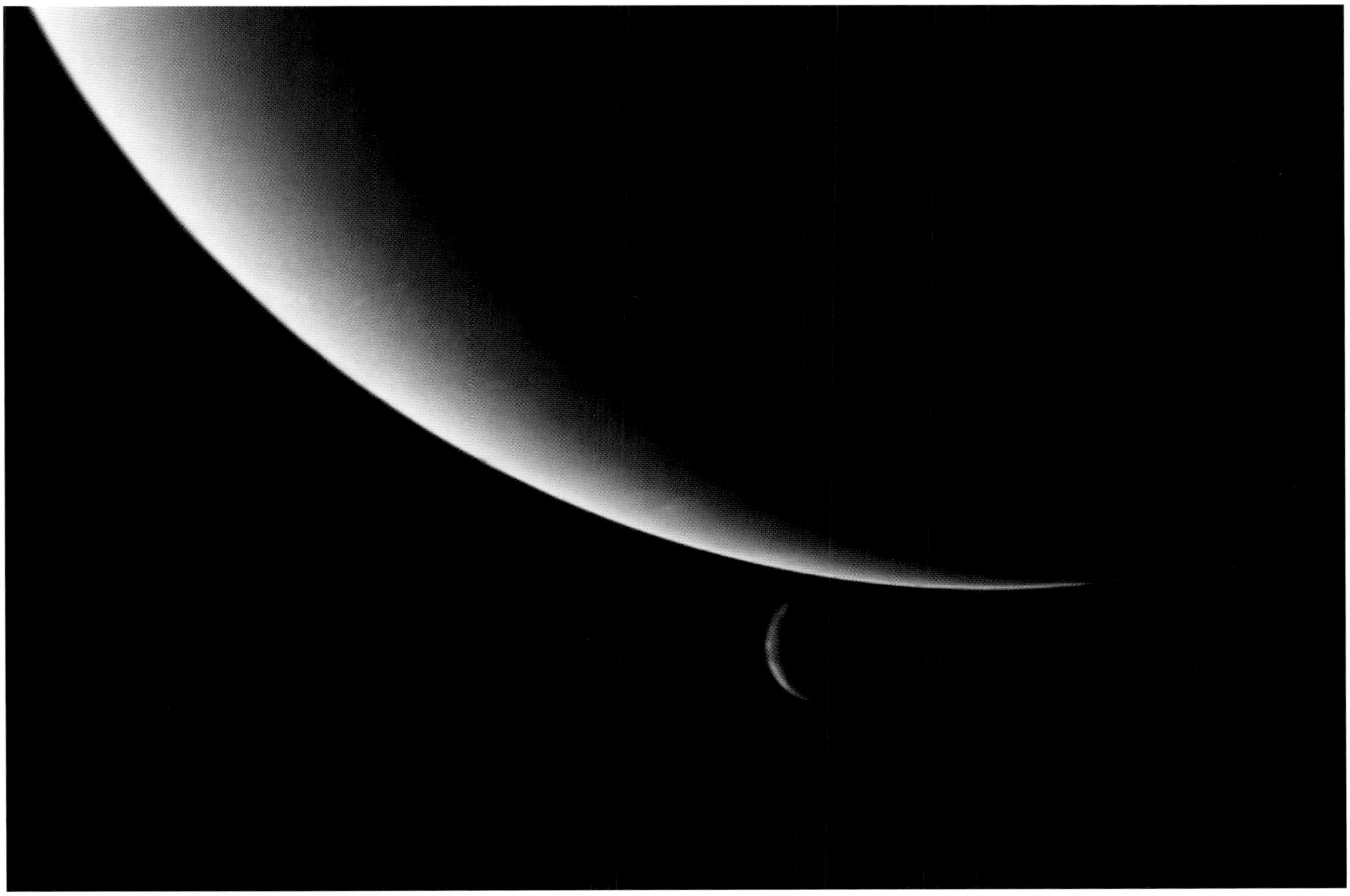

DISSECTING NEPTUNE

The distant but clear images of Neptune already revealed a large, Earth-sized dark oval in Neptune's southern hemisphere. The Voyager imaging team were quick to name the feature the "Great Dark Spot," taking inspiration from the similar looking Great Red Spot found on Jupiter. As the planet came into view, the close-up images revealed that the spot was made from a number of swirling clouds. Like on Jupiter, the dark spot found on Neptune was an anticyclonic storm, but unlike its red counterpart, this Great Dark Spot was relatively cloudless. It also showed that the winds found on Neptune are the fastest in the Solar System. Wind speeds were recorded at over 1,200 mph (2,000 km/h), nearly three times faster than the intense storm clouds found on Jupiter and nine times stronger than the most powerful hurricanes recorded on Earth.

Unlike Jupiter's spot, a storm that has lasted for many centuries, the lifespan of the Great Dark Spot of Neptune was short. Future observations of Neptune since the Voyager 2 flyby have revealed that the atmosphere changes quickly. The Great Dark Spot in the southern hemisphere had completely dissipated around five years later. Over the decades, several other dark spots have formed and disappeared, each lasting roughly two to five years. It is still unclear to planetary scientists how and why Neptune's atmosphere is so dynamic and why these spots emerge and disappear so quickly.

Voyager data had revealed to planetary scientists a new and unexpected subclass of planets within our Solar System. Voyager 2's analysis of Neptune's atmosphere and interior confirmed that Neptune was like Uranus, the second member of a newly identified subclass of the giant gas planet group. This new subclass became known as the name "ice giants" because of the larger amount of icy materials found within the planets compared to the more traditional gas giants, Jupiter and Saturn. The atmosphere of Neptune is similar to Jupiter, Saturn and Uranus, being composed primarily of hydrogen and helium, along with amounts of hydrocarbons. A higher proportion of ultra-cold methane in Neptune's atmosphere gives it its dark-blue color.

The mantle and core are more similar to Uranus, containing a higher proportion of water, ammonia and methane. It is predicted that at a depth of around 4,400 miles (7,000 km), the methane, under great pressure, breaks down into diamonds that rain down into a deep ocean of liquid diamond that contains solid diamond-like icebergs. The core of Neptune sounds slightly less exquisite, being likely to be composed of a number of metals and silicates.

Neptune has a similarly peculiar magnetosphere to the one found at Uranus. It was seen that the magnetic field was strongly tilted away from its rotational axis, and the magnetic center was offset by about 8,400 miles (13,500 km) from its physical center. The Voyager team had previously predicted that the strange tilted magnetosphere at Uranus was caused by the sideways rotation of the planet. Since the new data showed a similar magnetosphere at Neptune, the sideways rotation could no longer be considered the probable cause. With the new data, planetary scientists now think that the angled orientation of the magnetic field is caused by convective fluid motions within an off-centered thin spherical shell of electrically conducting liquids within the planet's core.

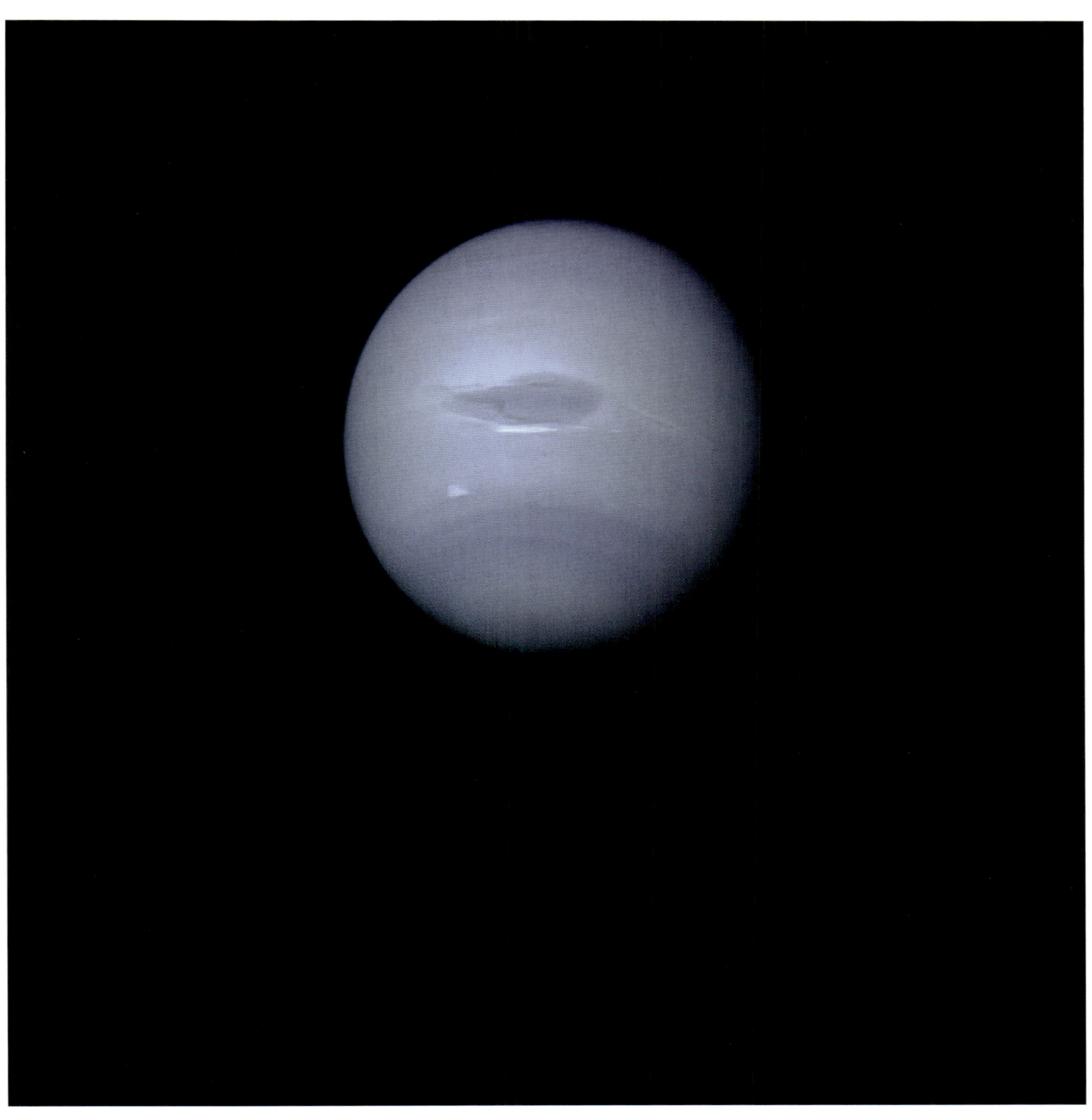

Depicted at the center, the Great Dark Spot (GDS) measures
8,100 miles (13,000 km) by 4,100 miles (6,600 km) in size. This
is roughly as large along its longer dimension as the Earth.

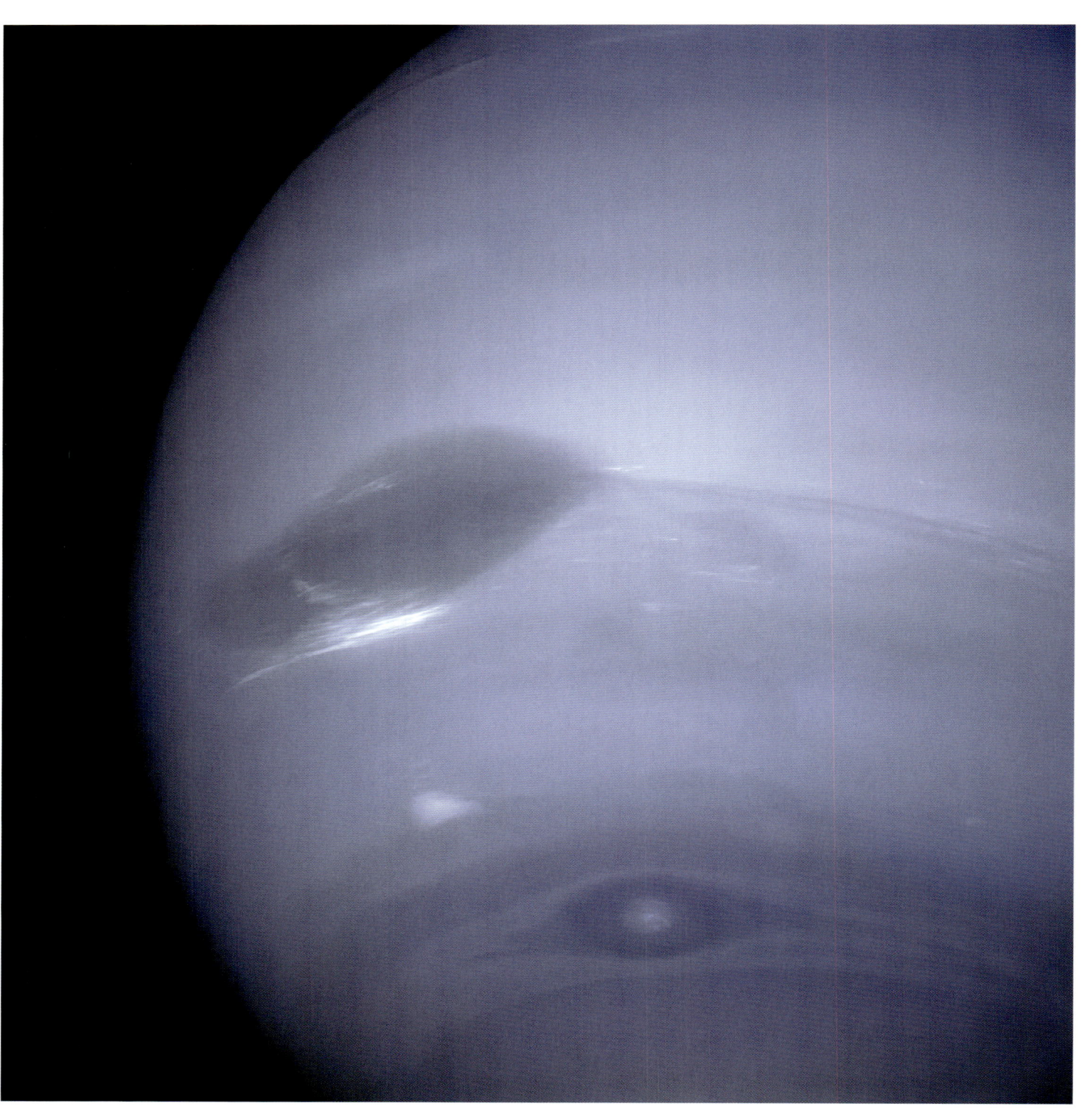

Bright white clouds, undergoing rapid changes in its looks, accompany the Great Dark Spot to the north. South of the GDS; are two features; one nicknamed 'Scooter' by the Voyager scientists and the other called Dark Spot 2.

Each feature moves eastward at a different velocity, so it is only occasionally that they appear close to each other, such as at the time this picture was taken.

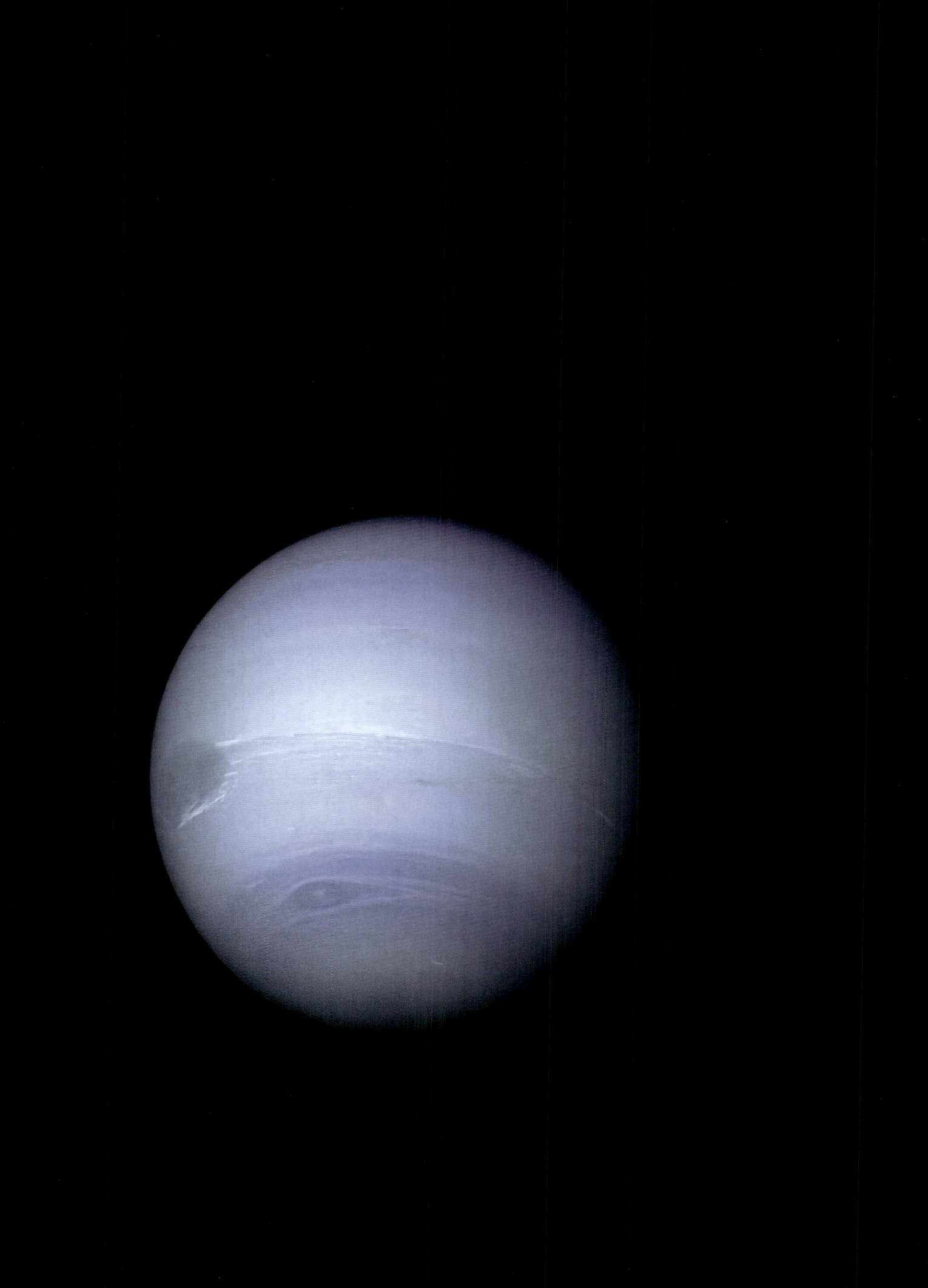

NEPTUNE'S RINGS

William Lassell, who discovered Triton in 1846, also reported the first possible mention of rings around Neptune. This claim was eventually dismissed as an observational error. In the following century, astronomers had invested vast amounts of time and money to see if they could find a ring system around Neptune. Many attempts did provide data to indicate that there was some kind of broken or incomplete system that orbited around Neptune. One of the most accurate descriptions was recorded by Patrice Bouchet, Reinhold Häfner, and Jean Manfroid from the La Silla Observatory (ESO) in Chile. They described five incomplete rings (arcs) that orbited around Neptune on July 22nd, 1984.

One of the goals from the Voyager team was to finally find out more about these rings. As like the previous flyby encounters, Voyager 2 waited until it had passed by the planet. Using the solar occultation trick once again, they pointed the camera platform to the area where the rings were believed to be. The imaging worked exactly as planned, answering a century-old unanswered question.

Voyager 2 showed a beautiful system of five separate rings that were not arced, but complete, albeit varying in thickness and density. The thinnest parts of the ring were made from dark dust-like material, impossible to see from Earth. It turned out that the Voyager flight path led it straight through the gap of the two outermost rings, and was hit by hundreds of thousands of tiny dust particles. Luckily, Voyager 2 emerged through the ring unscathed, and the particles were measured to be about as big as 1/100th the width of a human hair. The five main rings have since been named Galle, Le Verrier, Lassell, Arago and Adams. After the iconic astronomers who had made important discoveries within the Neptunian System.

It is still unclear why the rings around Neptune vary in thickness. Planetary scientists suggest that the ring clumps may be getting shepherded, like the rings around Saturn by tiny moons that were undetectable to Voyager's scanning platforms. During the solar occultation, the Voyager images also showed six new small moons that orbited around Neptune at distances between 30,000 and 73,000 miles (48,000 and 118,000 km) from the planet. In addition to Proteus, other moons were given the names Naiad, Thalassa, Despina, Galatea, and Larissa after mythological water deities.

Previous page: Neptune, Voyager 2, August 17th, 1989, 01:47:54.

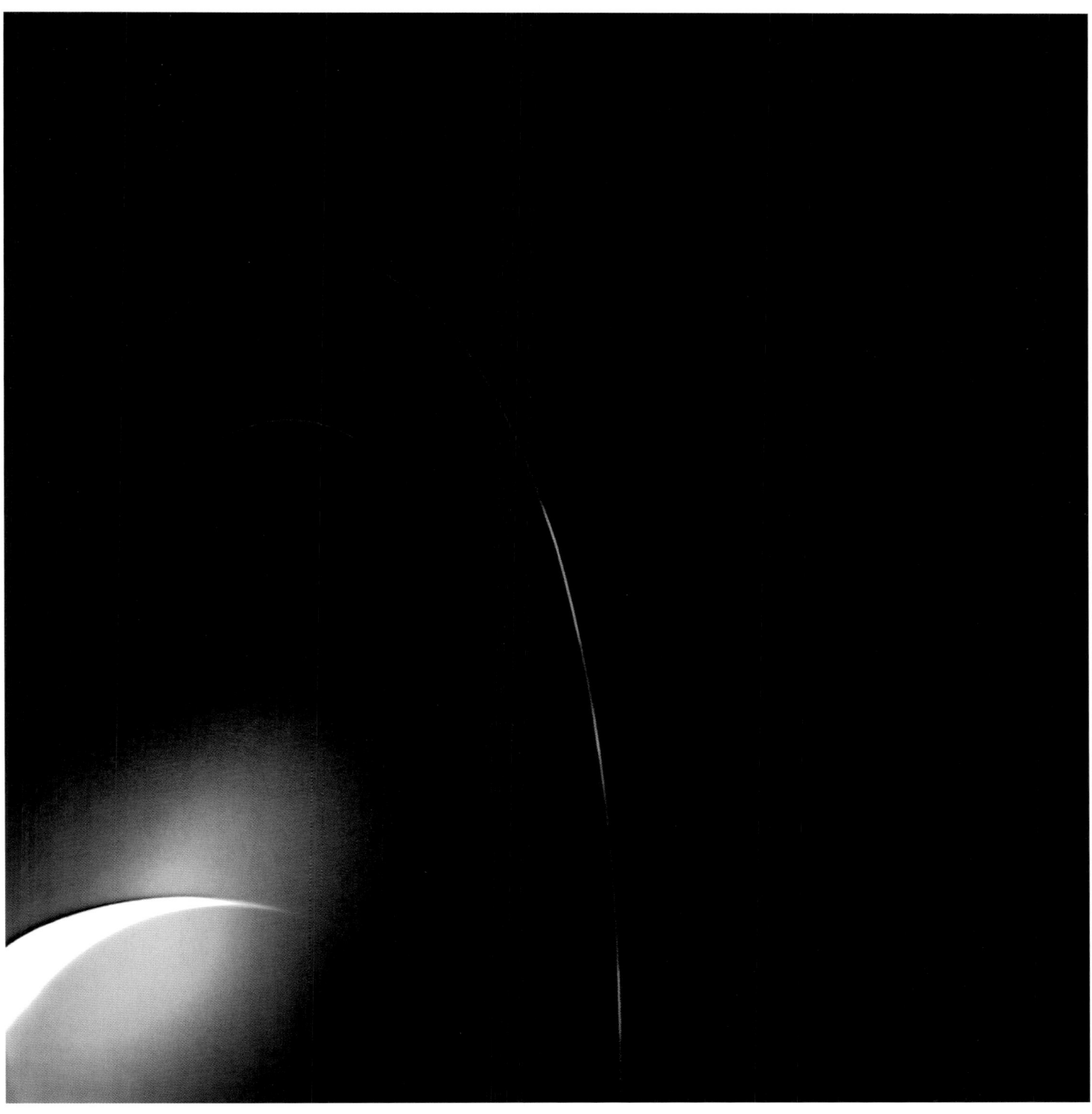

This image shows ring arcs, dense areas
within the otherwise faint rings. No planet
besides Neptune has these.

TRITON

Triton was discovered shortly after Neptune's discovery in 1846. Seventeen days after the Berlin Observatory had announced the new planet, Sir John Herschel—whose father, William Herschel, discovered Uranus—wrote to British astronomer William Lassell. The letter formed a rallying cry to search for moons orbiting Neptune. A mere ten days later, on October 10th, 1846, Lassell discovered Triton, which was not the hardest discovery in astronomy. Being a large and highly reflective moon, Triton is easy to spot if you know where to look.

Astronomers have been pretty puzzled by Triton's behavior since it was discovered. Voyager 2 confirmed during its flyby that the diameter of the moon is 1,680 miles (2,710 km), making Triton the seventh largest moon, and the tenth largest object in the Solar System. They observed that Triton performs a retrograde orbit around Neptune. That means that the moon travels around in the opposite direction to the spinning of Neptune on its axis. A strange and unique phenomenon, being the only large moon in the Solar System to orbit this way. It was hoped that the close flyby of Voyager would help explain why this occurs.

Earth-based spectroscopic studies had suggested that Triton would be similar to Saturn's largest moon, Titan. This would make it similarly composed of a mix of potentially life supporting organic chemicals such as ammonia, nitrogen, hydrogen and oxygen. Planetary scientists also predicted that its retrograde orbit could easily create the huge gravitational forces to cause tidal kneading to heat the moon's interior to form a subsurface ocean.

Voyager 2 took more than three years to travel from Uranus to Neptune. Every now and again, it would take distant shots of Neptune and Triton, to see if any valuable data and science could be obtained that might benefit and improve the close flyby mission. As the sets of photographs of Triton were sent back to Earth, it showed Triton to be smaller than the original calculation made in 1954 by Dutch astronomer Gerald Kuiper, and the surface to be significantly brighter. This vital information allowed the Voyager imaging team to shorten the exposure time to ensure quality imagery would be captured.

Five hours after Voyager 2 made its closest approach to Neptune on August 25th, 1989, the spacecraft passed the surface of Triton from 25,000 miles (40,000 km). The close flyby allowed detailed images of the moon's surface. The images that were brought back showed a relatively flat surface, atypical of an icy moon's surface. With the lack of impact craters, planetary scientists suggested the surface was very young, with regions varying between 6 to 50 million years old. A variety of strange types of terrain were seen, a fusion of bright and smooth plains, scattered with terraced depressions that most resemble frozen lakes. Other areas showed a darker surface that featured countless pits and ridges, like the skin of a cantaloupe melon.

Much of Triton's surface is covered in translucent reddish surface layers of nitrogen ices, as well as nitrogen snow and frost layered over much of the observed surface. The clean icy surface explains the reason for Triton's high visibility and brightness. The surface shows deposits of Tholins, which is made from a wide variety of organic compounds such as carbon dioxide, methane and ethane that has been formed by solar and cosmic rays. Tholins are considered to have been the precursor chemicals to the origin of life. Data scans also revealed that Triton had one of the coldest natural surfaces in the whole Solar System. Understandably from being so far away from the Sun. Average temperatures of just 38 degrees above absolute zero: -391 °F (-235 °C).

The strange retrograde orbit that Triton travels with around Neptune, proves that Triton did not originate in the Neptunian System. Instead, it was captured and pulled in from elsewhere, either from a collision with another object or that the object that Triton originated from had slowed. This caused its kinetic energy forces to become less than the gravitational forces from Neptune and thus was swallowed up by the planet. This capture of Triton may help to explain why there are very few large moons orbiting Neptune compared to the other gas giants. This suggests that the introduction of Triton to the system would have disrupted objects in regular orbit, such as seen with the highly bizarre orbit from Neptune's third largest moon Nereid, and that Triton collided and pushed away the other smaller moons that may have been present before its capture.

It is believed that Triton originated from the Kuiper belt, a ring of small icy objects that orbit from just within orbit of Neptune. The Kuiper belt is understood to be the origin of most of the short period comets we see passing through the night's sky. It is also home to the dwarf planet Pluto. Triton is slightly larger than Pluto and nearly the same in composition, which further provides evidence of Triton's origin. From newer observations and data, it is now understood Triton was part of a group of objects that formed a "binary system" that eventually encountered Neptune. This crossing broke up the binary system, slowed Triton down and brought it into Neptune's orbit. During that movement, it is considered that at least one collision occurred between Triton and another moon.

Right: Triton, Voyager 2, August 25th, 1989, 10:15:08.

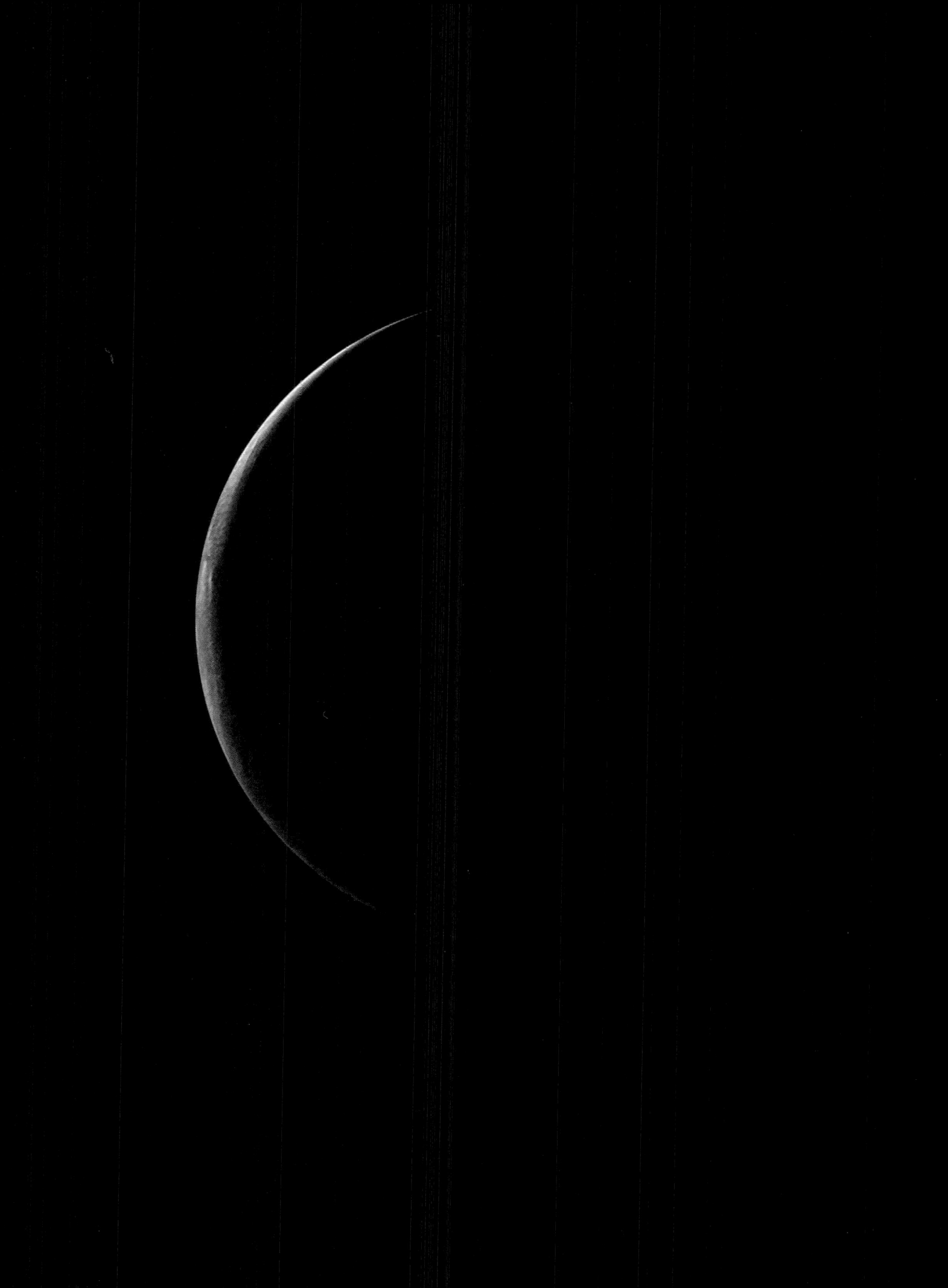

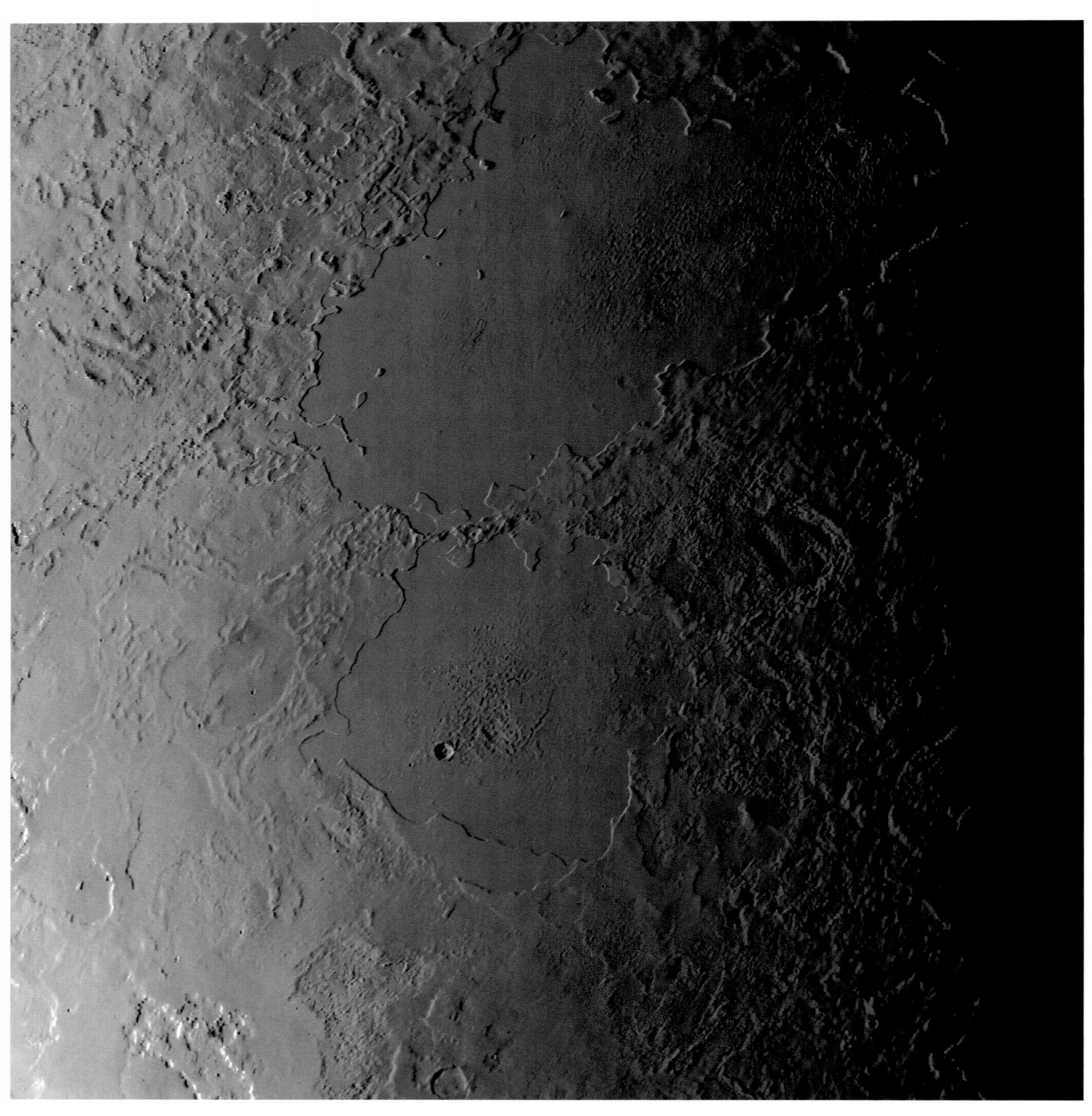

Tuonela Planetia (top) and Ruach Planitia
(bottom) are frozen nitrogen lakes with a
"land bridge" between them.

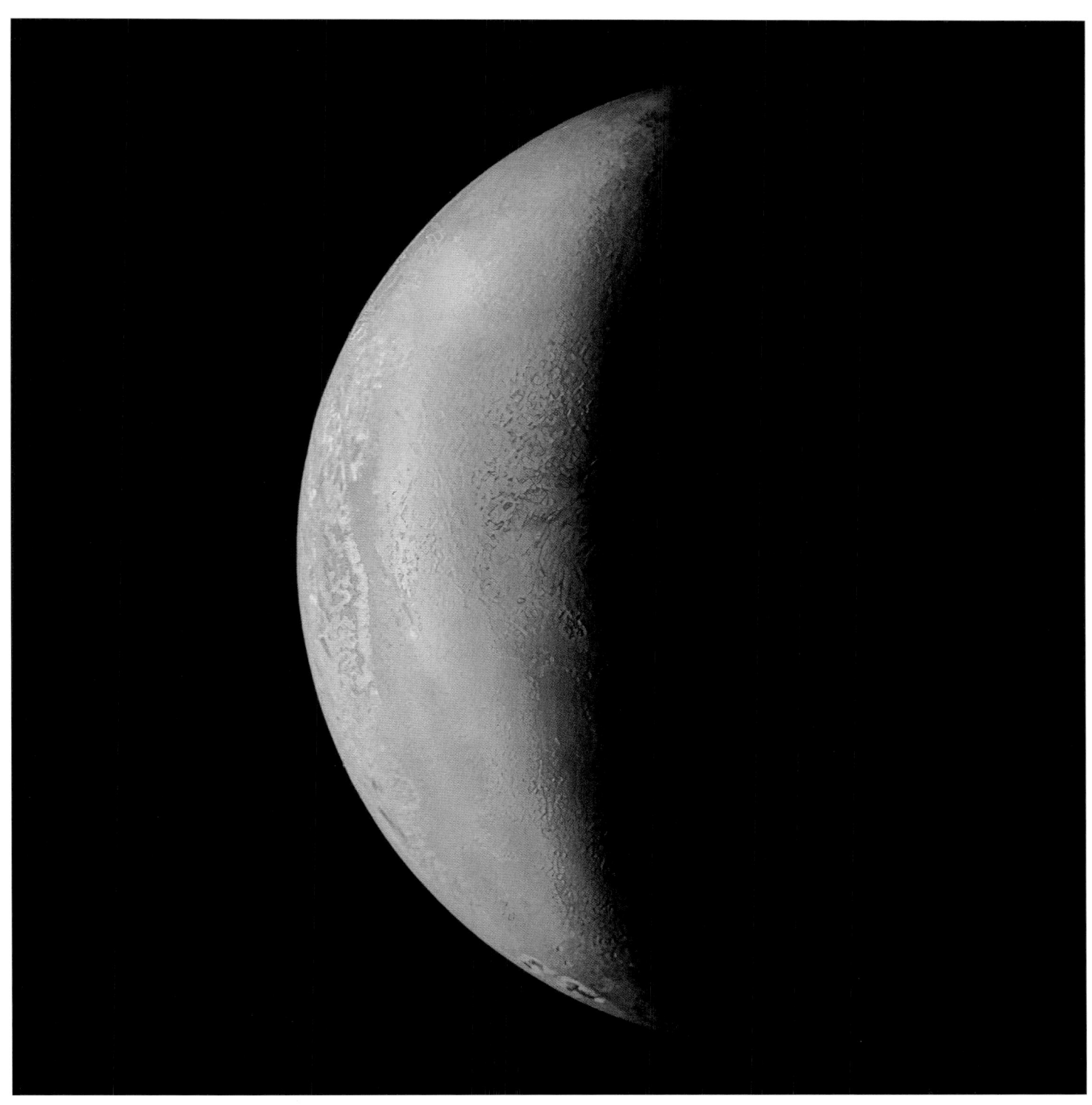

TRITON'S GEYSERS

One of Triton's biggest discoveries came from an unexpected source. Larry Soderblom, part of the Voyager imaging team, was making some short time-lapse movies of Triton. As he was compiling the different time images, he noticed something a little out of place. He saw dark and bright streaks that turned out to be plumes rising more than 5 miles (8 km) above Triton's surface of Triton, and spreading out more than 60 miles (96 km) downwind. Voyager 2, had, by pure chance, flown by and captured clear images of four erupting geysers of nitrogen gas and entrained dust and minerals on Triton's surface. Triton joined Earth and Io (and would later be joined by Europa and Enceladus) as one of the few places in the Solar System where active eruptions have been observed. The geysers were all spotted in an area between 50° and 57° south of Triton's equator. This is close to the subsolar point, where the Sun's rays strike Triton nearly perpendicular to the surface. It suggested to the Voyager team that, even with the great distance from the Sun, solar heating was playing a crucial role on Triton's surface. They proposed that the solar radiation would pass through the thin ice sheet on the surface, with a kind of "greenhouse effect" that would slowly heat and vaporize the subsurface nitrogen. Eventually, the gas pressure became too great and broke through the ice surface, creating the geyser, blasting up nitrogen gas and debris 5 miles (8 km) high. It has since been observed that each eruption may last for a whole year, each depositing large amounts of nitrogen ice.

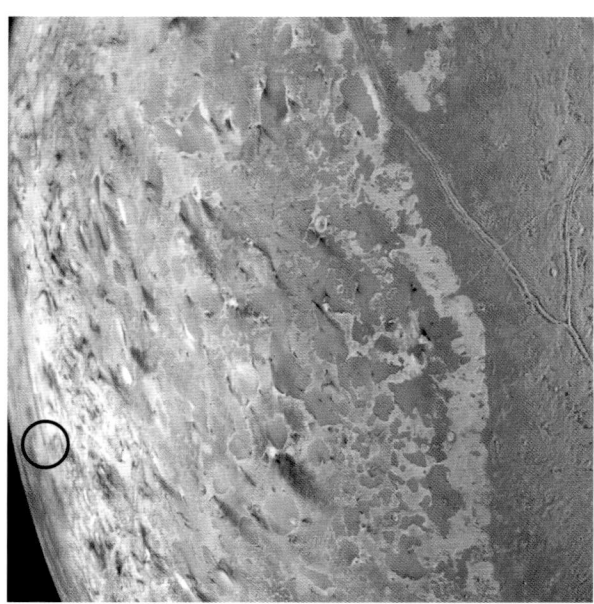

Around 50 dark plumes (one circled for reference), thought to be ice volcanoes, can be seen in the image of Triton's south polar terrain.

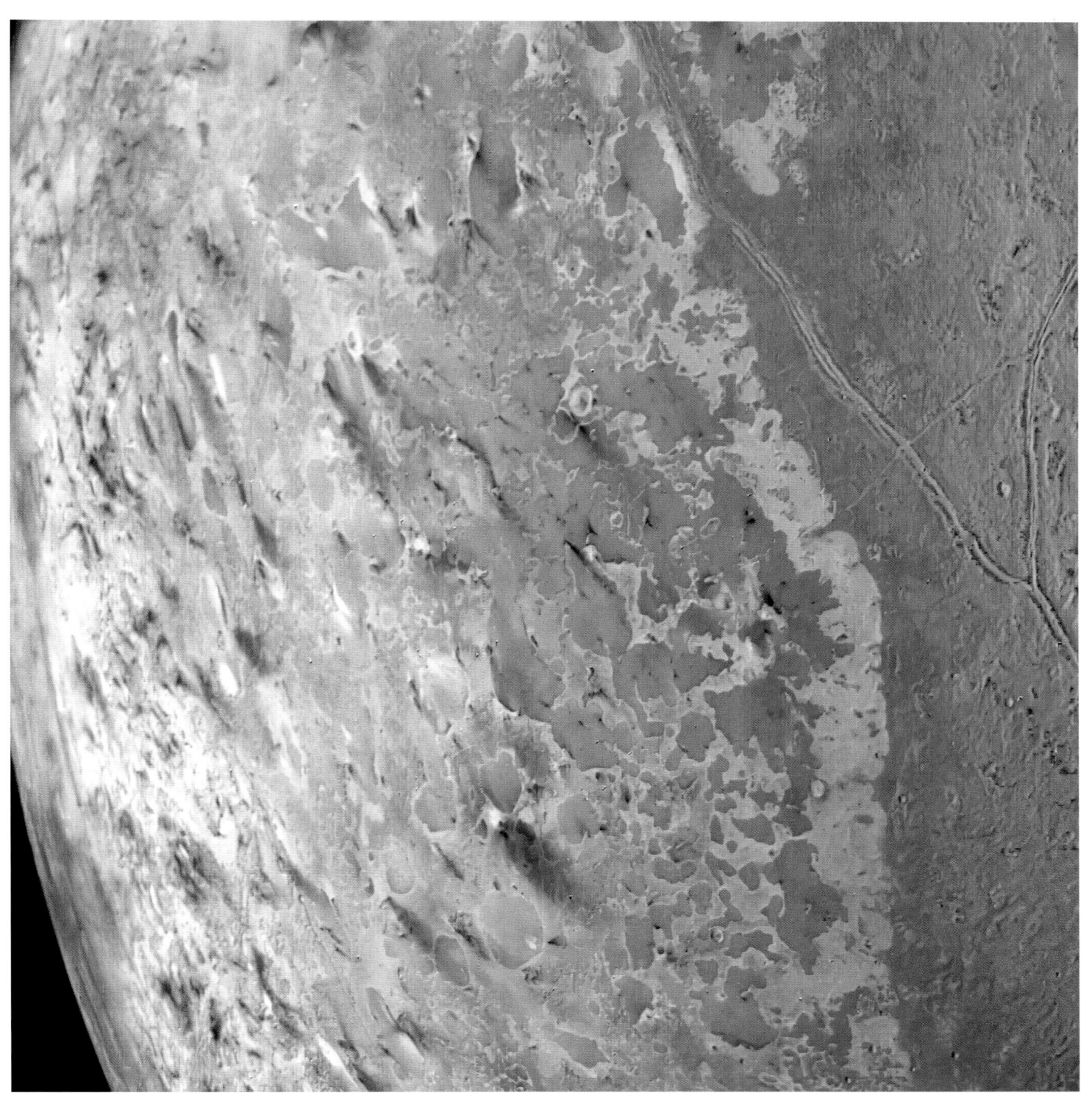

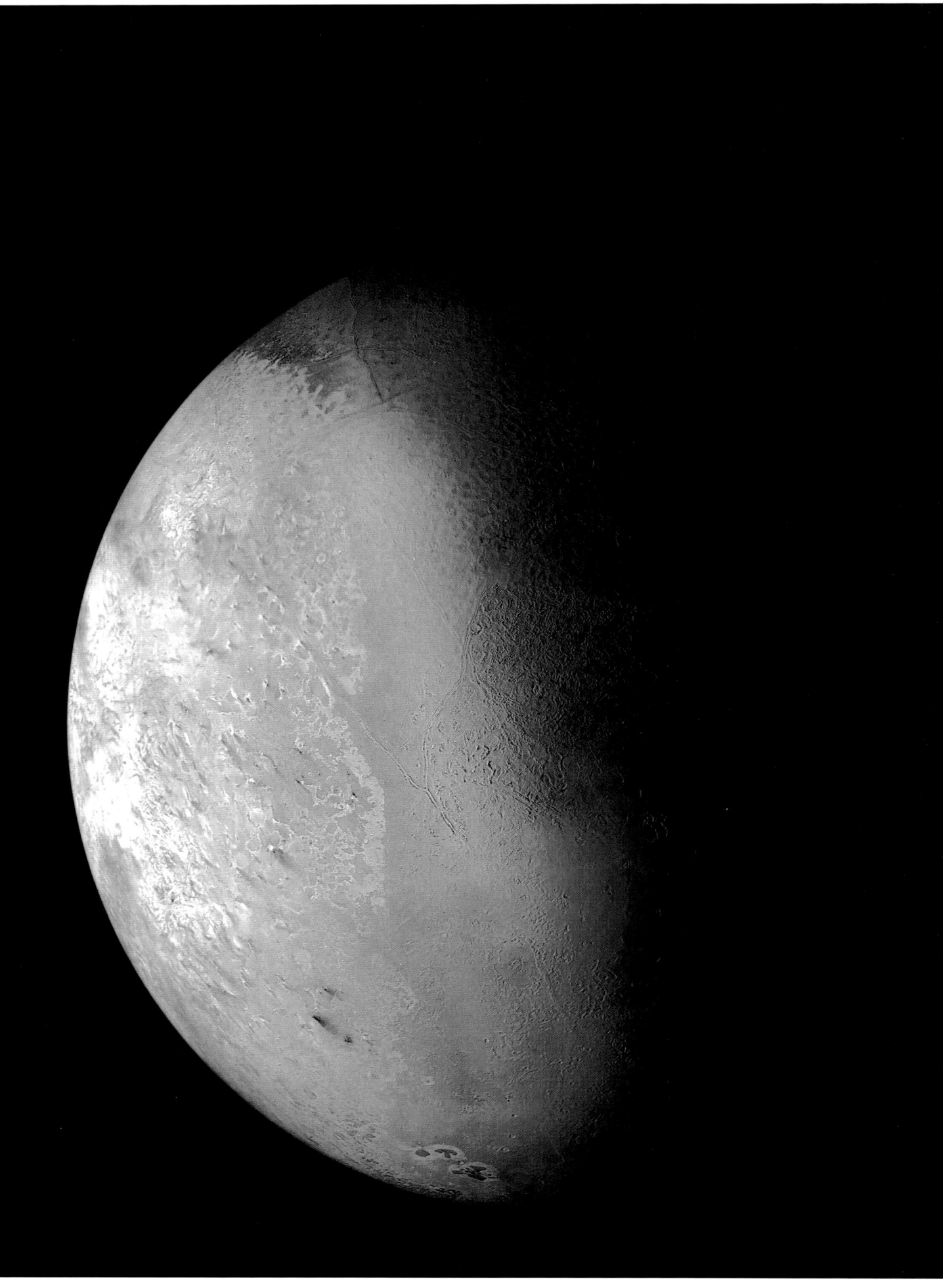

Triton's infamous cantaloupe terrain is shown here at the foreground. Forming large numbers of rising b obs of ice, this rugged terrain is believed to be formed when the icy crust of Triton underwent wholesale overturn.

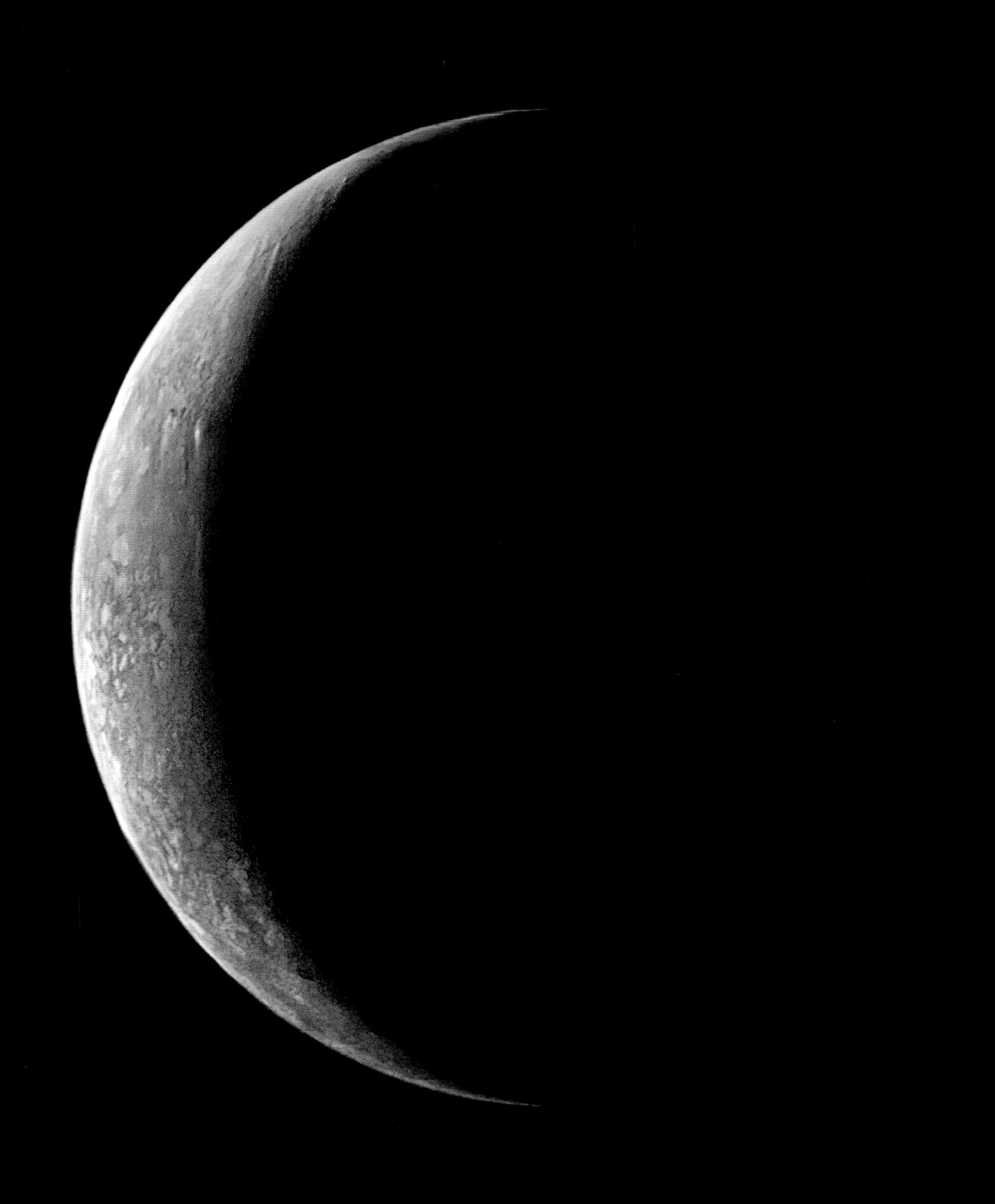

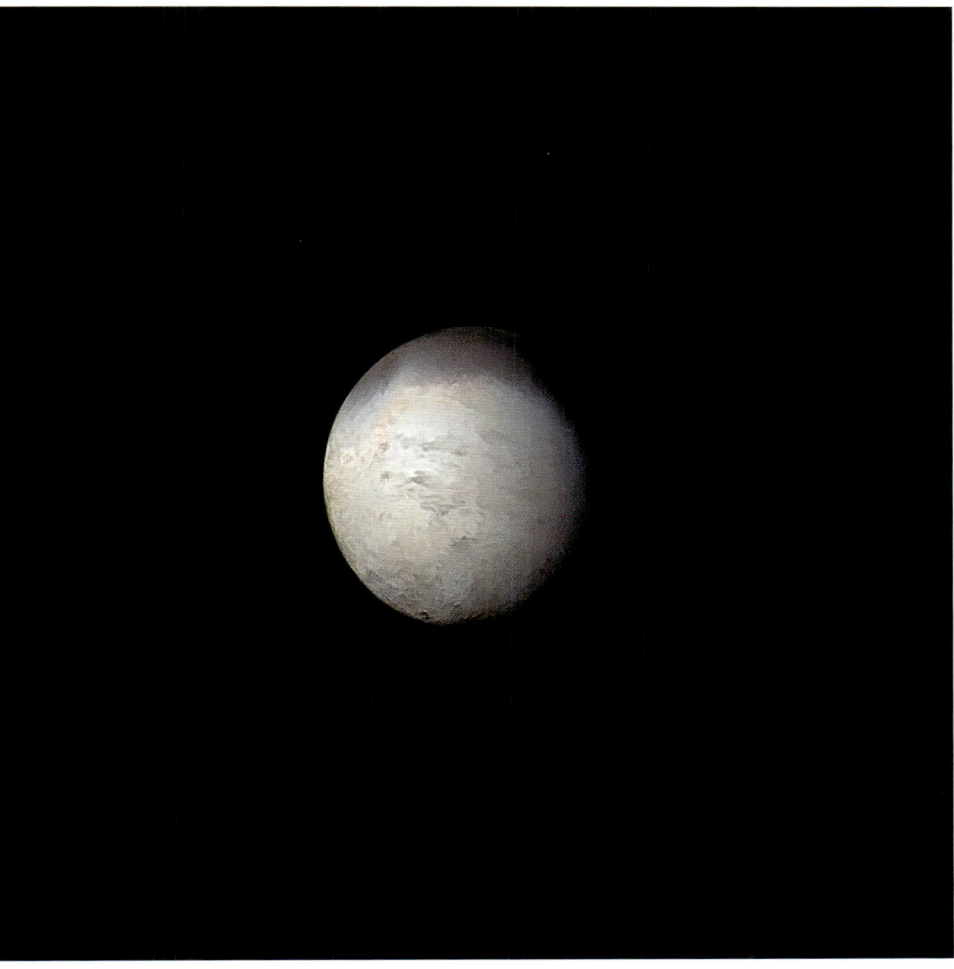

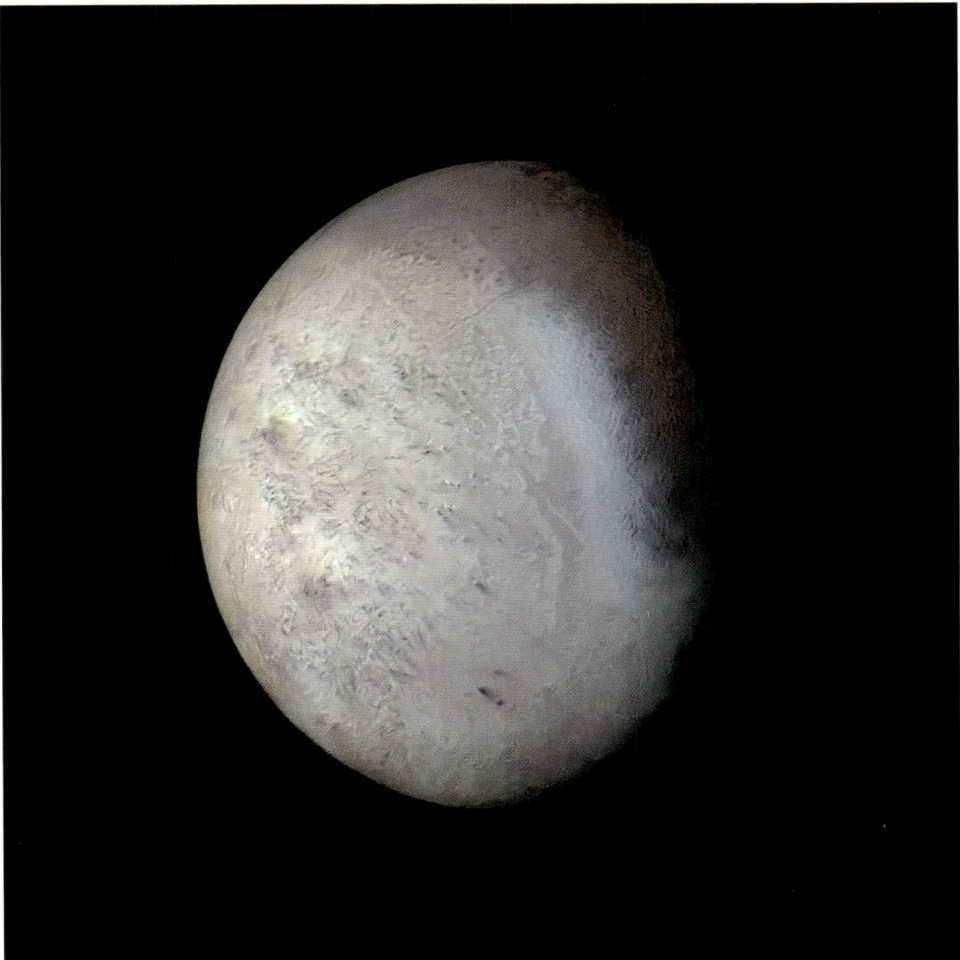

Left: Triton, Voyager 2, August 25th, 1989, 16:55:54.

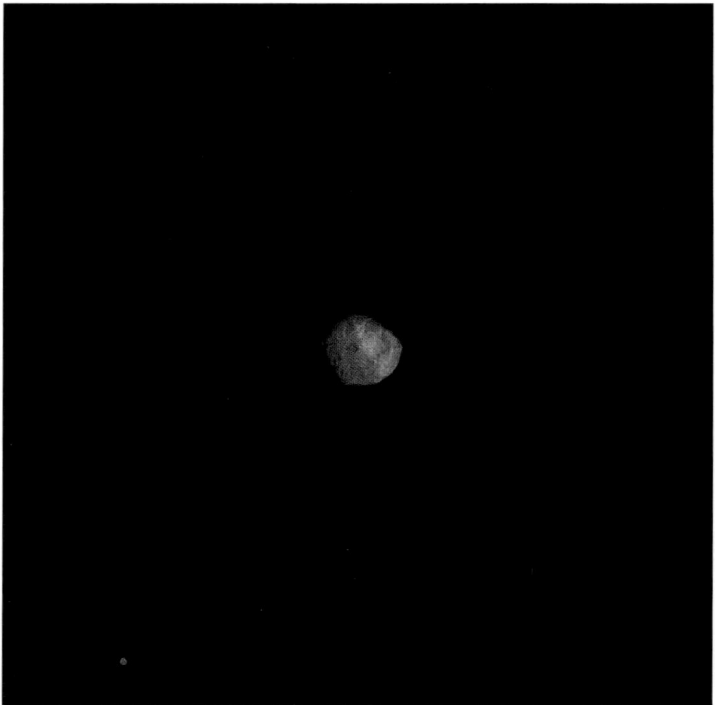

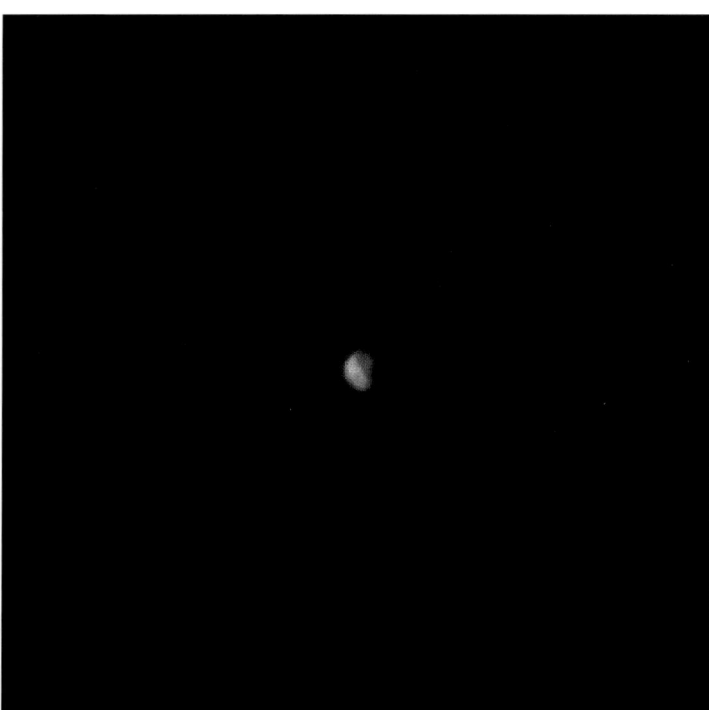

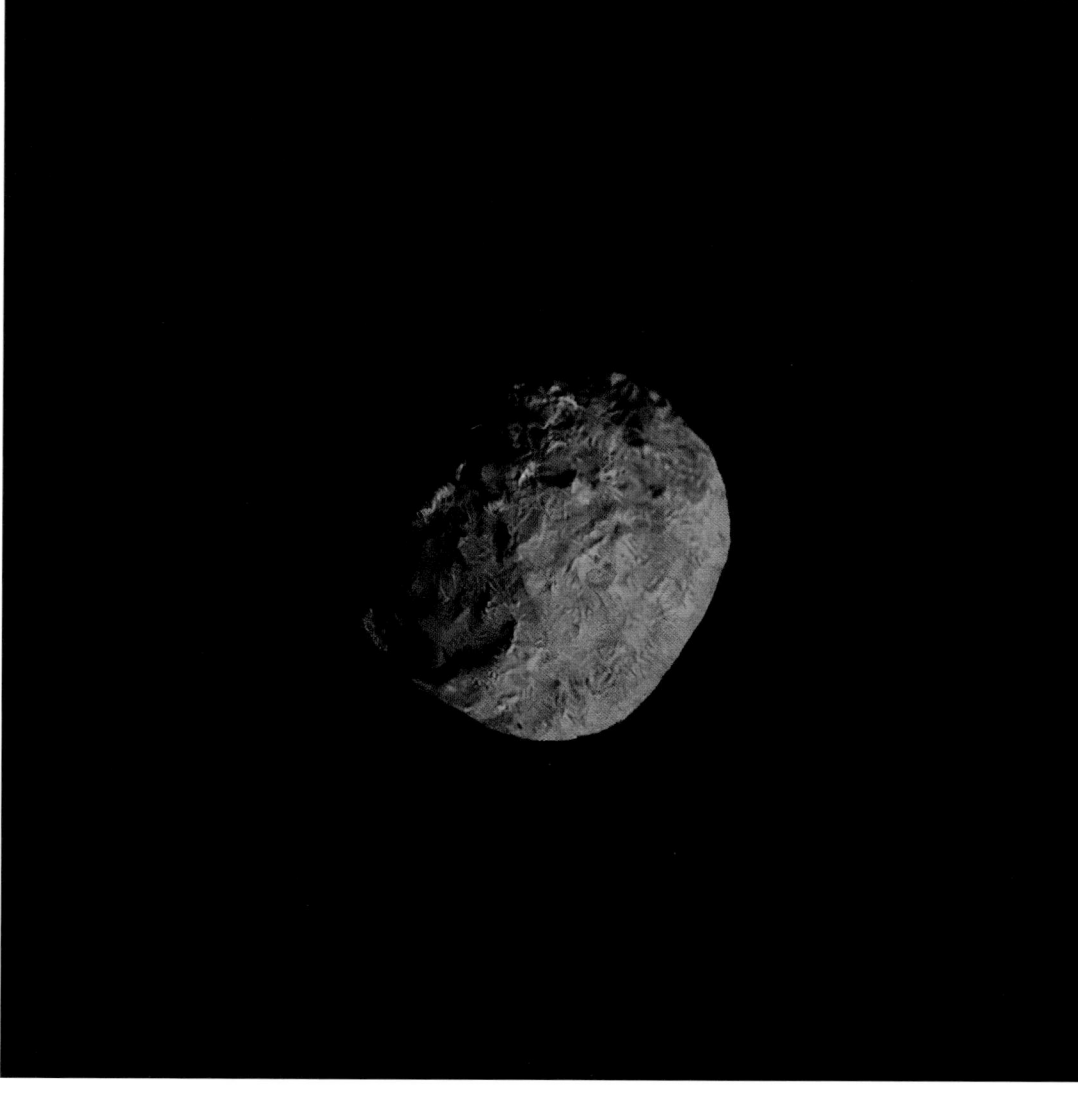

The three small inner moons of Neptune, captured for the first time. Nereid (top right), the largest, spans only 211 miles (340 km) wide. Proteus (bottom) isn't much smaller than Enceladus and Miranda, yet it is extremely primitive and isn't even round.

ONWARDS

The flyby of Neptune provided humankind's
first real encounter with the farthest planet in
our Solar System. The never-before-seen images
of an entirely new world were broadcast all around
the world for the first time. A beautiful result of the
remarkable feat of engineering and science that led
to one of the greatest voyages of exploration we have
ever embarked on.

Voyager 2 revealed a surprisingly active world
given how far away it was from the Sun's solar
energy. Storms the size of Earth were seen.
The fastest wind speeds in the Solar System
were recorded, faster than 1,000 mph (1,600 km/h).
The mysteries behind Neptune's ring system were
unveiled, revealing a faint and in places almost
invisible ring system. Six new moons are now known
to orbit around Neptune. The flyby revealed cold,
relatively small Triton to be an incredibly active world.
Amazing geyser plumes were spotted, so active that
they constantly changed the appearance of the
moon's surface.

It was as if the flyby of Triton signaled an end to
the interplanetary mission. On August 25th, 1989,
the Voyager team back home received the signal
confirming that Voyager 2 had passed the last of
Neptune's moons and was now dropping out of
the bottom of the Solar System. Time to join
Voyager 1 on its eternal interstellar mission,
among the vast emptiness of space between
the stars.

Right: Neptune & Triton, Voyager 2, August 31st, 1989, 07:35:40.

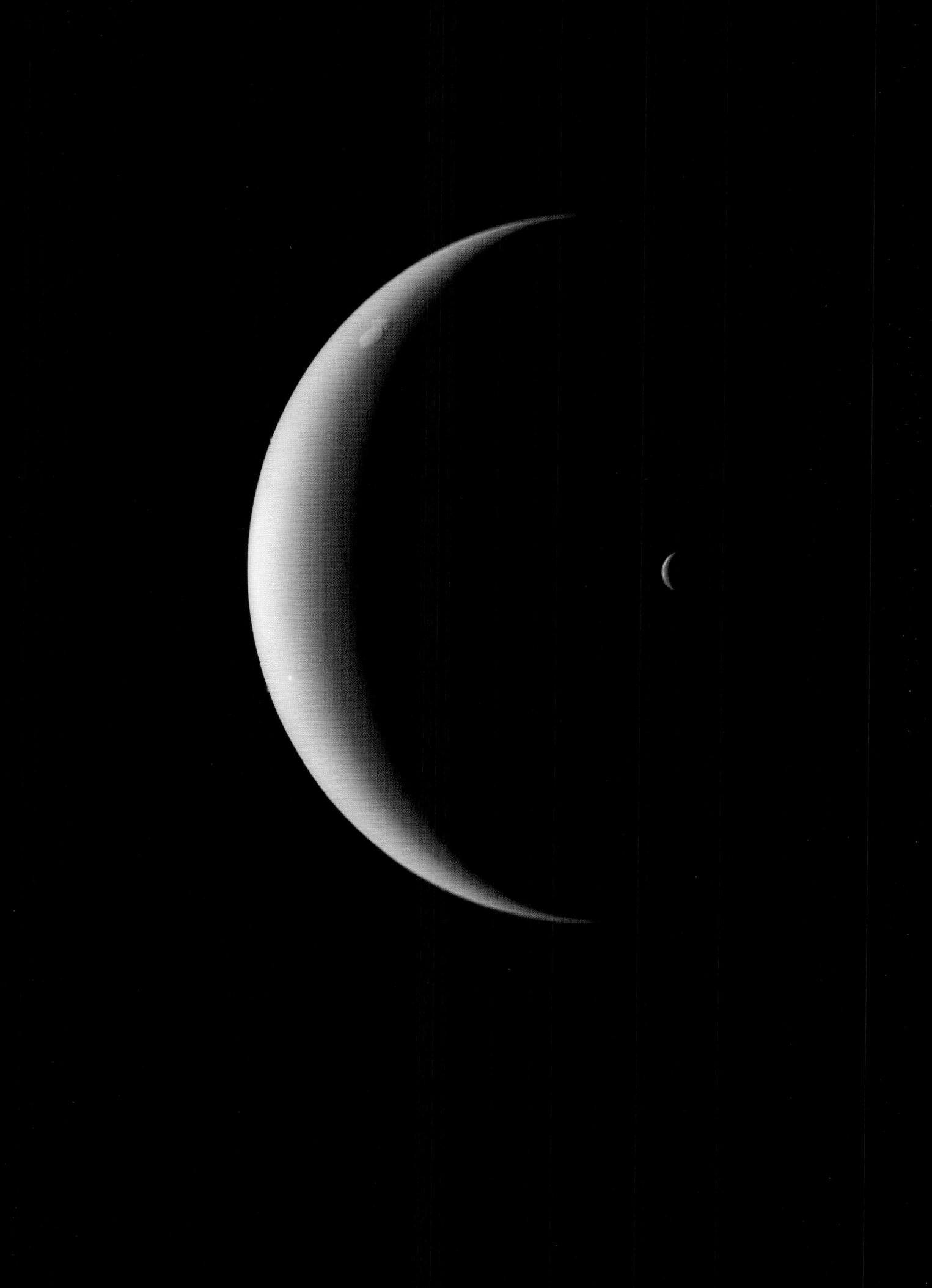

ALL OF PLANETARY EXPLORATION TO ME IS A STORY ABOUT LONGING. IT'S A LONGING TO KNOW OURSELVES. IT'S A LONGING TO UNDERSTAND THE SIGNIFICANCE OF OUR OWN EXISTENCE. IT'S A LONGING TO SAY TO THE UNIVERSE, WE'RE HERE. KNOW US.

Carolyn Porco

ARE WE THERE YET?

A PALE BLUE DOT

A farewell party

To finish off the Neptune encounter, to celebrate a great planetary journey, and to send Voyager off into the unknown, a big party was organized at JPL. The star of the party, who emerged from behind the wall of building 180, stepped onto the stairs of the administrative building and did a duck walk, was none other than Chuck Berry. A few minutes later the likes of Ed Stone and Carl Sagan were dancing to Johnny B. Goode being played live by the man himself. A piece of what was there on that festive night was simultaneously making its way out of our galaxy. At that moment, it was also time itself that appeared to be dancing.

After leaving behind a beautiful crescent of Neptune and Triton, the last port of call for this mission was passed. The countdown clock that was running went from counting down to counting up. The realization that the final thing the Voyager would see from the Solar System was now behind them, struck a chord with the whole team. A hugely successful mission ended, with enormous amounts of energy put into it over the years, while at the same time a new mission just started. Voyager began its interstellar mission.

One more historic first

With the end of the planetary mission also came the end of the photographic exploration. Although the spacecraft would continue on their way, higher and higher above the rest of the planets, their cameras would be turned off. This had nothing to do with a malfunctioning, but rather with the fact that the cameras used up a large part of the plutonium power supply's electricity. With the main science of the mission shifting to fields and particles experiments, this power was badly needed for the coming years. Besides, what fun is it anyway to take photographs of an endless stretch of darkness? There was simply not a photograph that needed to be taken anymore by Voyager.

Of course Carl Sagan would not have been Carl Sagan if he agreed with that. Already way before the end of the Neptune saga, when Voyager 1 completed its planetary mission in 1980, Sagan and his team started thinking about one more historic first that the cameras on the craft could achieve. A perspective they felt should not go to waste. The farther Voyager 1 moved out of the ecliptic plane, where the planets orbit, the better its view of the Solar System behind became. This quickly led to the realization that it would not only become possible to take a portrait of Earth, but actually one from the entire Solar System family.

The opposition

Although it seems like an instantly appealing idea, the opposition to this Solar System selfie was strong. A first reason for this was something that is taught in your first photography class: don't point your camera directly at the Sun. The cameras on Voyager utilized telescope optics to focus their images. If a beam of sunlight manages to pipe down, it could heat the photo detectors and fry the system. "So what?" others would say, "We don't need them anymore now that Voyager 1's planetary mission is over." While the cameras did not have a lot more pictures to take, calibration or diagnosis of certain potential kinds of problems could still serve as a good backup functionality to test for the twin craft, Voyager 2.

With Neptune in the rearview mirror ten years later, that argument did not hold up anymore since the flybys were over and done with. So Carl Sagan, Candy Hansen and the other imaging team members made their case again for the Solar System family portrait. The debate continued, however, turning into a more fundamental discussion this time around. With budgets being cut, some within the organization were questioning the worth of spending time and money on what they saw as a superfluous stunt. This hard-boiled science-only mentality was prevalent throughout NASA during the 70s and the 80s. Although scientists today must demonstrate the value of their work to the general public, this was absolutely not the case back in the day. The "soft science" that was directed at education and public outreach was frowned upon, and if there was one face they attached to this notion it was the face of Carl Sagan.

Green light

Luckily, there were a few key officials within NASA that shared Sagan's vision of the historic, cultural and aesthetic value of the desired photograph. Associate Administrator for Science Len Fisk as well as Administrator Richard Truly were charmed by the idea, and when Carl Sagan saw Fisk in the room during a Planetary Society dinner in 1989, he went over and shared some truth with the man. Not much later, Fisk and Truly stepped in and guaranteed to the team that the needed people as well as the resources would be made available to take the family picture. So, on Valentine's Day 1990, the planets all put on their happy face and said "cheese".

The updated version using the latest image-processing
software of the Pale Blue Dot released by NASA to celebrate
the 30th anniversary of when the original image was captured.

Look again at that dot. That's here. That's home. That's us. On it, everyone you love, everyone you know, everyone you ever heard of, every human being who ever was, lived out their lives.

Carl Sagan

Of course, as with any big family, getting them in line was not that easy. It actually took more than just one photograph. Voyager was commanded to snap pictures of one planet after another so that in the end a mosaic could be formed. Although some had saturation from the blinding gaze of our parent star, and some were just filled with vastness of pitch-black, Candy Hansen knew this camera like the back of her hand and was therefore appointed to locate the family members. Neptune? Check. Saturn? Check. The picture where Earth should have been captured was filled with scattered light that made our marble hard to mark down. Then it was spotted a "pale blue dot," right within a ray of scattered sunlight.

The only home we have ever known

Voyager 1 and Voyager 2 did unthinkable discoveries on their planetary journey and captured high-resolution images of the four gas giants and their moons for the first time in human history. Still, that distant photograph taken on that Valentine's Day is the thing for which this mission is most remembered. Not for its aesthetic appeal, but more so for its meaning. A perspective concerning the fragility of our planet and our existence, in the face of that tiny dot, that rings even more true and relevant today as ever before. A perspective that was most eloquently formulated by Sagan himself, in his book Pale Blue Dot from 1994:

"Look again at that dot. That is here. That's home. That is us. On it, everyone you love, everyone you know, everyone you ever heard of, every human being who ever was, lived out their lives. The aggregate of our joy and suffering, thousands of confident religions, ideologies, and economic doctrines, every hunter and forager, every hero and coward, every creator and destroyer of civilization, every king and peasant, every saint and sinner in the history of our species lived there – on a mote of dust suspended in a sunbeam. The Earth is a very small stage in a vast cosmic arena. Think of the rivers of blood spilled by all those generals and emperors so that, in glory and triumph, they could become the momentary masters of a fraction of a dot. Think of the endless cruelties visited by the inhabitants of one corner of this pixel on the scarcely distinguishable inhabitants of some other corner, how frequent their misunderstandings, how eager they are to kill one another, how fervent their hatreds.

Our posturing, our imagined self-importance, the delusion that we have some privileged position in the universe, are challenged by this point of pale light. Our planet is a lonely speck in the great enveloping cosmic dark. In our obscurity, in all this vastness, there is no hint that help will come from elsewhere to save us from ourselves. The Earth is the only world known so far to harbor life. There is nowhere else, at least in the near future, to which our species could migrate. Visit, yes. Settle, not yet. Like it or not, for the moment the Earth is where we make our stand.

It has been said that astronomy is a humbling and character-building experience. There is perhaps no better demonstration of the folly of human conceits than this distant image of our tiny world. To me, it underscores our responsibility to deal more kindly with one another, and to preserve and cherish the pale blue dot, the only home we've ever known."

The original Pale Blue Dot image released
by NASA taken on February 14th, 1990.

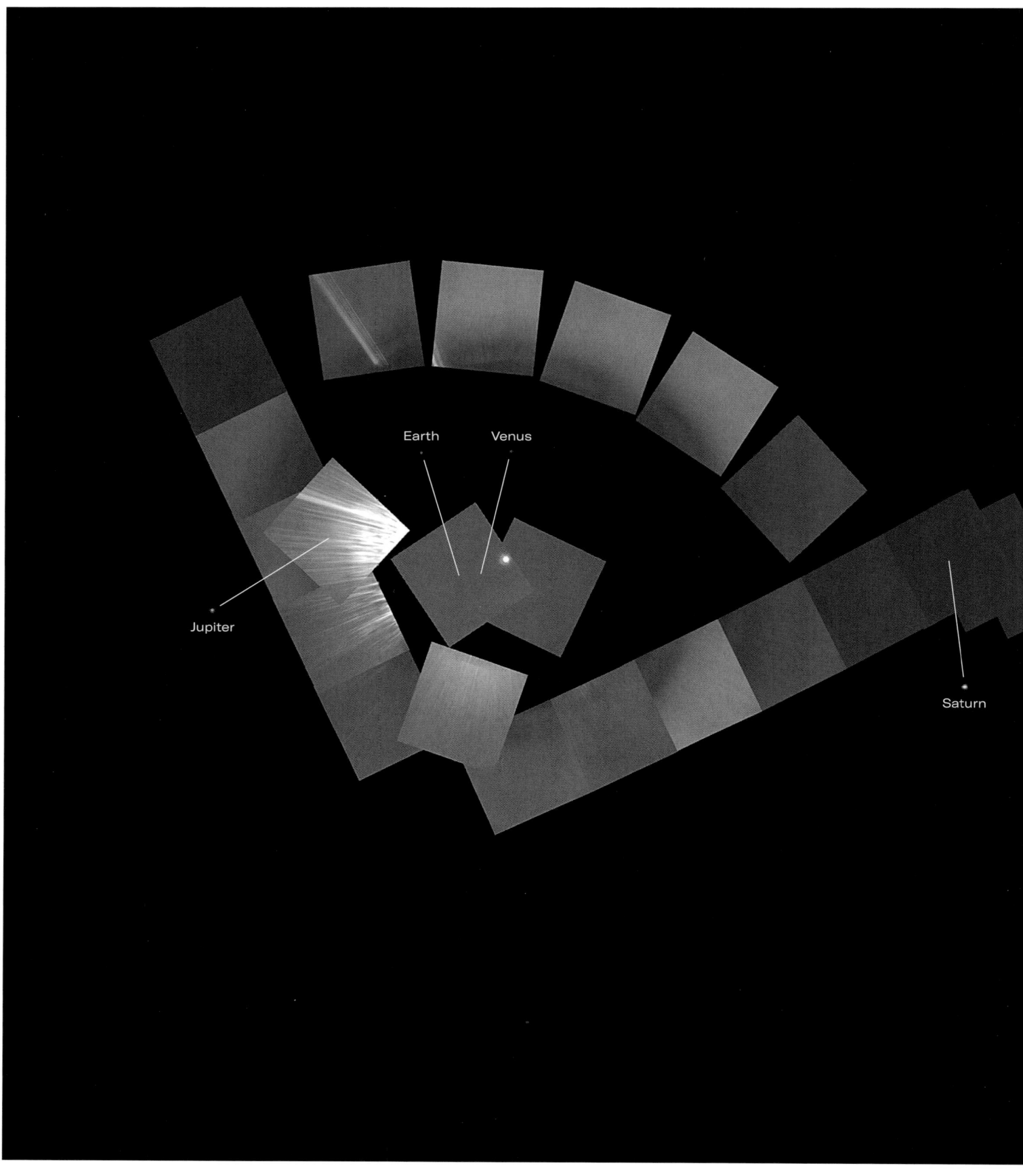

Uranus

Neptune

Earth
Venus
Sun
Jupiter
Mercury
Mars
Saturn
Uranus
Neptune

The Family Portrait was taken by Voyager 1 on February 14th, 1990. About 3.7 billion miles (6 billion km) from Earth, Voyager 1 took sixty individual frames that mosaic together to show Jupiter, Earth, Venus, Saturn, Uranus, Neptune and the Sun in one long image.

THE EDGE

Drawing the line

How do we determine when an object leaves our Solar System? Not an easy question. There is no border post at the end of the line, and there is no Google Maps Solar Edition that is able to draw a straight line and say "you have arrived at interstellar space." In fact, what that border looks like and where it is, was not common knowledge for people. For a long time the "communis opinio" was that when you got to Pluto, that was about it. When the "Voyager Planetary Mission" became the "Voyager Interstellar Mission," its main goal was to find out what exactly lies behind the planets. A search through uncharted waters for the edge of our Solar System.

That edge is naturally determined by the magnetic field of the Sun. This boiling hot star produces energy by fusing four hydrogen nuclei into one helium nucleus. In its core at temperatures of millions of degrees. This releases photons and subatomic particles that bounce around inside, and they eventually slip out into space. Sitting outside on a sunny afternoon, the sunlight that hits your face is made out of those photons and probably created inside the core of the Sun about 50,000 years ago. The solar wind is a flow of charged particles, primarily protons and electrons, that streams outward from the Sun. To determine the edge of our Solar System, this solar wind is crucial.

The gateway

The reason for this importance is that the solar wind creates a gigantic and spherical bubble around the Sun that is called the heliosphere. Every star has these cocoons that they blow around themselves. In our case, the heliosphere extends far beyond the orbit of Neptune, until it merges with the hydrogen and helium gas that permeates interstellar space. Simply said, inside the bubble resides the solar wind, outside the bubble is the territory of the interstellar wind. Finding the edge between them, is finding the gateway into interstellar space. Since this could tell us a lot about our own Solar System, as well as those of other stars, the next goal for Voyager was to find, discover and learn about this gateway.

Despite the fact that no one on the Voyager team had a faint clue on where the boundary would be, they extended the mission to find out. After Voyager 2 passed by Saturn, the mission was renamed "Voyager Uranus-Interstellar Mission," and after that it became the "Voyager Neptune-Interstellar Mission." "Interstellar" was always there.

With the launch and giant-planet gravity assists, both spacecraft had built up the speed to escape the Sun's gravity and pop out. But how long would it take? Ten years? Twenty? Thirty? Point of concern was whether the spacecraft would last long enough for the team to witness it at all. Voyager's plutonium power supplies would not last forever, so projections were that if the edge would not be found before 2020 kicks in, it might become a difficult feat to pull off.

Solar wind

Perhaps more importantly however, was making sure the team was able to detect and measure when the edge is found. As astronomers have discovered, solar wind streams are constantly in motion due to slow and fast winds. While the slow wind, which is an extension of particles created in the Sun's upper atmosphere, "only" hits a speed of 900,000 mph (144,840 km/h), the fast wind, which streams off the Sun's visible surface, blows more than 1.7 million mph (2.7 million km/h). Acting out their own weather systems these slow and fast winds can collide, fueled by particles called coronal mass ejections, this forms electromagnetic shock waves and sprays of ionizing radiation. These confrontations can lead to both gloriously beautiful aurora displays, as well as wreak havoc for orbiting satellites and surface power grids.

Why is this important? Well, the Voyager spacecraft have an instrument on board that was developed by Ed Stone himself, called the Cosmic Ray Subsystem. It is specifically designed to measure the high-energy particles from the Sun, and other stars, that we just mentioned. By identifying the difference between high-energy particles from the Sun (nuclei) and those from outside of our Solar System, this instrument can be used to map out the magnetic field. It is so called "squiggly line science," in that the instrument provides streams of data that need to be read. This illustrates the scientific change that came along with becoming an interstellar mission. Learning about the geological diversity of a planet is a job for a camera, but measuring energies and particles in space takes something else. It takes squiggly lines. Moreover, the fact that this instrument is on the Voyager spacecraft, is a beautiful testament to the perseverance of Ed Stone. It means that he had his eyes on interstellar discovery right from the get-go in the early 70s.

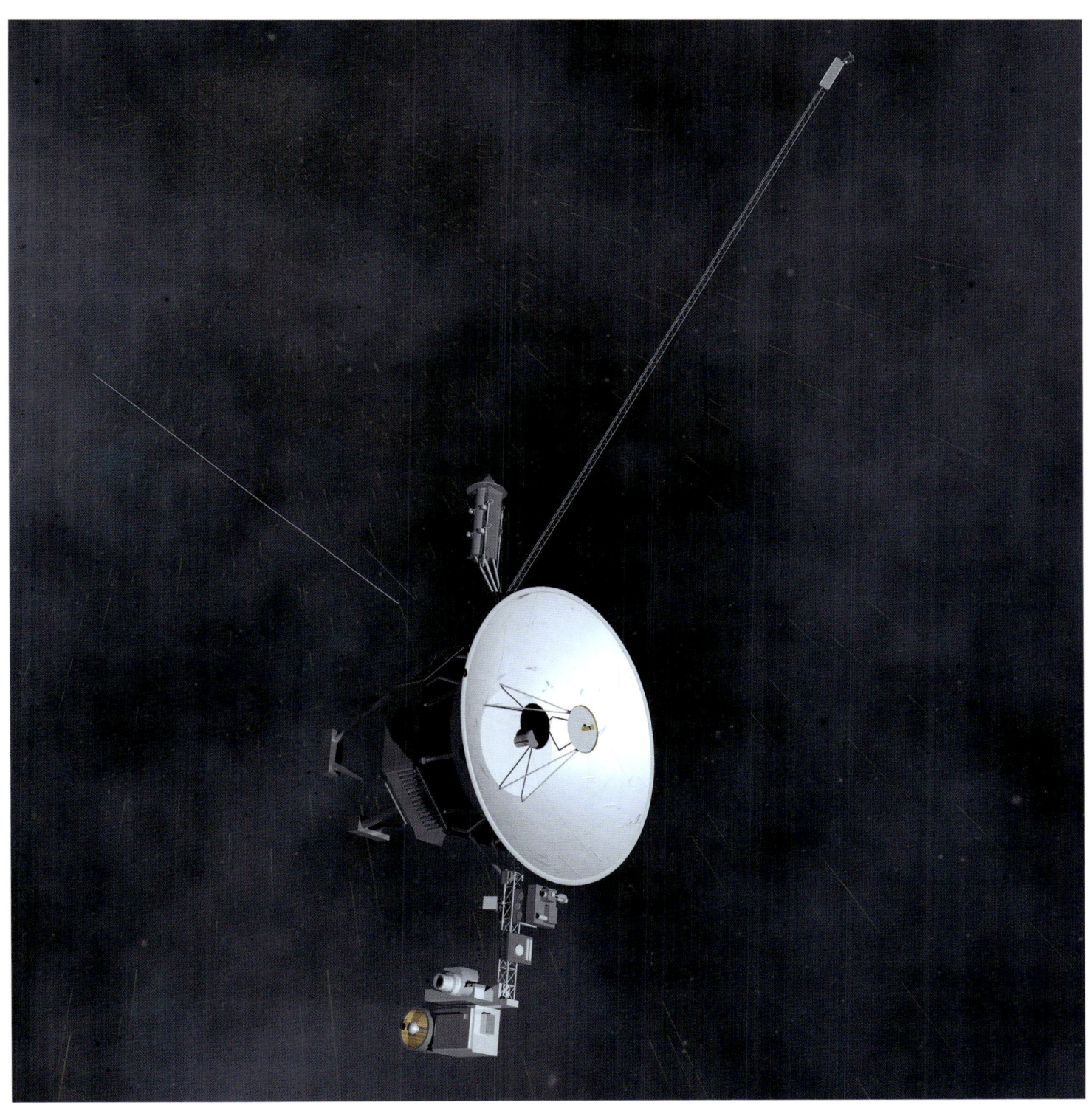

This artist's impression of the Voyager 2
spacecraft traveling out of the Solar System
at 34,000 mph (54,700 km/h). The solar wind
can be seen streaming past the craft four
times faster.

Dialing back

So as both Voyagers sailed on, it became a matter of waiting. This transitioned the science and operations teams into a different kind of mission. The imaging team, the ultraviolet spectrometer team and the four remote sensing teams were all disbanded, which helped to save the dwindling power and increase the chances of getting to the edge. The power is used to heat the computer, radio transmitter, receiver electronics, as well as the other instruments that were still operating. If it is left on its own in the deep cold space, possible cracks and breaks may cause the spacecraft to kick the bucket.

Dialing back on the equipment and the teams also had a budgetary motivation. The total amount it takes to keep the Voyagers running was around five million dollars per year. While that is still a great deal of money, it is a reasonable investment, given NASA's annual average budget of 17 billion dollars per year. That being said, Voyager went through eleven major reviews through the years to determine whether it would not be a better idea to pull the plug. Although the mission continued every time, scaling it back had a big impact. By 1990, those still on the team had been there for most of the ride since the beginning, when they were still youngsters. Now, they grew older and had kids on their own. The majority of their time, for about a full decade, was devoted to routine spacecraft maintenance. Although the mission originally occupied three full floors of Building 264, the one place for JPL's most high-profile projects, now they were being demoted into the former offices of a computer company in Sierra Madre. Suzy Dodd, JPL Project Manager is clear why they kept calm and carried on: "the Voyagers belong to all of us, they represent all of us, they will speak to the ages for all of us."

The remaining instruments

So if all of these teams were disbanded, and some of the equipment was turned off, what was still running? Five different science instruments became the main lifeline for the Voyagers. They touch, feel and smell their outer environment and measure the heliosphere since the beginning of the interstellar mission. Specifically, the instruments are used to detect the plasma ions in the solar wind, the force of the solar or the interstellar magnetic field, the directions and energies of the particles and the power of the radio waves possibly emerging from interstellar regions. The more years go by, the longer it takes this data to find its way back home to us. Since 1989, the NASA Deep Space Networks radio telescopes in the U.S., Australia and Spain have been listening and receiving the numbers carefully and faithfully. The signals are sent off by the Voyagers with a confident 23 watts. However, by the time it has traversed its way across the unfathomable distance to Earth, the strength of the signal has died down to a minuscule 0.0000000000000001 watts. When the data is received, the tape recorder is rewound and the whole trick is repeated. Year after year after year. Until we reach the gate.

The steps to the exit

Before Voyager was set to pop out of the Solar System, it was believed to transition through a number of distinct parts of the heliosphere. The first area that should show a shift is the boundary that is known as the "termination shock." This is where the speed and the direction of the solar wind is altered, caused by pressure from the interstellar wind outside the heliosphere. One-step further takes us to the "heliosheath", which can be seen as the skin of the heliosphere. Here, like water around the bow of a ship, a gush of interstellar ions slide around the outside of the sphere. Since no one knew how much pressure would be coming from outside the bubble, no one really knew how thick that skin is until the Voyagers reach the end of that part, the actual edge of our Solar System, which is called the "heliopause."

By 2004, at a staggering distance of nine trillion miles (14.5 trillion km) from the Sun, the models that were derived from the data started to show subtle changes. Voyager 1's magnetometer detected an abrupt increase in the strength of the surrounding magnetic field and a drop in the strength of the solar wind, suggesting that it had crossed the termination shock and entered the more turbulent heliosheath. A few years later, in 2007, Voyager 2 followed that path. Naturally, after a decade of routine spacecraft maintenance, these changes caused a ripple of excitement through JPL and the team. While the spacecraft push themselves forward with a speed of 10 miles per second (16.1 km per second), it still took 15 years to reach this point. Along the way, team members had left or retired. Those still on board fought with everything they had to keep this mission alive, even though sometimes the technical and the financial odds were stacked against them. Nevertheless, it was worth the wrangling, because they were getting close now. Heliopause, we are coming for you!

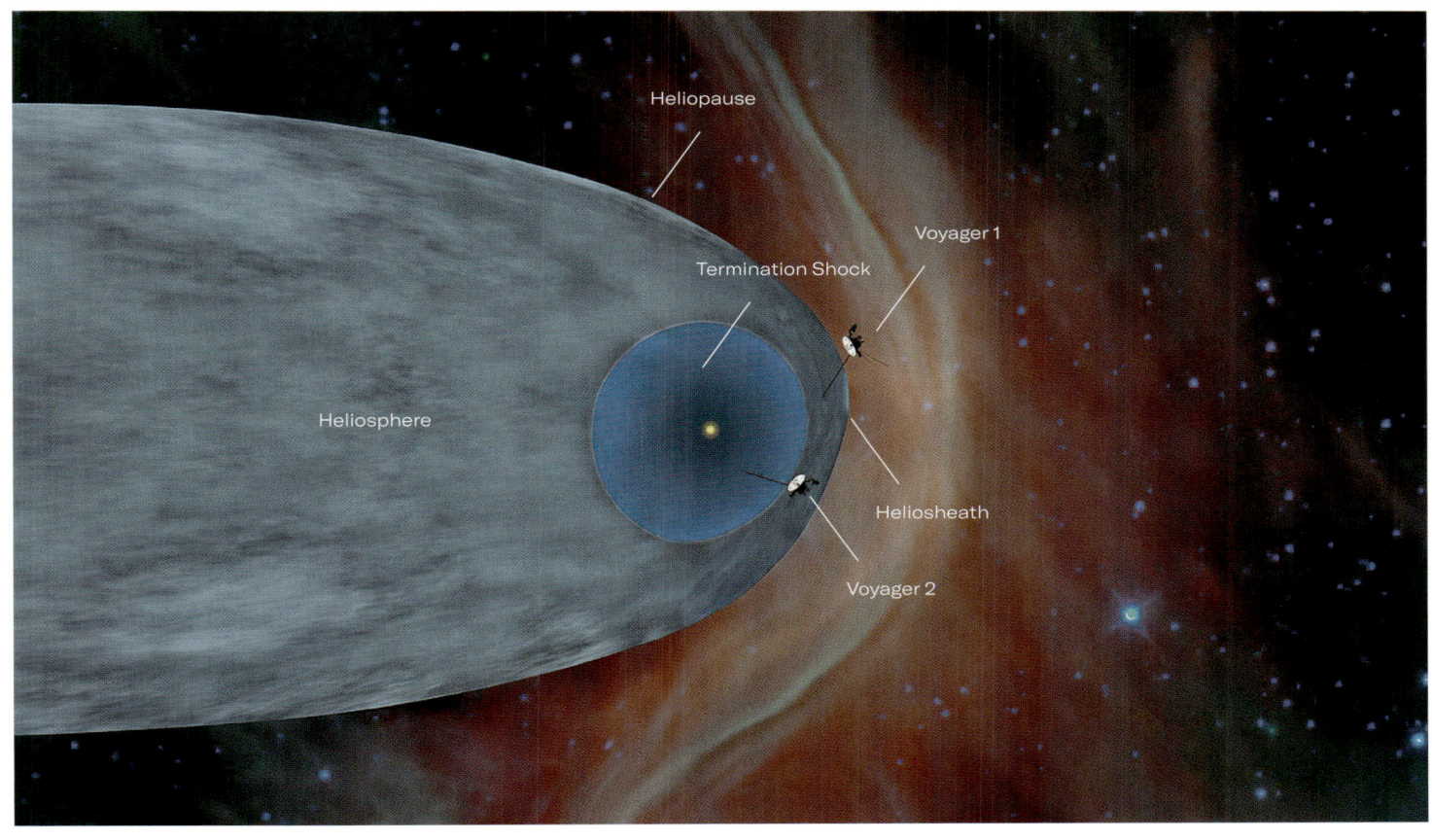

This illustration highlights the position of both Voyager spacecraft outside of the heliosphere, a protective bubble created by The Sun. Voyager 1 crossed over the edge of the heliosphere in August 2012. Voyager 2 followed later, in November 2018.

When it was just recently announced that Voyager One was in interstellar space, it was like humanity had just become an interstellar species. It was like knocking on eternity's door.

Carolyn Porco - Imaging Science

Are we there yet?

On July 28th, 2012, things started to get awry. A sudden and dramatic drop of 50% in the solar energetic particles was witnessed on Ed Stone's Cosmic Ray Subsystem (CRS) instrument. Right at the same moment, another instrument measured a big increase in the cosmic ray particles from outside the heliosphere. Just when the team started to figure it out and were ready to start drawing first conclusions, everything switched back in the blink of an eye and went back to normal. It was as if the universe was literally playing a trick on them.

Finally, on August 25th, 2012, the density of particles around the spacecraft precipitously increased, as though it had plunged from sky to sea. Now, it had dropped to nearly zero, and stayed there. Are we there yet? Just like before, the outside particles saw a steep increase as well. All solar energetic particles were now replaced by interstellar cosmic ones. For Ed Stone, he recalled it was like standing on the shore of a particle beach, your feet being hit by waves of water that constantly recedes, until at some point the tide changes and your feet are in the water indefinitely.

The smoking gun

It was not hosanna immediately. Being scientists, real proof they had entered interstellar space was still needed. The plasma and the magnetic field had not been measured yet, so there was no certainty. Perhaps the heliosphere itself was in motion or maybe had they entered an unexpected depleted region of the heliosphere? In search for more clues, they turned to the right direction of the magnetic fields. Based on their conceptual cartoons and computer models, by entering the heliopause the direction of the magnetic field would suddenly change. Unfortunately, the direction of the magnetic field had not changed at all, indicating that they were still in the heliosphere. At this point, from a particle point of view, they had broken away from the heliosphere. They just did not have the magnetic fields direction or the plasma to back it up, and the outside scientific community was building

the pressure for a smoking gun. One man grew determined to give it to them: Don Gurnett. As a professor at the University of Iowa and leader of the Voyager Plasma Wave Subsystem (PWS), Gurnett knew there was another, more indirect, way to measure the density of plasma. However, in order to do so, he needed the radiant power of the Sun. His instrument measures the size of waves that travel through the ionized atoms and molecules, providing information on the density and temperature of that place. Those waves become detectable through a disruption, which during the planetary mission was mainly caused by the magnetic field of the planets. However, at the edge of our Solar System, those planets are long gone. The only alternative, Gurnett and his colleagues knew, was a burst of energy from the Sun in the form of a solar flare. These were known to fire through the Solar System and make waves in the plasma.

Leaving the bubble

In May 2013, the Sun performed its duty. A solar flare flew by Voyager 1 and its particles caused a strong and detectable wave in the surrounding ionized gas. The electrons were moving back and forth with a frequency that fitted a region of space with 80 times the density of ionized particles as the normal heliosphere would have. Since the dictionary definition of heliopause is such a sudden peak in density, this finding was hailed as a eureka moment. According to Gurnett, it took the team ten seconds to figure it out. Yes, we are there. Although a debate still goes on within the scientific community, the results were officially published in Science in September of 2013. In addition, just like with crossing the termination shock a few years before, Voyager 2 followed its twin in 2018.

For the first time, humanity exited the bubble of our Solar System. Every second, the twin spacecraft are passing through a place that we have never been before. A momentous moment in the history of humankind. The objects that were launched back in 1977 to explore the planets and their moons are now set off on a journey into the truly unknown: interstellar space.

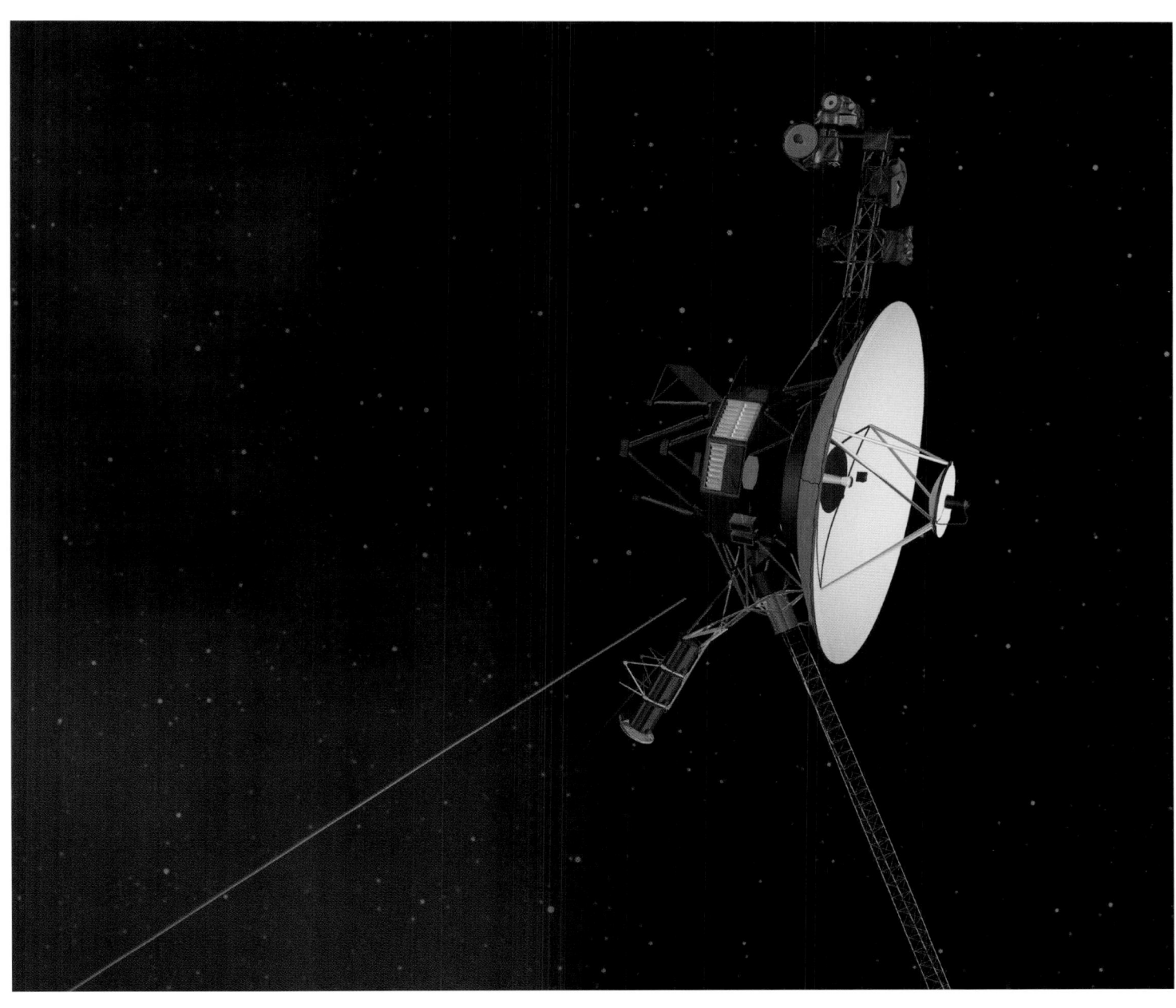

An artist's impression of Voyager 1 entering interstellar space. The interstellar plasma is shown by the orange background glow, which was first observed by NASA's Hubble Space Telescope. The Orion Nebula can also be seen in the background.

A COSMIC ODYSSEY

THE CONTACT

A New Frontier

The Voyagers are the first human-made objects to enter the space between the stars. It carries a little bit of us into the galaxy. Interstellar space is a new frontier that could be as novel and relevant for us as the Pacific was for the Europeans 500 years ago. Not for exploration's sake, but for the knowledge it gives us in return. The probes are collecting and sending back data that challenges fundamental rules of physics, and perhaps offers answers to even more fundamental questions that astronomers, philosophers and physicists have pondered throughout the ages. Why did our Sun give birth to life exclusively here on Earth? Where else might we find evidence that we are not alone?

Dwindling Power

Of course, these spacecraft will not be working forever. The plutonium power supplies, responsible for generating electricity, have carried the Voyagers along the gas giants and through the edge of our Solar System, but have been dwindling ever since. In this process, the power levels dropped from 470 watts in the beginning to about 160 watts now (as of 2024), as the radioactive plutonium-238 gradually decays, producing less heat and electricity over time.

This drop in the plutonium is already starting to have its effects. For example, if for some miraculous reason one of the craft passes by something useful to photograph, the power limitations will make it impossible for the camera to be turned back on without it gobbling up too much power and ruining other functionalities. And although the five instruments still in use today use up significantly less power than the camera did, the transmitters and the heaters are still eating it up at quite a pace.

How cold is too cold?

It is predicted that the power and thruster fuel of the spacecraft will be sufficient to stay in communication with the towers until sometime around 2025. Either way, engineers have to decide what to turn off and for how long, in order to save power on the Voyagers. This means an estimation needs to be made on how cold each component can get. At some point, the levels will drop to values where the team is forced to turn off the heaters and other subsystems. Still, the wattage level is unforgiving. The lower it drops, the faster parts will freeze and die out.

Luckily, the team is learning tricks along the way. One of the most valuable insights so far: spinning the wheels of an eight-track tape recorder — the spacecraft's only data-storage option — generates a bit of additional heat. Suzy Dodd states that her team is also looking into enabling only an engineering signal, with a minimum power utility, just to know where it is. When it comes to the time frame, she has a specific goal: "We launched in 1977, and so if we can keep in contact until 2027, that would be fifty years. That's my goal."

Losing touch

It is clear each spacecrafts' heaters, computers, and instruments have an expiration date and it is getting closer and closer. At some point, the DSN antennas in California, Spain or Australia will be looking for a signal and it will not be there. The day each Voyager shuts down will be sad for the team and those that have worked on the mission throughout the decades, similar to losing a grandparent or close relative who has lived a full and rewarding life without the opportunity to say goodbye.

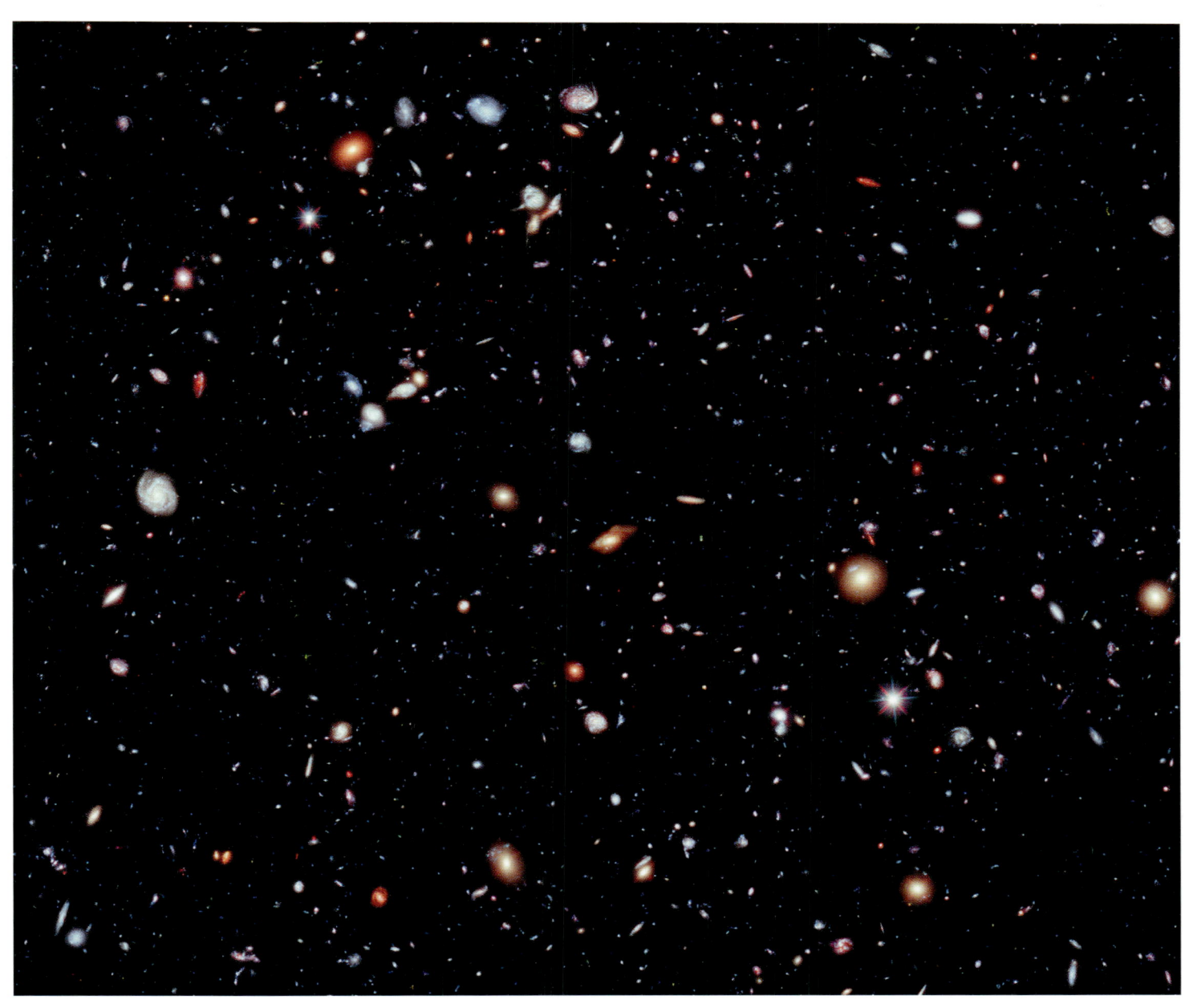

The eXtreme Deep Field (XDF) is the deepest image of
the universe ever to have been taken. It took 10 years of
images captured by NASA's Hubble Space Telescope to
create the image that contains around 5,500 galaxies
within this one frame.

THE JOURNEY

A hypothesized adventure

Although it might be the end of human contact with the Voyagers, it is just the beginning of a gigantic journey for each spacecraft, a cosmic odyssey that challenges our imagination. While these mechanical troubadours continue in their pursuit of limitless expanse of interstellar space, at some point not only the realm of the solar wind but even that of the Sun's gravity will be left behind. Both have picked up the pace to do so. The gravitational influence of the Sun is predicted to extend one-third to halfway to the nearest stars. From there, it becomes a hypothesized adventure.

After the spacecraft shut down, they will gently turn into a long and silent deep freeze. Although an occasional micrometeorite or cosmic ray might hit them, causing a minor dent, the Voyagers are solidly protected. Because of the low density of particles in interstellar space, the golden records and the spacecraft are predicted to only accumulate two percent damage in a period of about 100 million years. Set aside the highly unlikely event of a collision or burn-up by another star, there is not much that could destroy the Voyagers. Will they travel forever? What will their journey look like?

Oort Cloud

The first likely destination is a crowd of comets and asteroids, cast out of the inner Solar System through encounters with other objects in their lifetime. Every once in a while a new comet is discovered, like the Comet Hale-Bopp in 1995, Comet Hyakutake in 1996, and Comet NEOWISE in 2020, which has their ice boiled by the Sun, producing a majestic display of gas and dust turned into graceful arcing trails. Tracing their origin led Dutch astronomer Jan Oort to believe that our Solar System is wrapped with a spherical shell, filled with an estimated trillion or more asteroids and comets, extending out to the edge of the Sun's gravitational influence. Hence, we call it the Oort Cloud.

This cloud will be surrounding the Voyagers for quite a bit of time. Although they will arrive at the Oort Cloud in about 300 years, it is estimated it will take about 30,000 years to arrive at its outer edge. While "a trillion asteroids and comets" sounds like a lot, it doesn't even compare to the magnitude of the distances between them. The Voyagers passing anywhere near them is likely never to occur.

Aim for the stars

Once we escape the cloud and fast-forward about 10,000 years, the first stars will be on the horizon for the spacecraft. Voyager 1 is expected to pass only 100,000 AU from the red dwarf star Gliese 445,

while Voyager 2 is calculated to skim by another red dwarf star, Ross 248, with a distance of approximately 110,000 AU. Although both stars are relatively small and have a faint output of energy, there is a potential for inhibited planets around them. Carl Sagan and his Golden Record team have speculated on giving the crafts a push in the right direction to thrust them into the Solar Systems, using one final "empty the tank" thruster firing command. However, the chances that Voyagers will be ploughing through interstellar space even further is more likely.

The next star on the itinerary could very well be the famous, hot and young Sirius. In a little under 300,000 years, Voyager 2 is set to cross paths with it at a distance of 270,000 AU. Why is it famous? Mainly because it is the brightest star in the sky aside from the Sun. It is not a coincidence the Egyptians, Greeks and Polynesians used it to figure out what time it was. After Sirius, and about 100,000 years later, the craft is finding itself at a relatively close distance to Delta Pavonis and GJ 754.

Space is really empty

After that, it becomes more of a guessing game. Although we call it an adventure, it will mainly be a lonely trip through sheer volume. "Space" as Ed Stone says, "is really empty." The distances between the stars and other objects are unfathomable, which makes the probability of a visit or a collision close to zero. It is like being on a lifeboat in the Atlantic at night. Maybe you see a distant ship pass, maybe you do not. That is the future of Voyager for billions of years to come. That means, with our own Sun having a life expectancy of some five billion years, Voyager will outlive humanity.

The Milky Way Ferris Wheel

They will become Earth's ambassadors to the Milky Way, because what is certain is that the Voyagers will be taking stretches of 250-million-year-long orbits around the center of the Milky Way galaxy. Although Voyagers have become interstellar travelers, they are not intergalactic travelers. In order to escape the gravity of our galaxy, the speed of the Voyagers would have to be about fifteen times faster than they are now.

Along the way, they will be jostled around in different directions. The exact trails that the stars travel in their orbit around the center of the galaxy is not completely known to us, as the Voyagers can be flung around by the gravitational force of roaming planets without a home star. They will poke around the galaxy like billiard balls, changing their trajectory, with Chuck Berry and his colleagues on the Golden Record as the only remaining part of the mission.

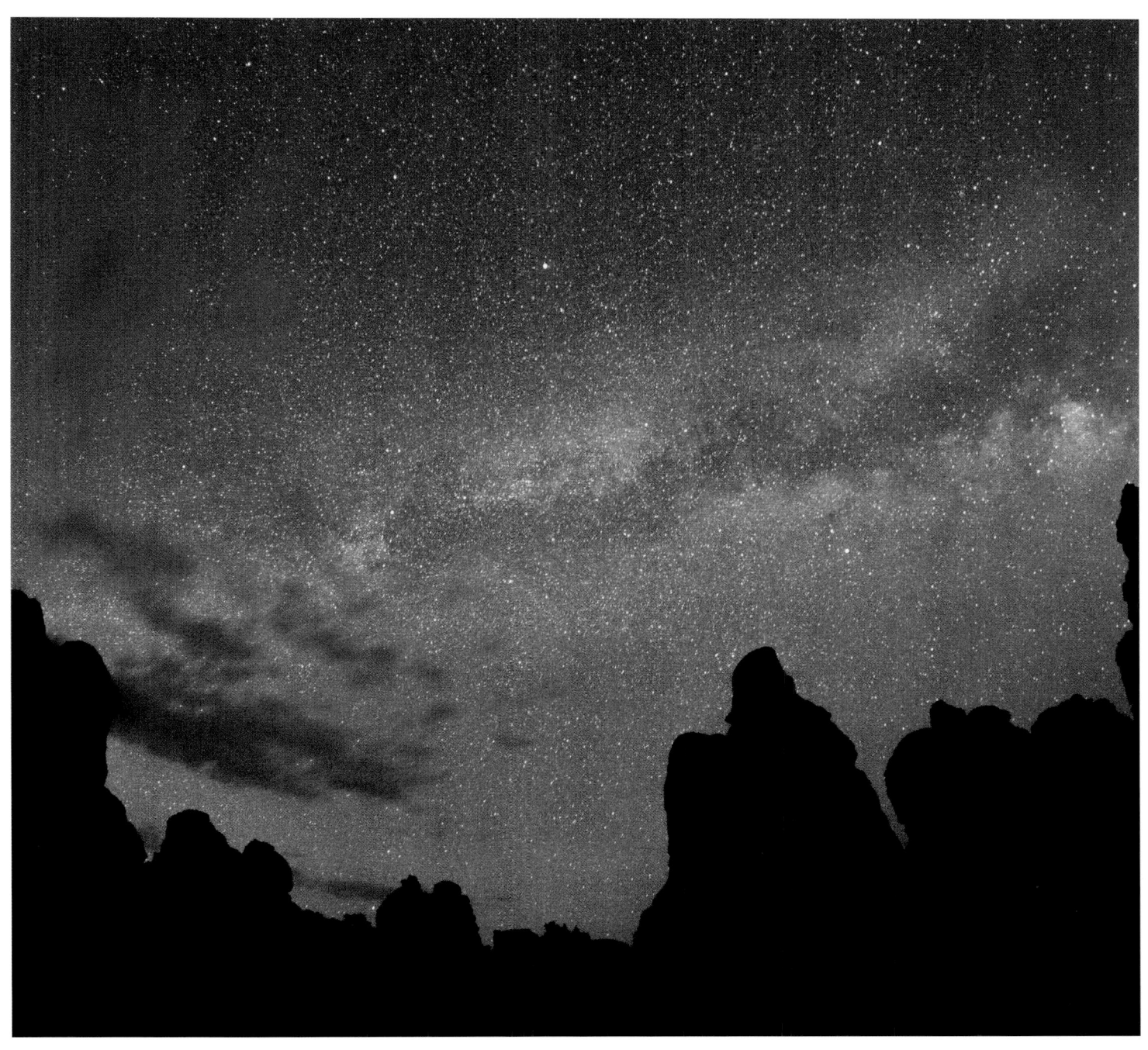

Looking up at the stars from our galaxy, the Milky Way,
which can be clearly seen in the night sky from the
Dollhouse Granaries in the Canyonlands National Park,
Utah. Taken on October 31st, 2019.

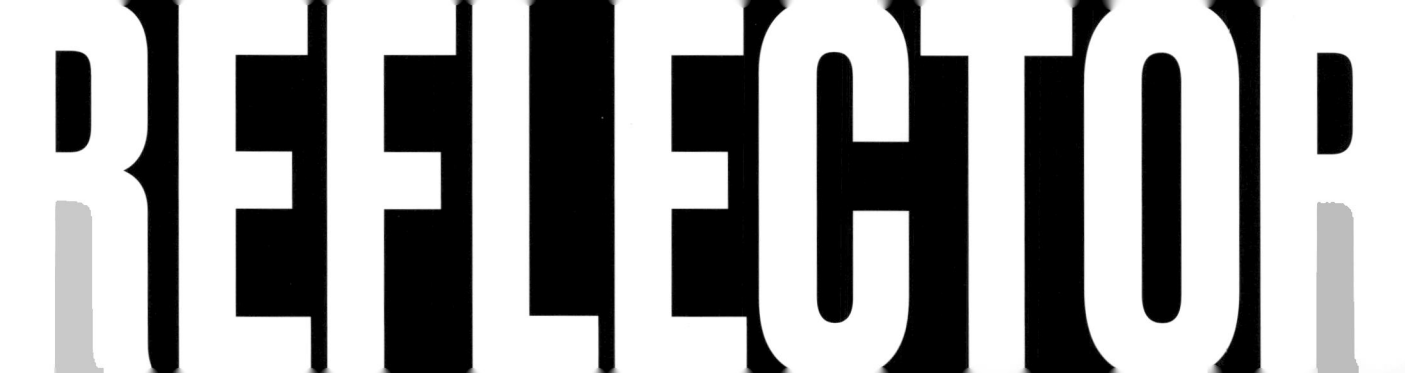

THE MEDIUM IS THE MESSAGE

Although it is the robotic exploration of space, Voyager is also very much a story of human drama. One of knowledge and discovery. Clashes and consensus. Triumph and sacrifice. A battle of perseverance, right down to the mundane. Men and women fighting for the existence of the mission, while their office was demoted and teams were disbanded. Astronomers, scientists, engineers, managers, technicians and students have designed the mission, processed the photographs, filled our textbooks, kept the craft alive and still to this day communicate with the ships as they sail through interstellar space. If Voyager is humanity's greatest journey, this might just make its nine flight-team engineers and everybody on the Voyager team, our greatest living explorers. There is no single 'Columbus' striking the honor of the discoveries. There is a group of heroes that have expanded our knowledge. They will always be remembered for doing so.

Several missions have strongly built on the legacy of Voyager. There is the Cassini-Huygens mission that explored Saturn and its moon to greater depth. And then there was New Horizons, that took up Voyager's mantle, flying by Jupiter on a trajectory that would take it by Pluto, the last remaining unexplored target of the original Grand Tour concept, before following the twin Voyagers on a path leading out into interstellar space. However, perhaps an even bigger legacy was realized within the technological realm. Besides the fact that this technology was hurled into space in the 70s and to this day flies itself, runs itself, and checks itself, it also inspired spin-off technologies that we still use today. Not only have the developed coding systems been applied in the development of cell phones and CD-players, the image processing technology in your smartphone evolved from systems first developed by Voyager engineers.

While the mission thus meant the beginnings of something new, it also quite possibly represents the end of something else. That something else would be space exploration for the sake of observation, rather than commercialization. In the U.S., Trump administration advisers have referred in internal memos to NASA's traditional contractors as "Old Space", while proposing to refocus budget to stimulate the growth and economic development of what they call "New Space". An attempt to turn space into an experience for those who can afford it. This "Westworld in the sky" goes against the very fabric of what the Voyager mission was about: expanding our knowledge instead of our wallets.

That expansion of knowledge did not limit itself to the planets, the edge of our Solar System or interstellar space; it also included the expansion of knowledge about ourselves. Voyager did not change the Solar System, but it did change us. Voyager historian Stephen Pyne insightfully notes: "Even as they are celebrated for racing forward, many of their most dazzling discoveries were the offshoot of staring back at what they passed in their slingshot flybys. Their trajectory is a triangulation of future and past." From the Pale Blue Dot perspective to the images of the planets, Voyager carried a cultural payload that increased the value of what it means to be human. Along with the Golden Record, it's taken that human culture to the stars.

Some are convinced that, with Voyager, we have moved into what can be called an Interstellar Age, where humankind is ultimately destined to leave our Sun's cradle to build a new future. This means we accept the fragility of our existence by adapting and moving on. To find a new home and become an interstellar species before the bubble bursts. Still, if there is one thing that Voyager has proven, it is the unique conditions Earth has been given. Finding another place to settle will not be easy. As Carl Sagan said, this should underscore our responsibility to preserve and cherish the pale blue dot. Perhaps this means coming to terms with our fate on our own planet and deal more kindly with one another for the time we still have. Even if that is "just" five billion years. Accept as a species that, not unlike individuals, life ends. It is a realization that can have a powerful and positive effect on how we cherish and enjoy life on Earth.

About thirteen years before the Voyagers launched, Marshall McLuhan popularized the phrase "The Medium Is The Message" in his influential work "Understanding Media: The Extension of Man". In it, he states that each new introduction of technology opens up a new dynamic that shapes the forms of human association and interaction. According to him, the nature of the medium is therefore more important than the message. So while it is tempting to think that, in the offset chance Voyager stumbles upon an alien civilization, the contents of the Golden Record represent our human message, it is worth the effort to reflect on what our presence there would mean? Perhaps even more than Chuck Berry or Mozart, it is the spacecraft itself that forms the deeper message. The two robotic traveling machines as an extension of humanity. The technology, the radioisotope power system, the communication system, the instruments... This is who we are. This is what we are capable of, and this is what has changed our understanding of ourselves, changed the way we interact with each other, and changed our outlook on the world. We hope that it does for you what it did for us.

References

There are a couple of 'stand out' books that helped us with the research and writing for the background information for this book. We highly recommend checking out the following books if you wish to find further, and more detailed information on the Voyager mission:

Bell, J. (2015). The Interstellar Age:
Inside The Forty-Year Voyager Mission. New York.
Dutton.

Riley, Dr C. (2015). NASA Voyager 1 & 2, Owners' Workshop Manual.
Sparkford, Yeovil. Haynes Publishing.

Other helpful and informative resources and images that have also been used and referenced in this book:

Hollingham, R. (2017, August 18). Voyager: Inside the world's greatest space mission. BBC. https://www.bbc.com/future/article/20170818-voyager-inside-the-worlds-greatest-space-mission

Hussey, J. (2014) Bang to Eternity and Betwixt: Cosmos. Amazon Kindle. John Hussey

Jet Propulsion Laboratory. (2019, April) https://voyager.jpl.nasa.gov/

Photo Journal, Jet Propulsion Laboratory. (2019, April). https://photojournal.jpl.nasa.gov/

Ring-Moon Systems Node voyager data. (2019, August). https://pds-rings.seti.org/voyager/data.html

Siy, A. (2017). Voyager's Greatest Hits: The Epic Trek to Interstellar Space. Watertown, MA. Charlesbridge

The Farthest. USA: Emer Reynolds, 2017. DVD

The History of Space Exploration. (2020, January 4). https://www.nationalgeographic.org/article/history-space-exploration/7th-grade/

Tingley, K. (2017, August 3). The Loyal Engineers Steering NASA's Voyager Probes Across The Universe. The New York Times.https://www.nytimes.com/2017/08/03/magazine/the-loyal-engineers-steering-nasas-voyager-probes-across-the-universe.html

Wikipedia, Voyager Program. (2019, April). https://en.wikipedia.org/wiki/Voyager_program

Wikipedia, Solar System. (2019, April). https://en.wikipedia.org/wiki/Solar_System

Photo & Image credits

All the raw 'dataset' images taken by Voyager 1 & 2 are courtesy of NASA/JPL/Caltech.

The raw Voyager dataset can be accessed and downloaded from www.opus.pds-rings.seti.org

The subsequent image processing and editing of the photographs was made by Ted Stryk, Joel Meter and Delano Steenmeijer.

All other photographs and images are referenced below:

Page 6:
The Blue Marble photograph taken from Apollo 17.
Courtesy of Apollo VII-XVII: ISBN-13: 978-3961711321

Page 11:
Rick Guidice depicting Pioneer 10 passing Jupiter, 1973
https://en.wikipedia.org/wiki/Pioneer_10#/media/File:Pioneer_10_at_Jupiter.jpg

Page 12:
Solar System diagram: Ian Moores
NASA Voyager 1 & 2: Haynes Publishing

Page 14:
Mariner spacecraft model presented to John F.Kennedy
https://www.nasa.gov/multimedia/imagegallery/image_feature_925.html

Page 14:
first 'image' of Mars sent from Mariner 4 in July 1965.
NASA / JPL / Science Photo Library

Page 15:
Voyager mission badge
https://www.jpl.nasa.gov/missions/voyager-1/

Page 15:
Mariner – Jupiter – Saturn – Uranus cover
NASA / JPL / Science Photo Library

Page 16:
Voyager 1 trajectory
NASA / JPL / Science Photo Library

Page 17:
Voyager 1 & 2 trajectory
https://www.jpl.nasa.gov/edu/news/2018/12/18/then-there-were-two-voyager-2-reaches-interstellar-space/

Page 18, 20, 21 & 22
Voyager Spacecraft & construction
https://voyager.jpl.nasa.gov/galleries/images-of-voyager/

Page 26:
First photograph from space
U.S. Army White Sands Missile Range/Johns Hopkins Applied Physics Laboratory

Page 28
Voyager imaging team
NASA/JPL-Caltech

Page 29, 31 & 32
Dr. Garry Hunt & Voyager imaging team
Courtesy of Dr. Garry Hunt

Page 33
Voyager imaging team
Photos by Roger Ressmeyer / Corbis / VCG via Getty Images

Page 34
AP Photo/Lennox McLendon

Page 37
Deep Space Network Antennas, Canberra
NASA/JPL-Caltech

Page 42 & 43
Golden Record
NASA/JPL-Caltech

Page 43
Carl Sagan
Photo by CBS via Getty Images

Page 44 & 45
Making of the Golden Record
NASA/JPL-Caltech

Page 46
Courtesy of Jon Lomberg

Page 47
Golden Record instructions illustration
NASA/JPL-Caltech

Page 48
Golden Record contents
NAIC / Jon Lomberg

Page 49
UN Photo/Yutaka Nagata
Underwater diver: David Doubilet

Page 50
Family portrait: Photo by Nina Leen / The LIFE Picture Collection via Getty Images)
Illustration: Jon Lomberg

Page 53
Golden Record Disk
NASA/JPL-Caltech

Page 56 & 57
Rocket launch
NASA/JPL-Caltech

Page 61
Trajectory diagram
NASA/JPL-Caltech

Page 66
www.gabrielevanin.it/Bertini.jpg

Page 67
Trajectory diagram
NASA/JPL-Caltech

Page 140:
wikipedia.org/wiki/Bestand:Huygens_Systema_Saturnium.jpg

Page 142:
Trajectory diagram
NASA / JPL

Page 166:
Saturn Ring System diagram
NASA / JPL

Page 207:
William Herschel & telescope
wikipedia.org/wiki/William_Herschel

Page 208:
Trajectory diagram
NASA / JPL

Page 212:
Trajectory diagram
NASA / JPL

Page 214:
Gerard P. Kuiper AObservatory & Trajectory diagram
NASA / JPL

Page 240:
John Couch Adams
https://commons.wikimedia.org/wiki/File:John_Couch_Adams_2.jpeg
Urbain Le Verrier
https://commons.wikimedia.org/wiki/File:Urbain_Le_Verrier.jpg

Page 244:
Trajectory diagram
NASA / JPL

Page 281:
Family Portrait diagram
https://nssdc.gsfc.nasa.gov/image/planetary/solar_system/family_diagram.jpg

Page 283:
Voyager craft render
NASA/Goddard Space Flight Center Conceptual Image Lab

Page 285:
Heliosphere diagram
https://www.jpl.nasa.gov/spaceimages/details.php?id=PIA22835

Page 287:
Voyager craft render
NASA/JPL-Caltech

Page 291
eXtreme Deep Field image
NASA; ESA; G. Illingworth, D. Magee, and P. Oesch, University of California, Santa Cruz; R. Bouwens, Leiden University; and the HUDF09 Team

Page 293
Milky Way
NPS/Kait Thomas

Making this book

The Voyager spacecraft returned thousands of images back to Earth over the span of thirteen years during which the cameras were operational. Naturally, it was quite difficult to select just a few of them for our book. With the 184 images that made the cut, we decided to showcase photographs based mainly on their photographic merit instead of purely selecting the ones that are well-known and find their way into publications often. This tough selection process meant we had to kill photographs we absolutely loved. Photographs that people know from education materials, candy packaging or magazines. But when deciding between appealing to people's recognition, or wowing them with something new, we gladly chose the latter.

A beautifully crafted photography book. That was our simple intention from the start. This meant really bringing out the very best in these incredible images, making them suitable to print in a large format for the first time. To do so, we have meticulously gone through the original Voyager dataset. Using the latest image-processing software and techniques available to us, we revisited and created new versions of the photographs that have more brilliance, clarity and information than the original images released by JPL / NASA. In the book, you might find some 'lower-resolution' images that look slightly blurred or pixelated. Since many of these images are the only known pictures of a certain distant moon or astronomical discovery, we decided to include them for their scientific importance.

There's a lively discussion on the 'correct' color appearance of all the planets and their moons and how it should be accurately reproduced for print. After extensive research on this topic, we have aimed to be as true to the color that the Voyager images captured. Because the Voyagers did not use color filters that precisely match what the human eye sees, we have made the intentional decision to not be 100 percent factually accurate (by human standards) with the coloring, but to focus more on highlighting the aesthetic beauty and to bring out the hidden details in each of these images that the Voyager imaging system was designed to capture. In some cases we have framed the image subject differently than the raw image data to highlight interesting aspects of the image.

A note of thanks

Dr. Garry Hunt for his time, sharing his personal stories from his time working on the Voyager mission, as well as the enormous wealth of knowledge and expertise on the subject matter. Thank you!

Similar thanks to Dr. Candy Hansen for her time, sharing with us her experiences on the vital work she undertook for the Voyager project, also for feedback and advice on the book.

Dr. Mark Showalter & Björn Jónsson for their help and advice to get this project off the ground.

Jon Lomberg for giving us a glimpse into his experiences relating to the creative and art direction work he did for the Golden Record, and for the very generous help, providing us with the original images and drawings that were included in the Golden Record.

And of course, to the many thousands of people who, past and present, worked on the Voyager project, who worked relentlessly in order for the rest of the world to get a glimpse and a better understanding of our neighbors.

IF THE DOORS OF PERCEPTION WERE CLEANSED, EVERYTHING WOULD APPEAR TO MAN AS IT IS, INFINITE.

William Blake, 1790

VOYAGER
Photographs from Humanity's Greatest Journey
Paperback Edition

1st printing, 2025

Published by gestalten, Berlin 2025
© 2026 Die Gestalten Verlag GmbH & Co. KG, Berlin

Copyright of the hardcover edition:
© 2020 teNeues Media GmbH & Co. KG, Kempen

Essay by Dr. Garry Hunt
Edited by John Doe
Editorial coordination by Roman Korn
Production by John Doe
Color separation by John Doe

Production by Alwine Krebber

Printed in the Czech Republic by PB Tisk a.s.

ISBN 978-3-96171-707-1

© teNeues, an Imprint of
Die Gestalten Verlag GmbH & Co. KG, Berlin 2025

For more information, and to order books, please visit
www.teneues.com and www.gestalten.com

Die Gestalten Verlag GmbH & Co. KG
Mariannenstrasse 9–10
10999 Berlin, Germany
hello@gestalten.com

Düsseldorf Office
Waldenburger Straße 13
41564 Kaarst, Germany
verlag@teneues.com

teNeues Press Department
press@gestalten.com

Bibliographic information published by the
Deutsche Nationalbibliothek. The Deutsche
Nationalbibliothek lists this publication in the
Deutsche Nationalbibliografie; detailed bibliographic
data is available online at www.dnb.de

https://instagram.com/teneuespublishing

www.teneues.com

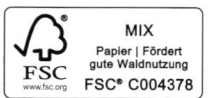